Analog Design and Simulation using OrCAD Capture and PSpice

Analog Design and Simulation using OrCAD Capture and PSpice

Dennis Fitzpatrick

ELSEVIER

AMSTERDAM • BOSTON • HEIDELBERG • LONDON • NEW YORK • OXFORD
PARIS • SAN DIEGO • SAN FRANCISCO • SINGAPORE • SYDNEY • TOKYO

Newnes is an imprint of Elsevier

Newnes

Newnes is an imprint of Elsevier
The Boulevard, Langford Lane, Kidlington, Oxford, OX5 1GB, UK
225 Wyman Street, Waltham, MA 02451, USA

First published 2012

British Library Cataloguing in Publication Data
A catalogue record for this book is available from the British Library

Library of Congress Control Number: 2011933640

ISBN: 978-0-08-097095-0

For information on all Newnes publications visit our website at www.elsevierdirect.com

Printed and bound in the United Kingdom

Transferred to Digital Printing in 2013

Contents

Contents

Preface

The Cadence/OrCAD family of Electronic Design Automation (EDA) software provides a complete design flow from schematic entry to circuit simulation through to PCB layout. The circuit is drawn using the Capture or Capture CIS schematic editor and circuit simulations are performed using PSpice. The schematic diagram is translated into a printed circuit board design using Cadence Allegro or PCB Editor which has replaced OrCAD Layout. The book has been written incorporating the features in the latest Cadence/OrCAD 16.6 software release and can be used with previous software releases and the latest demo **OrCAD PCB Designer Lite DVD (Capture & PSpice only)** software which can be downloaded from the Cadence website:

http://www.cadence.com/products/orcad/pages/downloads.aspx

This book will benefit anybody with an interest in using the Cadence/OrCAD professional simulation software for the design and analysis of electronic circuits. The book provides a practical hands on approach to using the software and at the end of each chapter there are exercises with step by step instructions to complete.

Thanks are due to the technical staff at the University of West London, Keith Pamment and Seth Thomas for reviewing the simulation exercises, Taranjit Kukal and Alok Tripathi from Cadence for reviewing the technical aspects of the book and Parallel-Systems UK for their support.

Instructions

Throughout the book, bold type will indicate tool specific keywords and also which menus to select, for example, the menu selection to create a new project is shown below.

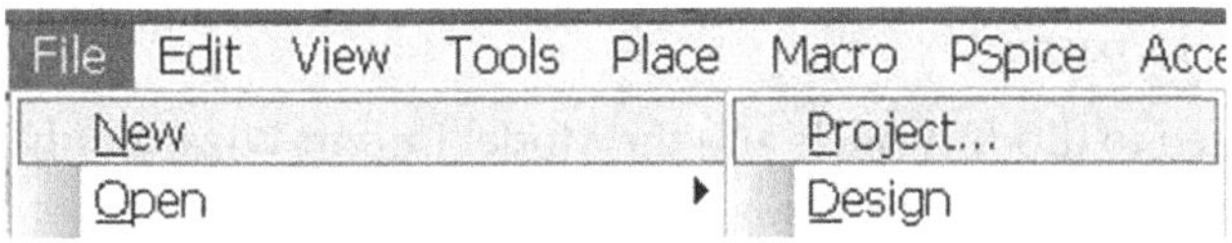

The instruction sequence will be, **File > New > Project**. The words in bold indicate which successive menus to select from the top toolbar, as shown above.

A right mouse button click will be written as **rmb**. For example, select the part in the schematic and **rmb > rotate**.

Bold type is also used to name any dialog box and windows that may appear. For example, the **Create PSpice Project** window is shown below.

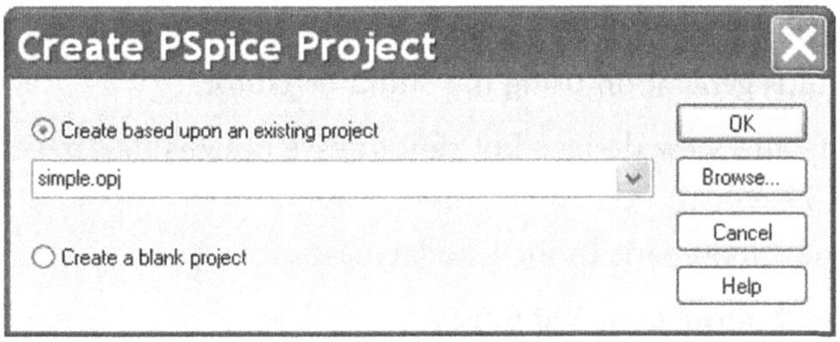

LIMITS OF ORCAD DEMO CD

The latest OrCAD demo CD, OrCAD 16.6 Lite, which is free to download or order from www.Cadence.com, is not time limited, it is only limited in functionality and size of the schematic. You can use the demo CD for most of the exercises in this book. However, the number of limits from 16.5 onwards has changed.

PRE-16.5

For PSpice simulation, you are limited to 64 nodes, 20 transistors, two subcircuits or 65 digital primitives and 10 transmission lines (ideal or non-ideal) with not more than four pairwise coupled lines. The maximum number of Tlines is limited to four.

The internal subcircuit nodes for opamps count towards the total node count. So use the uA741 from the eval library on the demo CD, which has 19 internal nodes. For digital circuits, the maximum number of nodes is limited to 250.

In Capture you can create and view designs but cannot save designs that have more than 64 nets or 60 parts. The Capture CIS demo database is limited to 10 parts. You cannot create parts with more than 14 pins and you cannot save part libraries with more than 15 parts.

The Model Editor is limited to diode models and the Model Import Wizard only supports two pin parts and models.

The Stimulus Editor is limited to sine waves and digital clocks.

Only simulation data created in the demo version can be displayed.

The Magnetic Parts Editor can only be used to design power transformers. The Magnetic Parts Editor supplied database cannot be edited and only contains a single magnetic core.

16.5 AND 16.6

From 16.5 onwards, for PSpice simulation, the number of nodes has increased to 75 nodes and there is no subcircuit limit.

There is no limit to stimulus generation using the Stimulus Editor.

In Capture you can create and view designs but cannot save designs that have more than 75 nets or 60 parts.

You cannot have more than 1000 parts in the CIS database.

You cannot create parts with more than 100 pins.

CHAPTER 1

Getting Started

Chapter Outline

Those of you who are familiar with setting up projects and drawing schematics in Capture may want to skip this chapter, as it has been written for those of you who have little or no experience of using Capture. This chapter will describe how to start Capture and how to set up the project type and libraries for PSpice simulation.

At the end of each chapter there are some exercises to do and as you go through the book, each chapter will build upon the exercises from previous chapters.

1.1 STARTING CAPTURE

Circuit diagrams for PSpice simulation are drawn in either Capture or Capture CIS schematic editor. The CIS option, which stands for Component Information System, allows you to select and place components from a component database instead of selecting and placing components from a library. For this book, it does not matter whether the circuits are drawn in Capture or Capture CIS.

If you have the OrCAD software installed, launch Capture or Capture CIS, by clicking on:

```
Start > Program Files > OrCAD xx.x > Capture
```

or

```
Start > Program Files > OrCAD xx.x > Capture CIS
```

Analog Design and Simulation using OrCAD Capture and PSpice. DOI: 10.1016/B978-0-08-097095-0.00001-5

where xx.x is the version number, e.g. 10.5, 11.0, 15.5, 15.7, 16.0, 16.2, 16.3, 16.5 and 16.6.

For example:

```
Start > All Programs > Cadence > OrCAD 16.6 Lite > OrCAD Capture
CIS Lite
Start > All Programs > Cadence > Release 16.5 > Capture
```

If you have the Cadence software installed, the tools are installed under the Allegro platform name. In this case, only Capture CIS is available and is branded as Design Entry CIS:

```
Start > Program Files > Allegro SPB xx.x > Design Entry CIS
```

1.2 CREATING A PSPICE PROJECT

New designs started in Capture will automatically create a project file (.opj) which will reference associated project files such as the schematics, libraries and output report files.

Before the circuit diagram is drawn, the project type and libraries required for the project need to be set up. First of all a new project is created by selecting from the top toolbar:

```
File > New > Project
```

In the **New Project** window (Figure 1.1), you enter the name of the project and then you have a choice of one of four project types:

- **Analog or Mixed A/D** is used for PSpice simulations.
- **PC Board Wizard** is used for schematic to PCB projects.

FIGURE 1.1
Creating a new project.

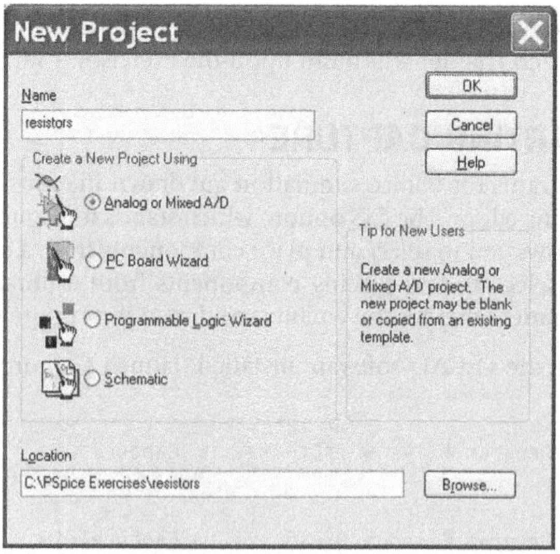

- **Programmable Logic Wizard** is used for CPLD and FPGA designs.
- **Schematic** is used for schematic and wiring diagrams.

When you select a Project type, the **Tip for New Users** gives a brief explanation of the project type. For PSpice projects, select **Analog or Mixed A/D**. This will activate the PSpice menu on the top toolbar in Capture.

It is recommended that a new directory location (folder) is created for each new project. This can be done by clicking on the **Browse**... button shown in Figure 1.1, which opens up the **Select Directory** window shown in Figure 1.2.

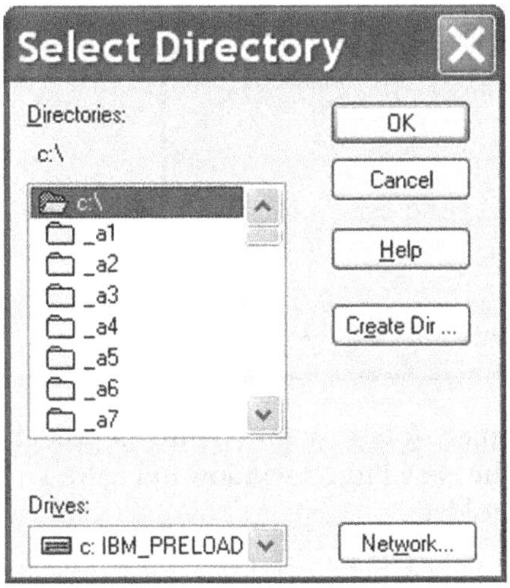

FIGURE 1.2
Creating a project folder location.

By selecting the **Create Dir**... button, the **Create Directory** window (Figure 1.3) appears, which allows you to name the directory (folder).

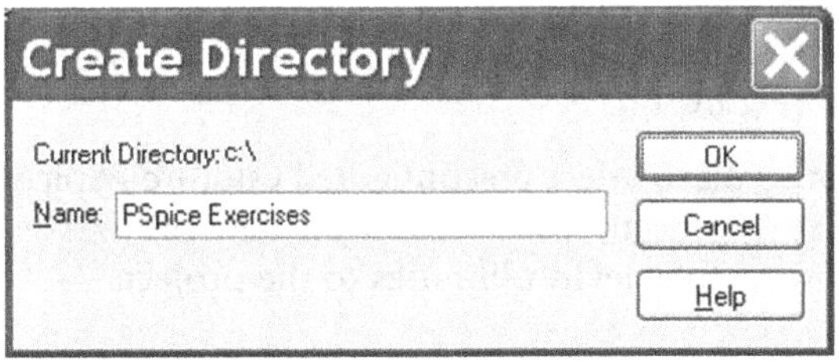

FIGURE 1.3
Creating the project folder.

The created folder, PSpice Exercises in this example, will appear in the **Select Directory** window. However, you must highlight and select the folder by clicking twice with the left mouse button, which will show the 'open' yellow icon as shown in Figure 1.4. A further subdirectory or folder can be created by clicking on the **Create Dir**... in the **Select Directory** window button and following the same procedure above.

The project folder location will then appear in the Location box of the New Project window (see Figure 1.1).

FIGURE 1.4
The project folder has been selected.

An alternative method of creating the project folder is to type in the folder location directly into the Location box in the New Project window in Figure 1.1 and Capture will automatically create the folder.

NOTE

It is a common mistake to create a project folder and not select the folder. Make sure you double click on the created folder name in the **Select Directory** window (Figure 1.4).

The next window to appear is the **Create PSpice Project** window, which sets up the project for PSpice simulation (Figure 1.5).

The pull-down menu option allows you to select preconfigured Capture-PSpice libraries for the project. The most commonly used option for new projects is **Simple.opj**, which adds the following five default libraries to the project:

Analog.olb
Breakout.olb
Source.olb
Sourcstm.olb
Special.olb

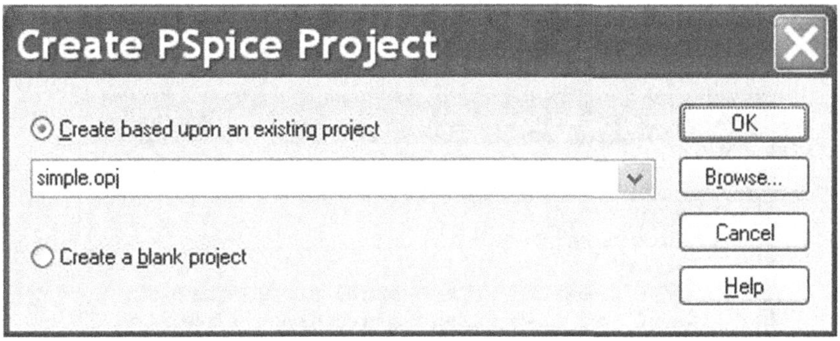

FIGURE 1.5
Create PSpice Project.

These libraries contain the most commonly used parts for PSpice projects and are recommended for new projects.

There is also an option to create updated versions of an existing project, i.e. to create a newer version 2 based upon the original version 1 project. In the Create PSpice Project Window (Figure 1.5), select the function **Create based upon an existing project** and then **Browse** to select an existing project. This will copy the existing project and all its associated files into the new project. This is similar to using the **File > Save As** function.

If the **Create a blank project** option is selected, then no Capture-PSpice libraries are added to the project. The libraries can be added later. This will be demonstrated in one of the exercises at the end of this chapter.

When a new project is created, a **Project Manager** window is created (Figure 1.6) which shows the absolute path to the libraries. Remember that these are Capture symbol libraries which define the graphics for the parts. They are not the PSpice model libraries. The Capture libraries are installed by default and can be found, depending on the OrCAD or Cadence software version you are using, for example, at:

```
<software install path> OrCad > OrCAD_10.5 > tools > capture > library
> pspice
```

or

```
<software install path> Cadence > SPB_16.3 > tools > capture > library
> pspice
```

Normally the **<software install path>** is the C: drive.

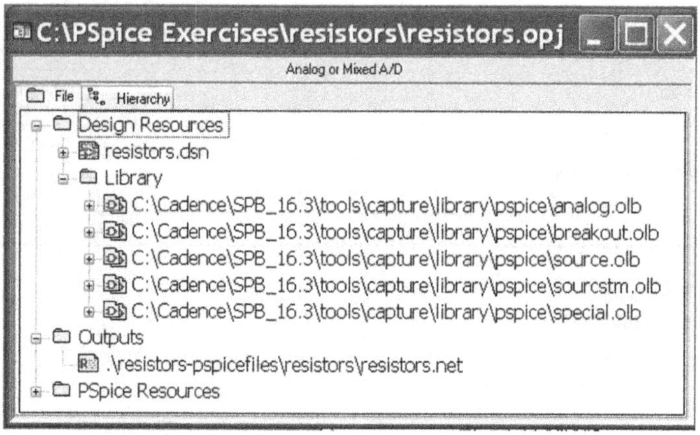

FIGURE 1.6
Project Manager showing the Capture parts libraries and their location.

TIP

If the **Project Manager** window is not displayed, select from the top toolbar, **Window** > <**project name**>**.opj** file (Figure 1.7). Here the project name is resistors. Note the project name file extension .opj.

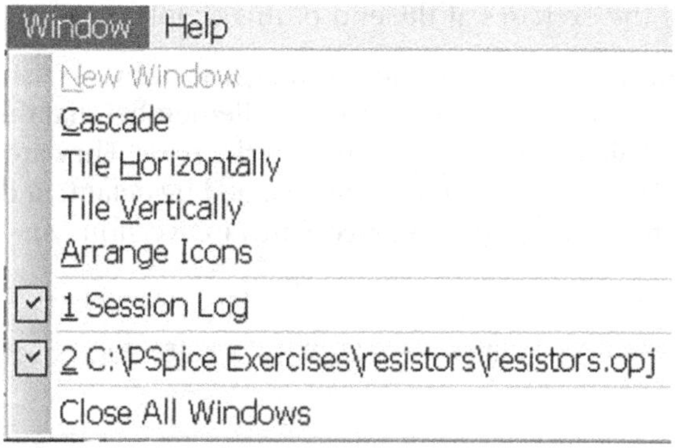

FIGURE 1.7
Displaying the Project manager window.

Alternatively, click on the Project manager icon or .

1.3 SYMBOLS AND PARTS

1.3.1 Symbols

Before drawing a schematic diagram, it is useful to know the difference between a part and a symbol. Symbols differ from parts in that they are not placed from the **Place Part** menu in Capture. You have to select the symbol from the **Place** menu (Figure 1.8).

FIGURE 1.8
Place menu.

The **Place** menu also shows the corresponding shortcut keys. For example, to place a Power symbol, press **F** and the **Place Power** menu appears as shown in Figure 1.9.

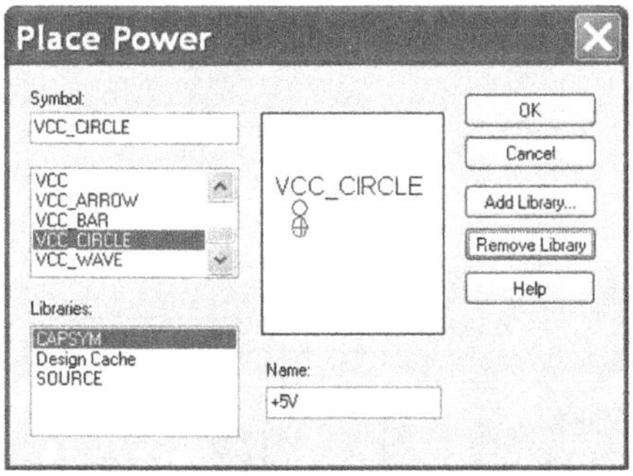

FIGURE 1.9
Place Power menu.

Wires connected to symbols take on the name of the symbol. For example, to define a wire to be connected to zero volts, you place a '0' symbol. To define a +5 V connection you can use a VCC_CIRCLE symbol and rename it +5 V. All wires connected to the +5 V symbol will take on a net name of +5 V. A net is a wire connection. There are many different symbols you can use to define the power and grounds connections and you can rename them accordingly.

In the **Place Power** menu in Figure 1.9, a **VCC_CIRCLE** symbol has been selected and its name has been changed to +5 V. Any wires (nets) connected to +5 V will take on the net name +5 V.

Other symbols include hierarchical ports and off-page connectors which allow signals to be connected together throughout the design. These will be discussed in Chapter 20.

There are two symbol libraries, **source** and **capsym**. **Capsym** contains all the analog ground and power symbols, while **source**, which also contains the analog **0V** symbol, contains the digital **$D_HI** and **$D_LO** symbols, which are used to set a digital level of 'hi' or 'lo' on a wire or pin of a digital device.

1.3.2 Parts

To place a part, select **Place > Part**. Figure 1.10a shows the Place Part menu for version 16.0 and Figure 1.10b shows the Place Part menu for version 16.3.

Although the two menus look different they have the same functionality in that they display the list of libraries available and the parts available in the libraries; and they both provide a part search function. In Figure 1.10a, only the **analog** library has been highlighted and so only those parts for that library are shown in the Part List.

FIGURE 1.10
Place part menu: (a) version 16.0; (b) version 16.3.

(a)

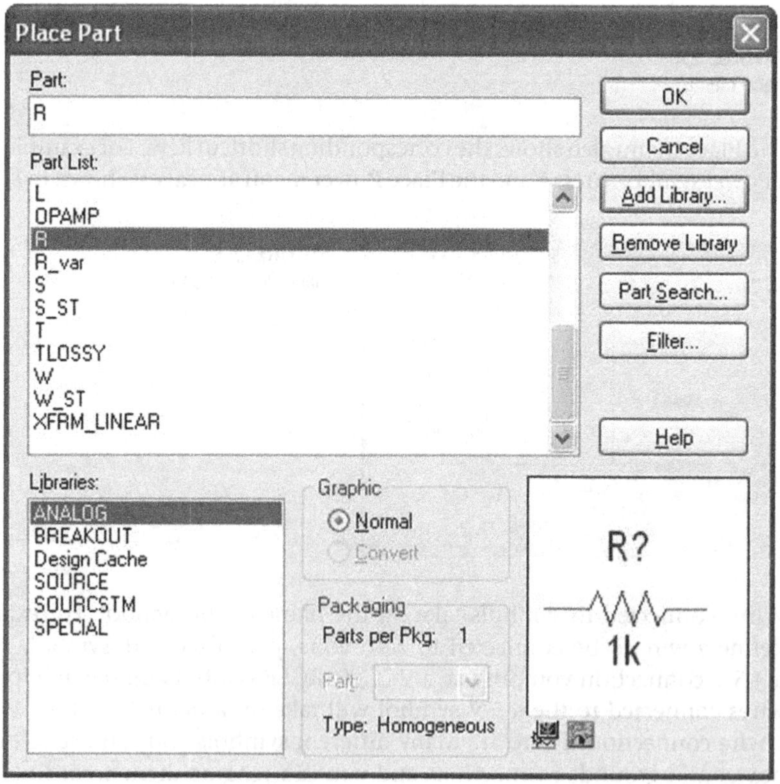

(b)

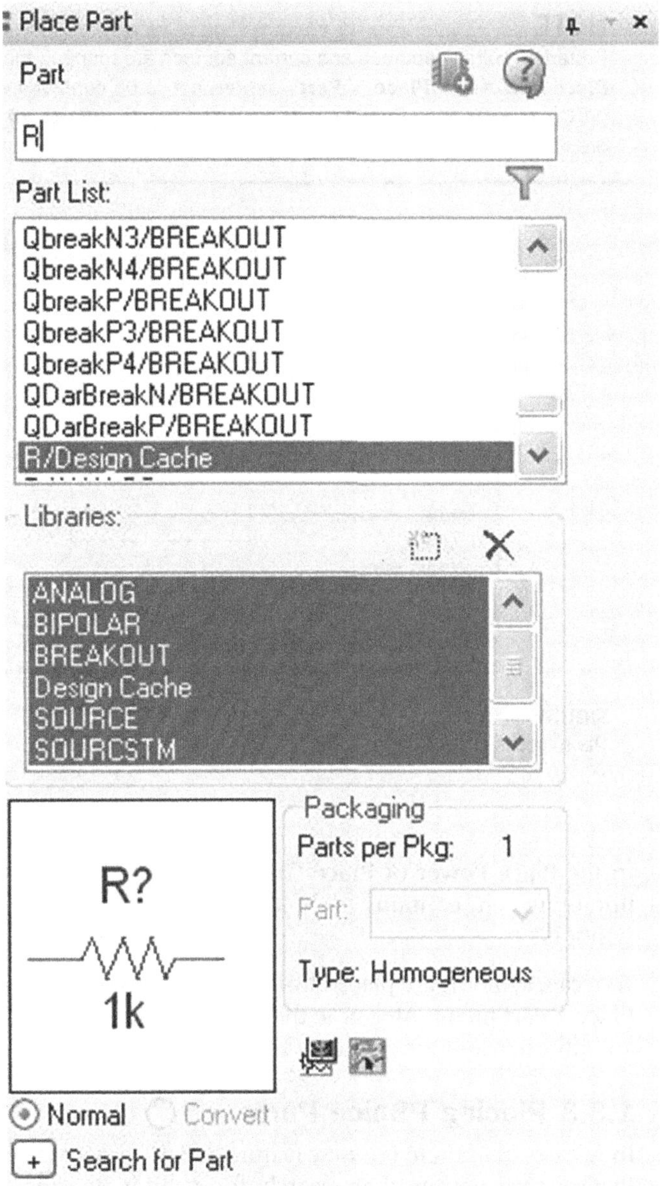

FIGURE 1.10
(Continued)

In Figure 1.10b, all the libraries have been highlighted and so you see the name of the part and which library it comes from. If you place the cursor over any part in the Part List, a tool tip rectangular bar appears showing the absolute path to the library part.

NOTE

Batteries, voltage sources and current sources are found in the **source** library from the **Place Part** menu (**Place** > **Part**) and are not to be confused with the power symbols (VCC_circle, 0V, etc.) from the **capsym** library (**Place** > **Power or Place** > **Ground**), which are effectively used to 'invisibly' connect wires with the same net name together.

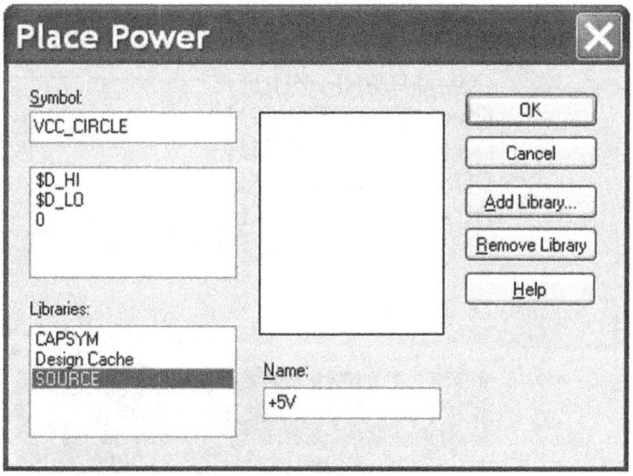

FIGURE 1.11
The source library for Place Power.

In the **Place Power** or **Place Ground** window (Figure 1.11) there is a **source** library which contains only the digital HI, digital LO and ground 0 V symbols.

To recap, symbols are placed from the **Place** menu and parts are placed from the **Place > Part** menu. Also note that both **Part** libraries and **Symbol** libraries have an .olb extension and are the Capture graphical parts.

1.3.3 Placing PSpice Parts

In release 16.6, there is a new feature which allows you to quickly place generic PSpice parts rather than search for vendor specific semiconductor device numbers. Parts are placed from the top tool bar **Place** menu as shown in Figure 1.12. Note that the **Passive** parts are from the **analog** library except the **Potentiometer** which is from the **breakout** library.

Figure 1.13 shows the selection of PSpice **Discrete** parts which are from the **Breakout** library. These are generic semiconductor parts with default names and parameters which can be customised using the PSpice Model Editor but the demo Lite version only allows you to edit diodes.

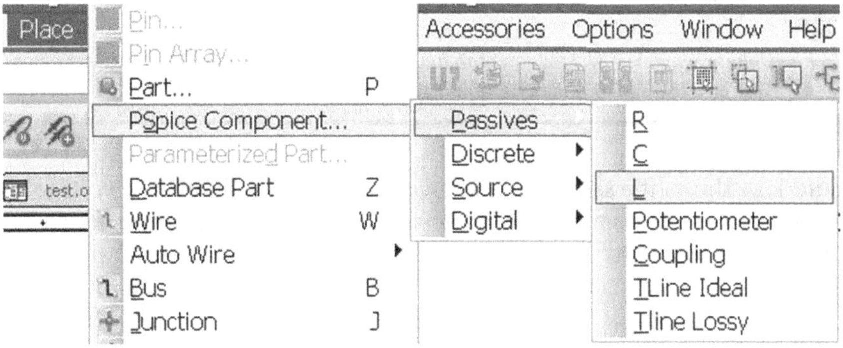

FIGURE 1.12
Placing a passive PSpice part.

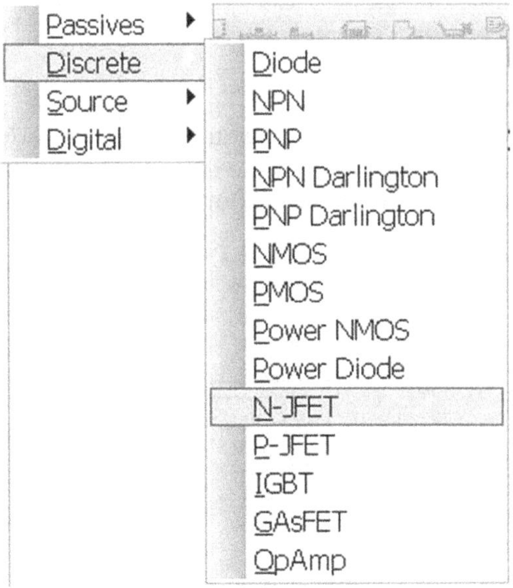

FIGURE 1.13
Placing a discrete PSpice part.

NOTE

If you know the actual semiconductor device number, then it may be best to find and use that known semiconductor device from the PSpice libraries. The generic parts are useful if you are new to electronics and you just want to get a circuit up and running quickly. Customising semiconductor parts from the Breakout library is mainly used for more advanced circuits where specific semiconductor characteristics are required.

TIP

The **eval** library which is available in both full and demo Lite software releases, contains general purpose standard analogue and digital semiconductors.

Figure 1.14 shows the selection of sources which are available. The voltage and current sources are from the **source** library whereas the controlled sources are from the **analog** library.

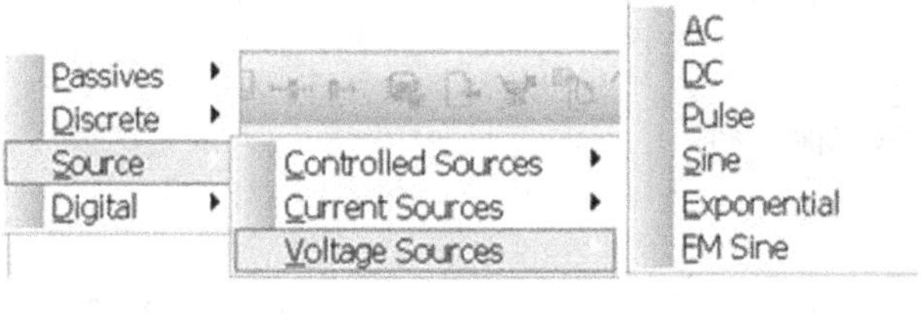

FIGURE 1.14
Placing a source PSpice part.

Figure 1.15 shows the selection of digital devices, if you have the full 16.6 software release, the gates and flip-flops are found in the dig_prim library and the ADC, DAC and Memory devices are from the **Breakout** library.

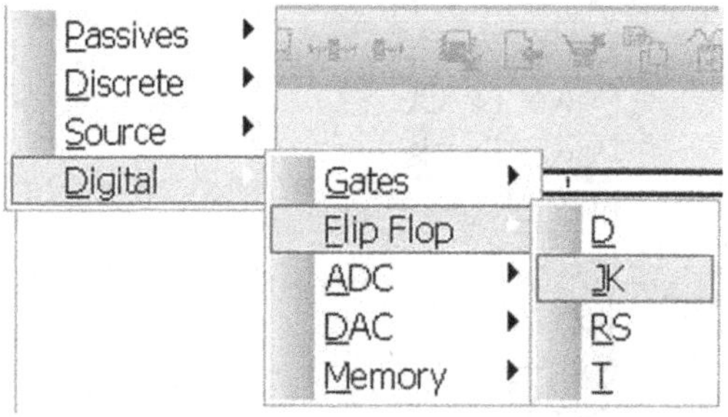

FIGURE 1.15
Placing a digital PSpice part.

At the time of writing, the digital gates and flip-flops are not currently available in the OrCAD Lite 16.6 version. They will be added in a later release. The ADC, DAC and Memory devices are from the **Breakout** library.

NOTE

If you are using the demo 16.6 OrCAD Lite version, be careful not to use the gates and flip-flops from the gate and latch libraries. These devices do not have PSpice models and therefore cannot be stimulated.

1.4 DESIGN TEMPLATES

From version 16.3 onwards, **Design Templates** have been added, which are complete electronic circuits and topologies including simulation profiles for analog, digital, mixed and switched mode power supplies. You can select any of these templates from the pull-down menu in the **Create PSpice Project** window when you create a new project (Figure 1.16).

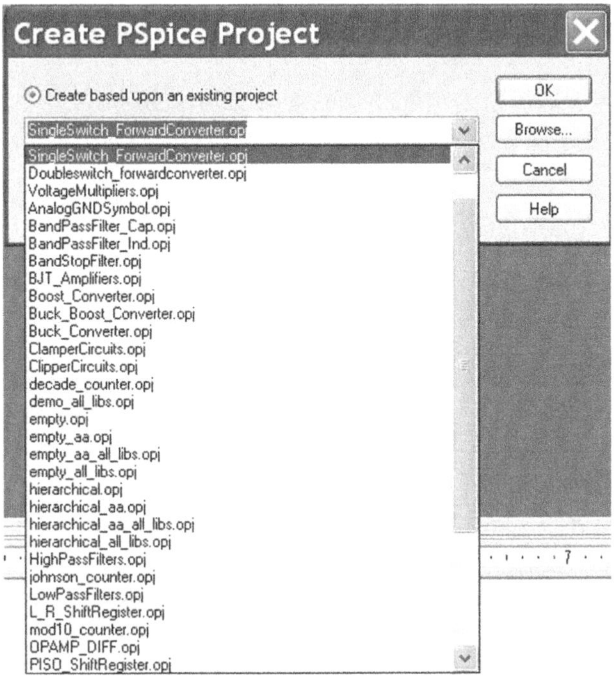

FIGURE 1.16
Available design templates.

Figure 1.17 shows the Design Template for a Single Switch Forward Converter which includes the schematic and explanatory text.

1.5 SUMMARY

Figure 1.18 shows the Project Manager created for a PSpice project and the checks that can be made to ensure that a PSpice project has been set up correctly. One common mistake is not selecting and highlighting the project folder that is created (see Figure 1.4).

Another common mistake when creating a project is that the wrong project type has been created; for example you see PCB instead of Analog or Mixed A/D in the Project Manager title. One way around this is to create a new project (of the

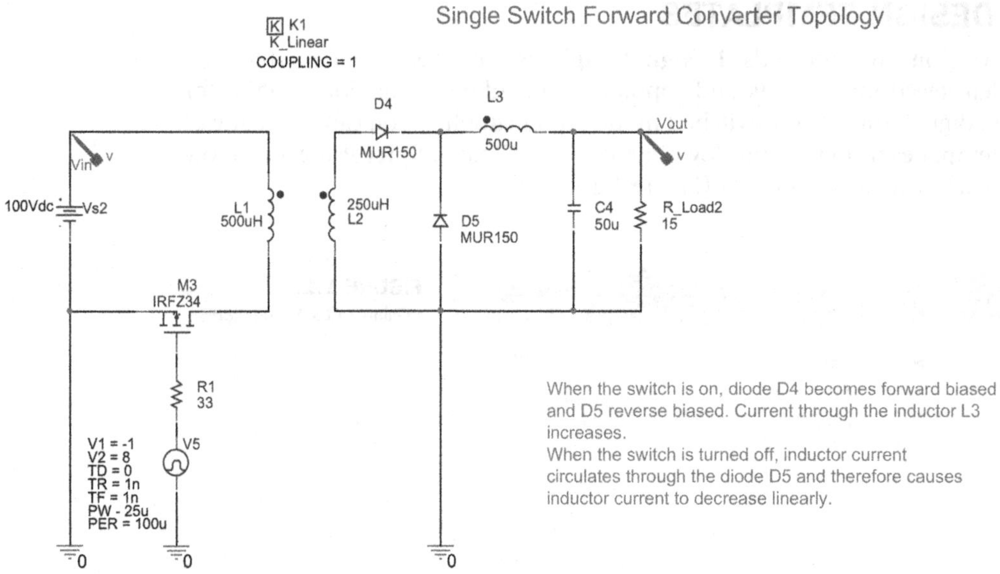

FIGURE 1.17
Design template for a single switch forward converter topology.

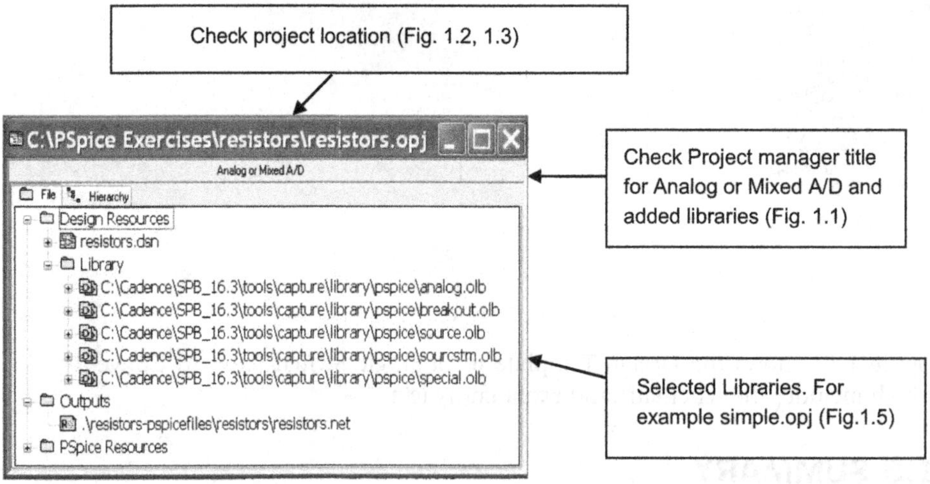

FIGURE 1.18
Project Manager setup for a PSpice project.

correct type) and copy and paste the .dsn file from the previous Project Manager into the new Project Manager. From version 16.3 onwards, you can change the project type by highlighting **Design Resources** in the Project Manager and **rmb > Change Project Type** (Figure 1.19).

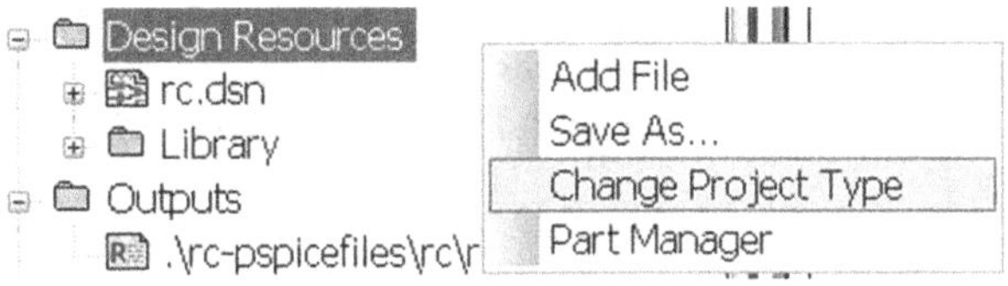

FIGURE 1.19
Changing project type.

1.5.1 Saved Designs

In version 16.6, pages that are not saved in the design are now marked with an asterisk. Figure 1.20 shows that Page 2 in the RC schematic has not been saved and hence the schematic **rc** folder and design file, **rc.dsn** are also shown as not having been saved.

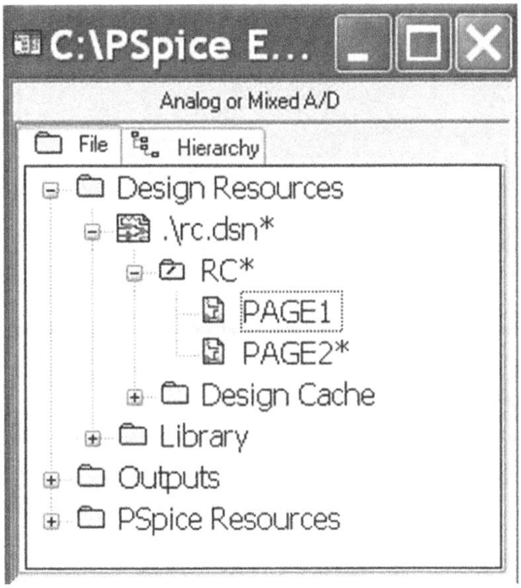

FIGURE 1.20
Unsaved pages are marked with an asterisk.

1.5.2 Opening Designs Created in Earlier Versions of OrCAD

In 16.6, you can open designs created in earlier versions of OrCAD without having to update the design. The designs will only be updated to 16.6 if you save the design.

1.6 EXERCISES

Exercise 1

You will create a new PSpice project as discussed in Section 1.2 and name it **resistors**. The project will be created in a folder called, for example, C:\PSpice\resistors and will be configured with the **simple** five default libraries.

1. Select **File > New > Project**. Enter **resistors** for the **Name** and select **Analog or Mixed A/D**. In **Location**, enter **C:\PSpice exercises\resistors** or you can use your own folder location. Check your entries with Figure 1.21 and then click on OK.

FIGURE 1.21
Creating a new PSpice project called resistors.

NOTE

You can also use the **Browse** button to create and name the project folder.

2. In the **Create PSpice Project** window, select **simple.opj** as shown in Figure 1.22 and click on OK.

FIGURE 1.22
Selecting the simple.opj project template.

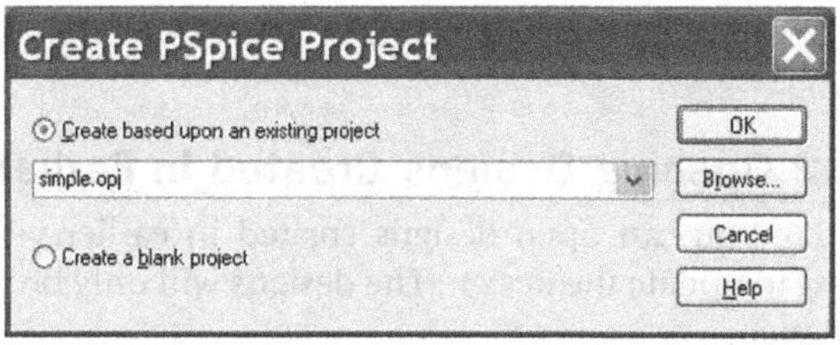

3. The Project Manager window will appear as shown in Figure 1.23.
4. Expand **resistors.dsn** by double clicking on it to open the SCHEMATIC1 folder (Figure 1.24).
5. Double click on SCHEMATIC1 to open PAGE1 and double click on PAGE1 to open up the schematic page (Figure 1.25).

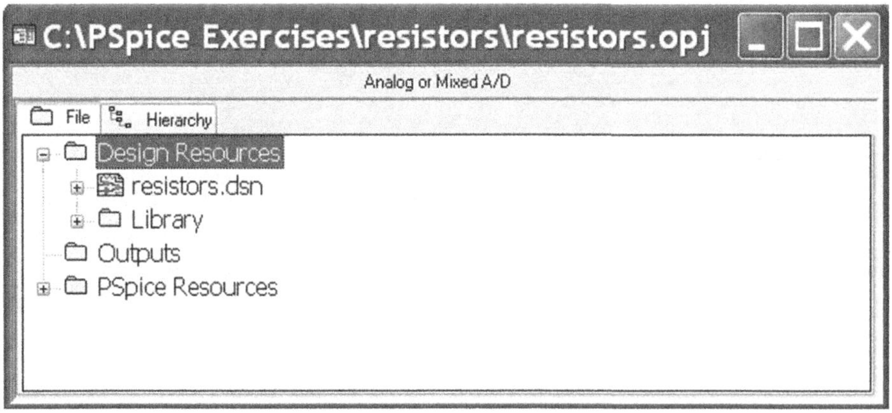

FIGURE 1.23
Project Manager window.

FIGURE 1.24
Schematic1 folder.

FIGURE 1.25
Page1 folder.

6. When you first open the schematic page, you will see some preplaced text and two voltage sources preplaced. Delete the sources and text by drawing a box around the sources and text and pressing the delete key.

Exercise 2

Draw the resistor network shown in Figure 1.26.

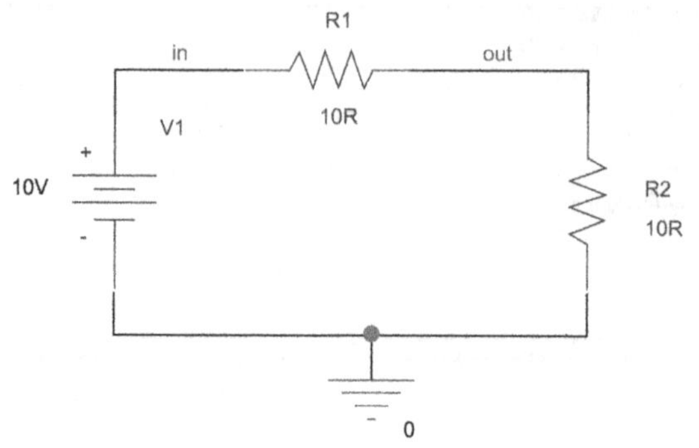

FIGURE 1.26
Simple resistor circuit.

1. To place a resistor, select **Place > Part** and select an R from the analog.olb library. Double click on the R and the resistor will attach to the cursor. In previous versions, click on **OK** and the **Place Part** menu disappears. From release 16.0 onwards, when you double click on a part or click on the **Place Part** , the menu will remain open.
   When you place the first resistor in the schematic, another resistor will be attached to the cursor; click **rmb > Rotate** or press R on the keyboard and place the second resistor. To exit place part mode, **rmb > End Mode** or press escape. Whenever a part is selected, there is an **rmb** context menu for place part options. P is the hotkey to place a part or you can select the Place Part icons, depending on which software release you have or .
2. For R1 and R2, double click on the default resistor value of 1k and change its value to 10 R.
3. Place the voltage source, which can be found in the **source** library. Change its voltage to 10 V.
4. To place a ground symbol, **Place > Ground** (or press G) or click on the icon or and select the 0 V symbol from the capsym.olb library (Figure 1.27).
5. To draw a wire, **Place > Wire** (or select the wire icon or or press W). You can always zoom in by pressing the 'I' key on the keyboard, or 'O' for zoom out.

NOTE

To exit wire mode, press escape (**Esc**) on the keyboard or press W on the keyboard, which toggles wire mode on and off. If you make a mistake you can always select the undo icon .

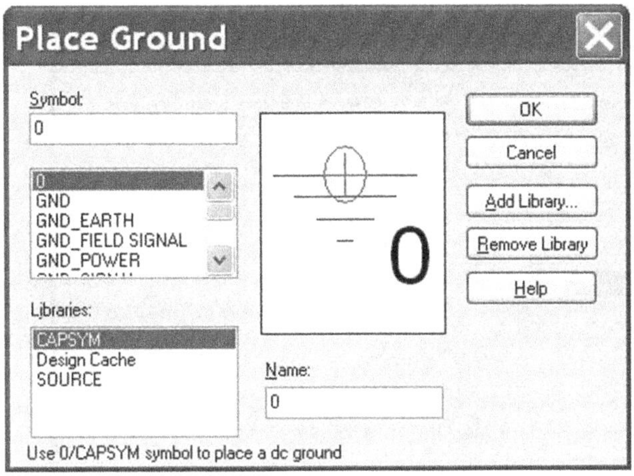

FIGURE 1.27
Placing a 0 V ground symbol.

NOTE

There are new features available from version 16.3 onwards to automatically connect wires to two or more points and a feature to automatically connect wires to a bus, which is described in Chapter 18 on Digital Simulation. To automatically draw a wire, select **Place** > **Auto Wire** > **Two Points** (Figure 1.28) or click on the icon. Click on the first wire point and then click on the second wire point.

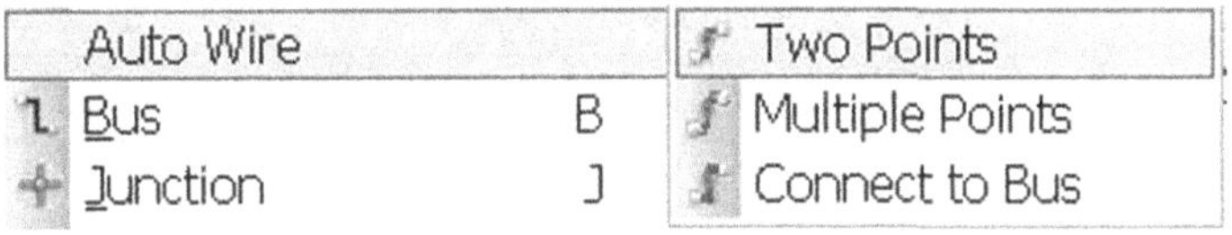

FIGURE 1.28
Auto Wire allows for the automatic connection of wires and busses.

6. Capture automatically labels each wire connection, also known as a node, with a node number, which by default is not displayed on the schematic. However, you can assign your own labels to wire nodes, which will give meaning to a node, i.e. **input** or **output**, and is useful when you want to analyze different nodes in a circuit. These labels are known as net aliases and are placed on a wire by highlighting a wire and then selecting **Place > Net Alias** (or selecting the net alias icon N1 or abc or pressing N).
7. Save the project by selecting **File > Save.**

NOTE

Parts can be pushed together such that when you move the parts away, wires are automatically drawn. Select **Options** > **Preferences** > **Miscellaneous** > **Wire Drag** and check **Allow component move with connectivity changes on** (Figure 1.29). This also allows parts to be moved onto a wire and to be connected.

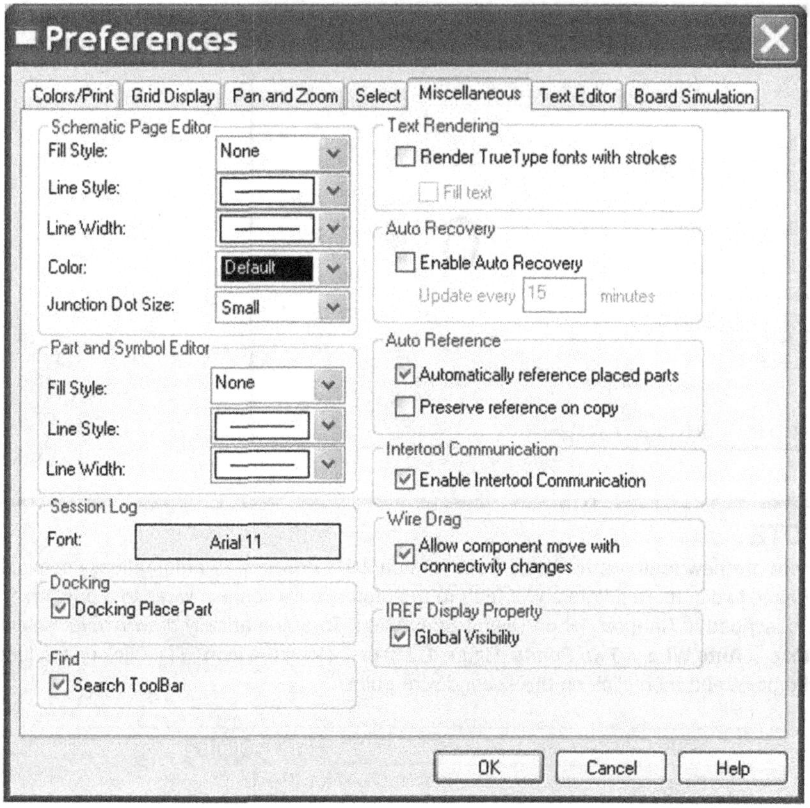

FIGURE 1.29
Enable Wire Drag to automatically connect wires to parts.

1.7 EXTRA LIBRARY WORK

1. Select **Place Part** to open the **Place Part** window and **Add Library** by clicking on the icon (Figure 1.30). In previous versions, just click on **Add Library...** (see Figure 1.10a).

FIGURE 1.30
Adding a library.

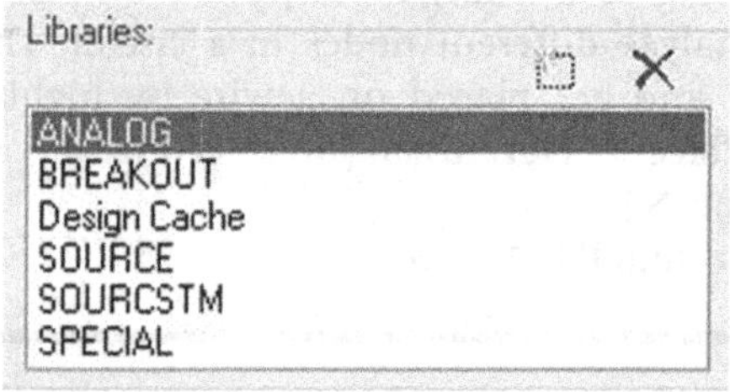

2. The **Browse File** window (Figure 1.31) will open. Make sure you are in **[install path] tools > capture > library > pspice** and select the **ana_swit.olb** library and click on **Open**.

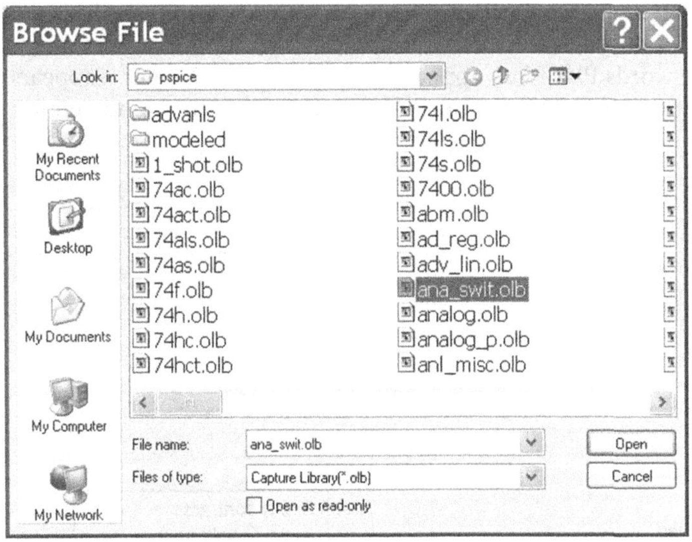

FIGURE 1.31
Browse library files.

3. Close **Place Part**. Has the library been added to the library list in the Project Manager? Expand the **Library** folder.
4. In the Project Manager select the **Library** folder and **rmb > Add File** (Figure 1.32). The **Browse File** window will open as in Step 2. Make sure you are in **[install path] tools > capture > library > pspice** and select the **1_shot.olb** library and click on **Open**.

FIGURE 1.32
Adding a library to the Project Manager.

5. Select **Place Part**. Has the **1_shot.olb** library been added to the list of **Libraries**?
6. In **Place Part** select all the libraries and click on **Remove Library**. Note which libraries are now available.

As you create new projects, any libraries added to previous projects in **Place Part** will be added to the list of available libraries. However, these libraries are not added to the Project Manager. Only libraries added via the **Library** folder in the Project Manager will be added to the configured list of Project libraries and these can only be deleted via the Project Manager.

From version 16.2 onwards, when you select **Place Part**, the menu appears on the right-hand side of the schematic and reduces the available size of the schematic page. However, at the top right of the **Place Part** menu, there is a thumbtack icon (Figure 1.33) which gives you the option to effectively hide the Place Part menu. When you select the thumbtack icon, the Place Part menu disappears and the words Place Part appear (Figure 1.34). The menu reappears when you move the mouse inside the Place Part window box and retracts when you move the mouse back to the left.

FIGURE 1.33
Hiding the Place Part menu.

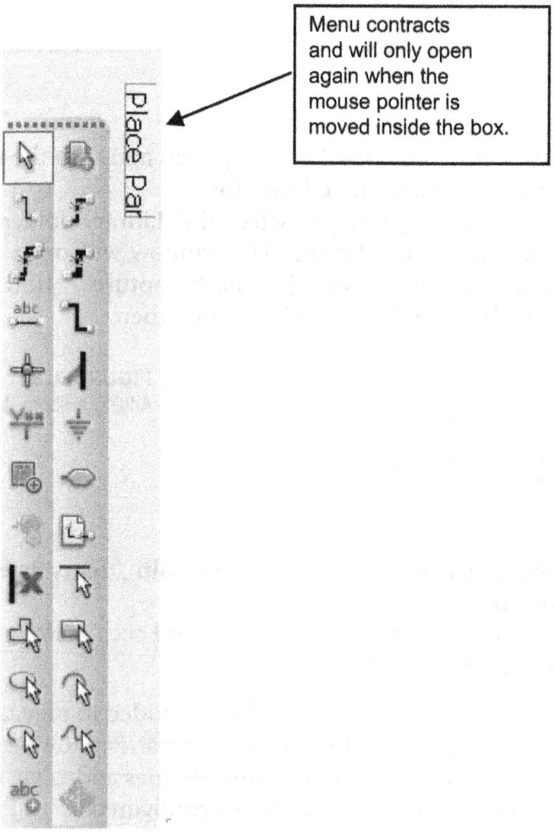

FIGURE 1.34
Place Part menu contracted.

CHAPTER 2

DC Bias Point Analysis

Chapter Outline

When you connect a battery or a power supply to a circuit, the circuit voltages and currents effectively settle down to what is known as a DC steady-state condition. This is also known as the operating point or bias point of a circuit under steady-state conditions. In PSpice, the bias point analysis calculates the node voltages and currents through the devices in the circuit. For example, for a simple common emitter transistor amplifier, the bias point analysis will calculate the base, emitter and collector bias voltages, and the base, collector and emitter quiescent currents.

Bias point analysis will also take into account any voltage sources applied to the circuit and any initial conditions set on devices or nodes in the circuit. For example, you may want to preset a capacitor to a known voltage or set an initial digital state, a logic '1' or '0' on the pins of a digital device.

The calculated bias point voltages and currents are also used as a starting point for the other circuit analysis calculations. For example, when you run a transient (time) or an AC (frequency) analysis, PSpice automatically runs a bias point analysis first. However, the bias point analysis can be turned off for special cases in which a DC steady-state solution cannot be found. This is especially useful in the case of an oscillator which relies on the fact that it has no steady-state condition.

With the bias point enabled, the output file will provide a list of all the analog and digital node voltages, the currents and total power of all voltage sources in

Analog Design and Simulation using OrCAD Capture and PSpice. DOI: 10.1016/B978-0-08-097095-0.00002-7

the circuit and a list of small signal parameters for all devices in the circuit. There is an option to suppress the bias information in the PSpice simulation profile.

NOTE

With a bias point simulation, all capacitors are implemented as open circuit and all inductors are implemented as short circuits in order to calculate the DC Bias Point.

The RC circuit in Figure 2.1 is based upon the resistor circuit in Chapter 1. A capacitor, C1, has been added in parallel with R2. With the circuit drawn, a PSpice simulation profile needs to be set up. The settings are accessed from the

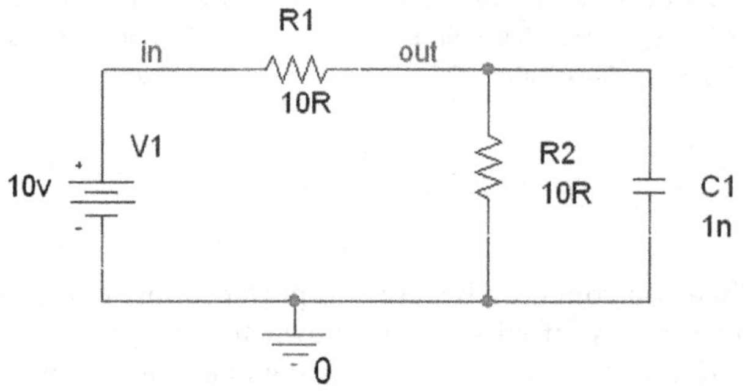

FIGURE 2.1
RC circuit ready for DC bias point analysis.

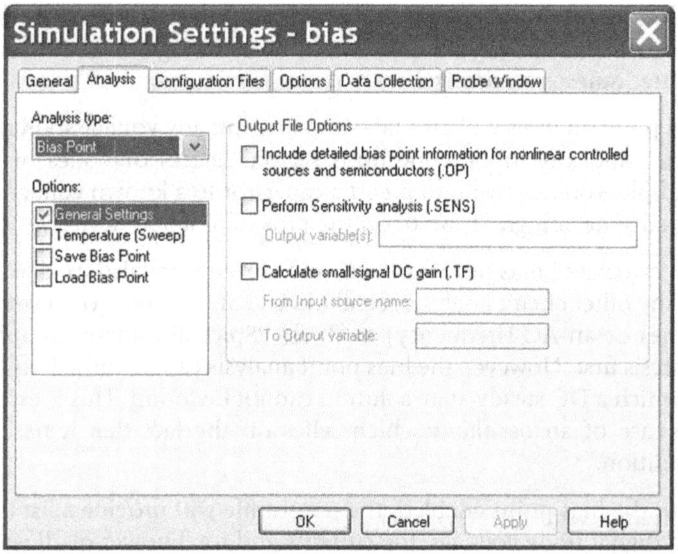

FIGURE 2.2
Simulation Settings windows with Bias Point analysis selected.

top toolbar, **PSpice > New Simulation Profile.** This is where the different analysis types for DC, AC, transient and bias point are selected. By default, **Bias Point** is selected.

Figure 2.2 shows the default PSpice simulation profile for a DC bias point analysis. For this example, the default settings are used. To run the simulation, select **PSpice > Run** or select the play button.

NOTE

In previous versions, when you run a simulation, the **PSpice Netlist Generation** dialog box shown in Figure 2.3 appears.

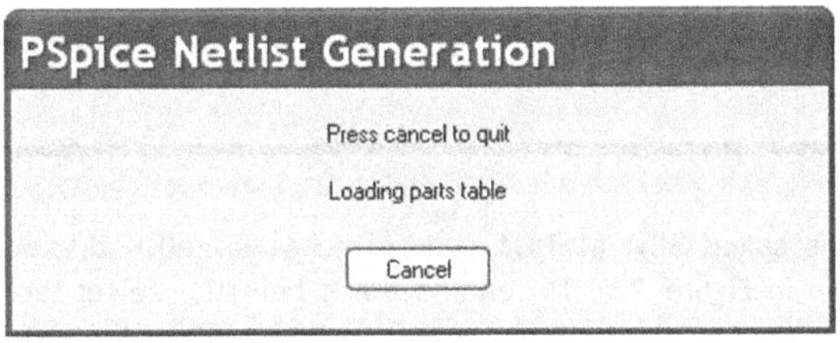

FIGURE 2.3
PSpice netlist generation.

From version 16.3, when you first run a simulation, the **Undo Warning** dialog box shown in Figure 2.4 will appear. This just states that you will not be able to undo or redo any previous actions. Just check the **Do not show this box again** (Figure 2.4) and click on **Yes**.

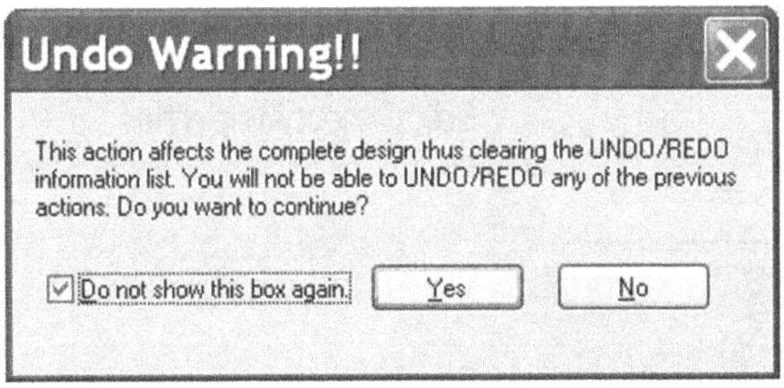

FIGURE 2.4
Undo Warning.

2.1 NETLIST GENERATION

The circuit diagram drawn in Capture is represented as a netlist of all the components and their respective connections to other components. This netlist is automatically generated when you run the simulation and can be seen in the **Outputs** folder in the Project Manager (Figure 2.5).

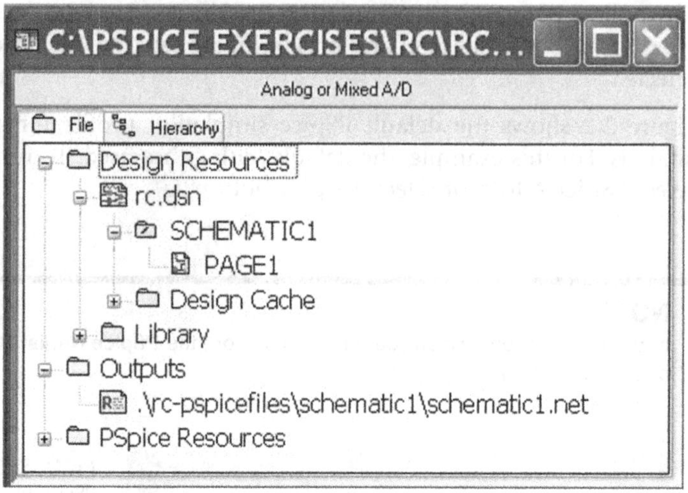

FIGURE 2.5 The Schematic1.net netlist file has been generated in the **Outputs** folder.

NOTE

By default the circuit diagram is called SCHEMATIC1 and so the generated netlist is called schematic1.net, as seen in Figure 2.5. To rename the schematic, select the **SCHEMATIC1** name, **rmb > Rename** as shown in Figure 2.6, entering **RC** as the new schematic name.

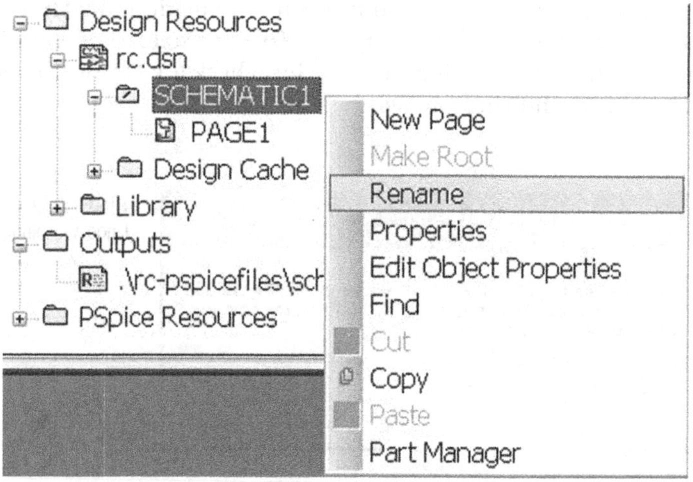

FIGURE 2.6 Renaming SCHEMATIC1 to RC.

When you run the simulation, an **rc.net** netlist will appear in the outputs folder. Figure 2.7 shows the original **schematic1.net** and the newly generated **rc.net**. Note that Capture is not case sensitive. Any schematics named in upper case will generate lower case named netlist files (Figure 2.7).

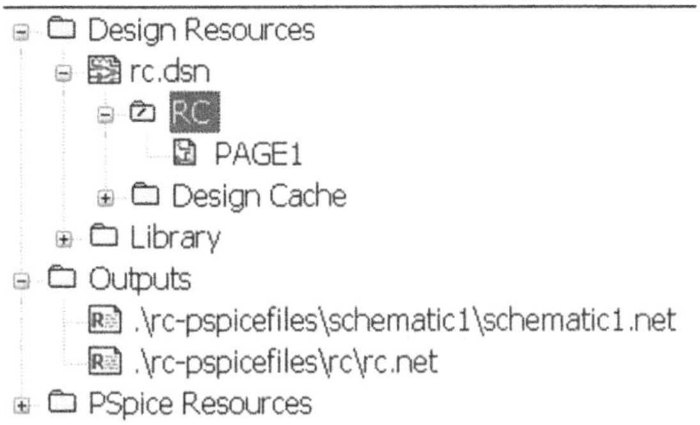

FIGURE 2.7
Renamed RC schematic and generated rc.net netlist.

To open the netlist file, double click on the name and the netlist will appear as shown in Figure 2.8.

```
* source RC
V_V1          IN 0 10V
R_R1          IN OUT  10R TC=0,0
R_R2          0 OUT  10R TC=0,0
C_C1          OUT 0  1n  TC=0,0
```

FIGURE 2.8
resistors.net netlist.

The netlist in Figure 2.8 shows that resistor R1 is connected between nodes, IN and OUT, and has a value of 10 ohms and a zero temperature coefficient (TC=0). The voltage source, V1, is connected between IN and 0 V and has a value of 10 V. The **R_** indicates that this is a flat netlist compared to a hierarchical netlist. Resistor properties and hierarchical circuits will be discussed in more detail in Chapter 20. For now, the netlist defines the net connections between R1, R2 and V1 and any defined net names.

When you run a PSpice simulation, PSpice is launched and the PSpice environment window appears. However, there is no graphical output for a DC bias point analysis as no waveform data are calculated and hence no traces are available to be plotted. Instead, you can look in the **Output File** to see the calculation results or view on the schematic, the bias voltages, currents and instantaneous power for the circuit components. From the top toolbar in Capture or PSpice, select **View > Output File** and scroll to the bottom of the file (Figure 2.9). Because the nets have been named, you can easily see the calculated output voltages and currents. Remember, during a DC bias point analysis, capacitors are open circuit; hence the RC circuit is behaving as a simple potential divider with R1 and R2 dividing the 10 V DC voltage down to 5 V on the '**out**' net.

FIGURE 2.9
Bias Point analysis Output file.

```
**** INCLUDING RC.net ****
* source RC
V_V1         IN 0 10V
R_R1         IN OUT  10R TC=0,0
R_R2         0 OUT  10R TC=0,0
C_C1         OUT 0  1n  TC=0,0

**** RESUMING bias.cir ****
.END

**** 05/15/11 11:03:59 ******* PSpice 16.3.0 (June 2009) ****** ID# 0 ********

 ** Profile: "RC-bias"  [ C:\PSPICE EXERCISES\RC\RC-PSpiceFiles\RC\bias.sim ]

 ****     SMALL SIGNAL BIAS SOLUTION       TEMPERATURE =   27.000 DEG C

******************************************************************************

 NODE   VOLTAGE     NODE   VOLTAGE     NODE   VOLTAGE     NODE   VOLTAGE

(   IN)   10.0000  (  OUT)    5.0000


    VOLTAGE SOURCE CURRENTS
    NAME         CURRENT

    V_V1         -5.000E-01

    TOTAL POWER DISSIPATION   5.00E+00  WATTS
```

NOTE

Capture is not case sensitive when naming nets.

A new feature (in release 16.3), as seen in Figure 2.10, is the use of colors in the output file to highlight the different syntax such as text, component values, comments, expressions and keywords. The default colors can be changed by selecting **Options > Preferences > Text Editor**, as shown in Figure 2.10.

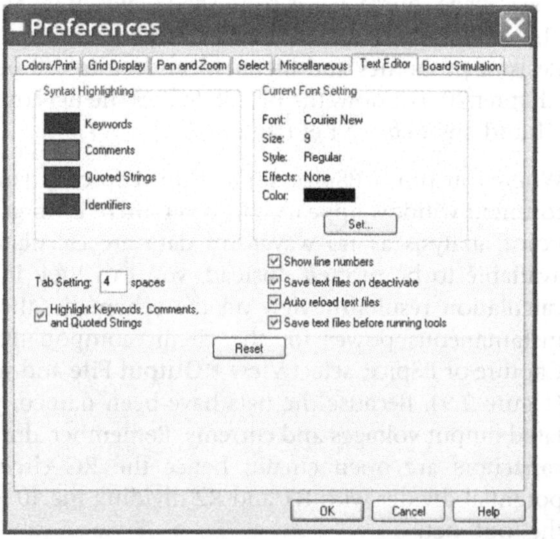

FIGURE 2.10
Text Editor default colors and fonts.

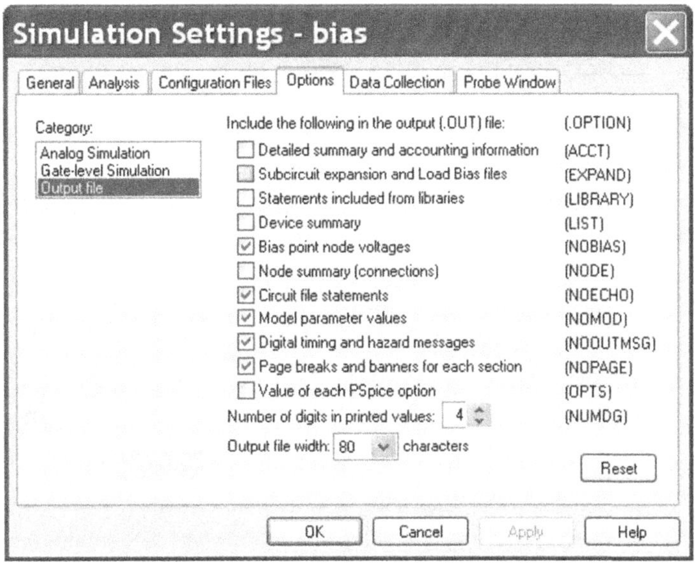

FIGURE 2.11
Output file options.

The **Output File** also highlights any errors and warnings that may have occurred and is useful to determine where the errors occur. See the exercises at the end of the chapter for an example.

As mentioned previously, the output file provides a report on the analog and digital node voltages and device currents together with a list of small signal parameters for the devices in the circuit. There is the option to suppress the bias information reported in the output file. Under the **Options** tab in the PSpice Simulation profile (Simulation Settings), select **Output** file in the **Category:** box as shown in Figure 2.11 and uncheck the **(NOBIAS)** option. There is also the option to list all the devices **(LIST)** in the circuit, summarizing their connecting nodes, values, models and other parameters. Figure 2.11 shows the other available options available for the output file. The **Reset** button resets all the options back to their default settings.

2.2 DISPLAYING BIAS POINTS

After a simulation is run, the bias voltage, current and power values can be displayed on the schematic. In Capture, select **PSpice > Bias Points > Enable** or select the bias display icons, which have changed in appearance in release 16.3. See Figures 2.12a and 2.12b.

Figure 2.13 shows the displayed bias voltages, currents and instantaneous power for the resistor RC circuit.

The number of significant digits displayed for bias points can be changed by selecting **PSpice > Bias Points > Preferences**, as shown in Figure 2.14. Up to 10 precision digits can now be displayed.

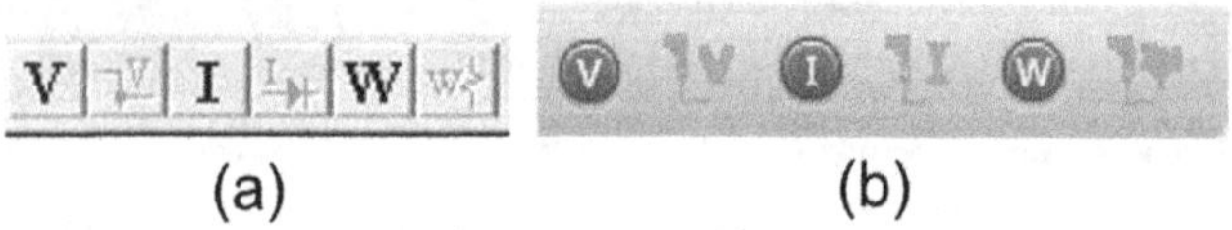

FIGURE 2.12
(a) Bias display icons pre-16.3; (b) bias display icons in 16.3.

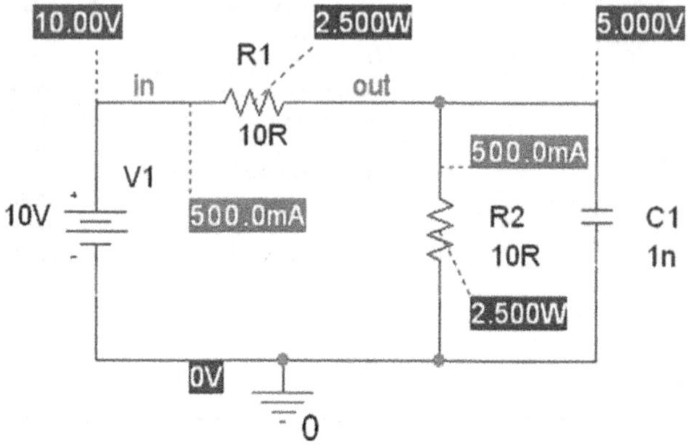

FIGURE 2.13
Displayed bias point voltages, currents and power.

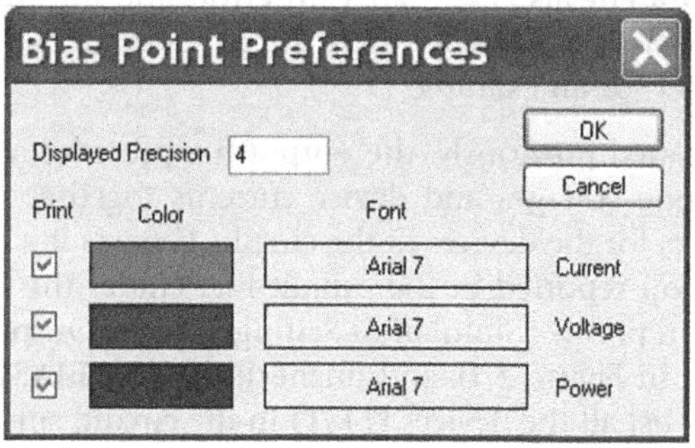

FIGURE 2.14
Changing the bias point display preferences.

Individual bias values can be turned on and off for voltage, current or power. For example, if you select a wire net, the voltage icon will be activated, which will enable you to toggle the value for bias voltage on or off for a selected wire.

If you select a component pin, the current icon will be activated, which will enable you to toggle the value for bias current on or off for a selected part.

If you select a component, the instantaneous icon will be activated, which will enable you to toggle the value for bias power on or off for a selected part.

TIP

You may need to press F5 to refresh the display after you turn bias display on and off.

2.3 SAVE BIAS POINT

You can save and reuse the bias point data from a simulation, which is useful if you have to run a number of simulations on a large circuit that has a long simulation runtime. This is assuming that the circuit netlist, i.e. the connectivity of the components, has not changed. Remember that other analyses use the calculated results from a bias point analysis, so when you resimulate the circuit, assuming that the netlist has not changed, the initial bias point calculations can be saved and reused, thus reducing the simulation runtime. Saving the bias point is also useful when a simulation fails to converge to a solution.

In the **Simulation Profile Settings** select **Bias Point** analysis and then select **Save Bias Point**. As you can see in Figure 2.15, the data from the bias point analysis are saved in a file called saved_bias_point.txt. You select the **Browse...** button to select or create the folder in which to save the file.

The bias point saved data contain node voltages and digital states for all the devices in the circuit, the total power and current supplied by any voltage sources, and a list of model parameters for the devices in the circuit.

NOTE

When saving bias point data, it is a good idea to add a .txt extension to the bias point data filename so you can readily open the file in a text editor such as WordPad or Notepad.

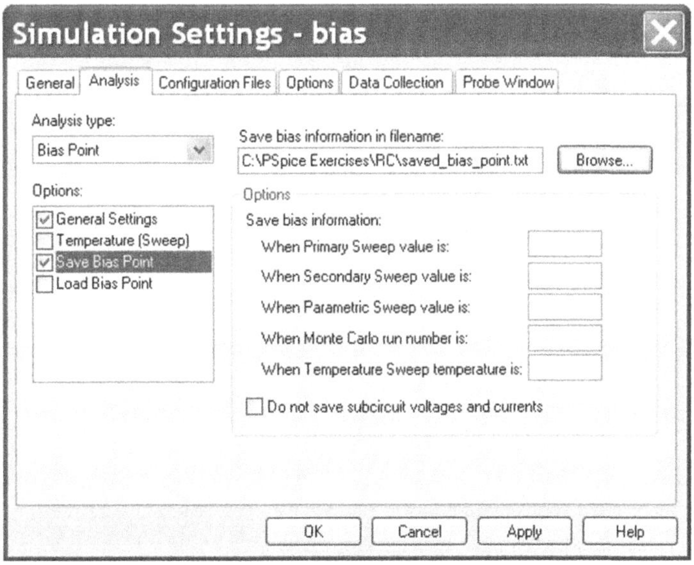

FIGURE 2.15
Bias point settings.

2.4 LOAD BIAS POINT

Saved bias point analysis data are loaded by selecting the **Load Bias Point** option in the simulation profile. Figure 2.16 shows a previously saved bias point data file being selected. Bias point information can also be saved and used for a DC sweep and a transient analysis.

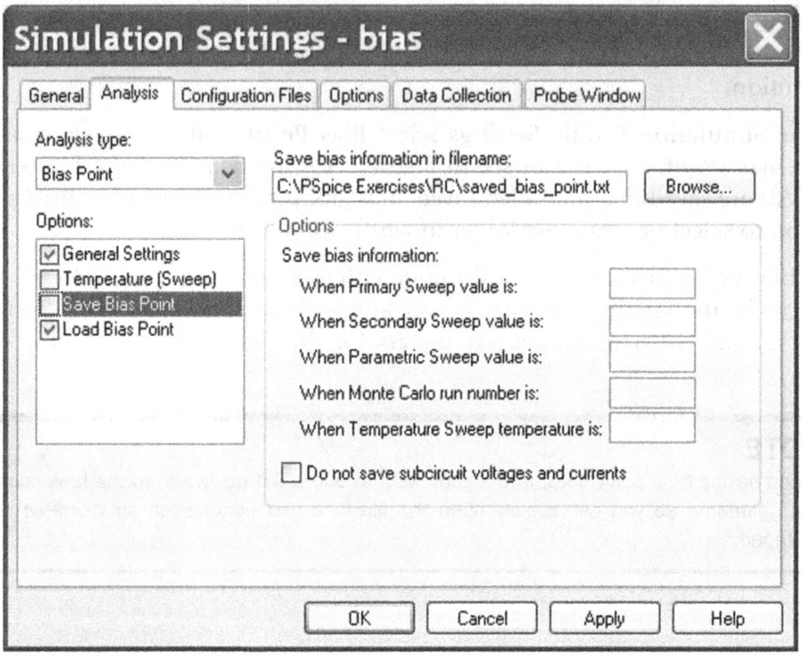

FIGURE 2.16
Loading saved bias point data.

2.5 EXERCISES

Exercise 1

1. The circuit in Figure 2.17 is based upon the resistor circuit in Chapter 1. Add a 1n capacitor, from the **analog** library, in parallel with R2.

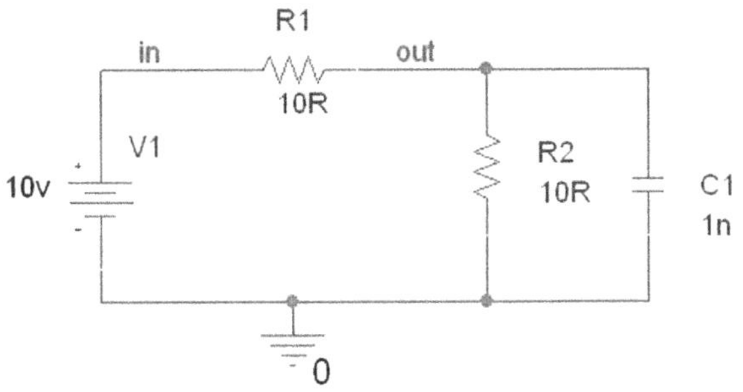

FIGURE 2.17
RC circuit.

2. Delete the 0 V volt symbol and resimulate. A warning message will appear (Figure 2.18) asking you to check the session log, which is normally open at the bottom of the screen in Capture.

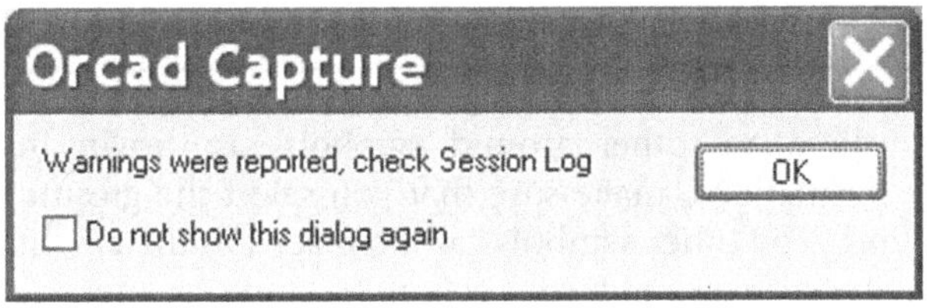

FIGURE 2.18
Warning message.

You may have to expand the window upwards to see the complete message. If the session log is not visible, it can be found from the top toolbar, **Window > Session Log** and should contain the following message:

```
WARNING [NET0129] Your design does not contain a Ground
(0) net.
```

Your reported net number shown will be different to the one above. The PSpice window will then open and display the output file, which will report that a number of numbered nodes are floating (Figure 2.19).

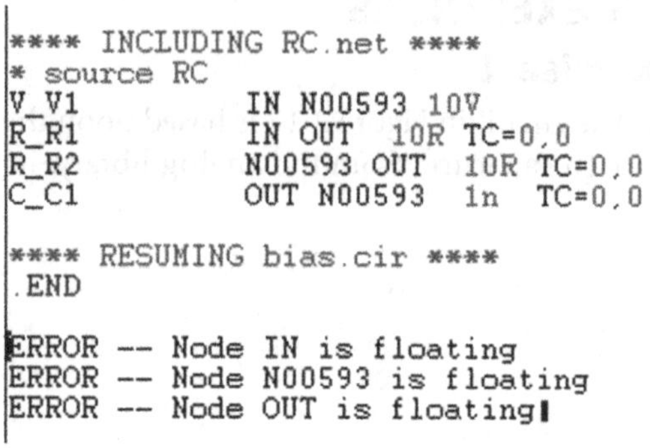

```
**** INCLUDING RC.net ****
* source RC
V_V1          IN N00593 10V
R_R1          IN OUT   10R TC=0,0
R_R2          N00593 OUT  10R TC=0,0
C_C1          OUT N00593  1n   TC=0,0

**** RESUMING bias.cir ****
.END

ERROR -- Node IN is floating
ERROR -- Node N00593 is floating
ERROR -- Node OUT is floating
```

FIGURE 2.19
Output file reporting floating node errors.

3. In the RC circuit, reconnect the ground symbol and resimulate. There should be no errors.

FIGURE 2.20
Ground symbols.

NOTE

PSpice automatically assign node numbers to wires in your circuit unless you assign a net name to them. In the above output file, every node is floating because node 0 has not been assigned. In PSpice as with other Spice simulation software tools, there must be a 0 V node in the circuit, otherwise nodes will be reported as floating in the output file.

In the **capsym** library there are other ground symbols, as shown in Figure 2.20. For PSpice simulations, make sure that you select the ground symbol with the **0** showing. The other symbols can be placed in the circuit to show the difference between ground connections as long as there is a 0 V node in the circuit.

4. Create a bias point simulation profile based upon the initial bias point: **PSpice > New Simulation Profile** or click on the icon and enter **bias** for the simulation **Name** (Figure 2.21) and click on **Create**.

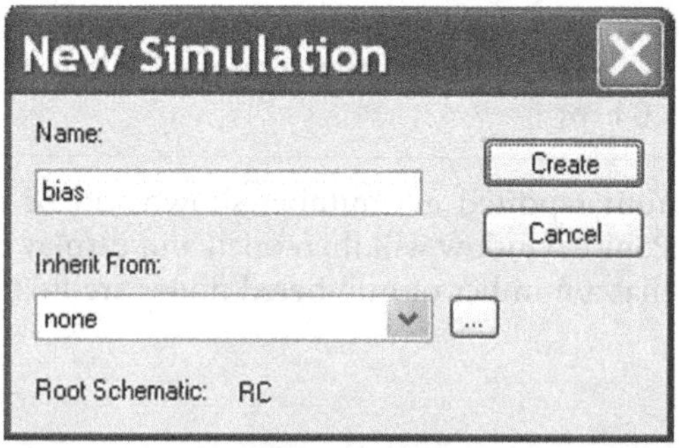

FIGURE 2.21
Creating a bias point simulation profile.

5. The dialog box in Figure 2.22 will appear telling you that a simulation profile of the same name already exists. New projects already include a default bias point simulation profile called **Bias**.

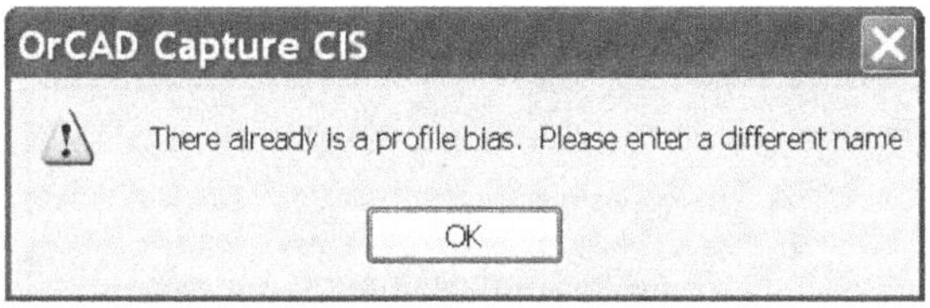

FIGURE 2.22
A simulation profile where the same name already exists.

Click on OK and Capture will automatically present you with a new name of **bias1** (Figure 2.23). Click on **Create**.

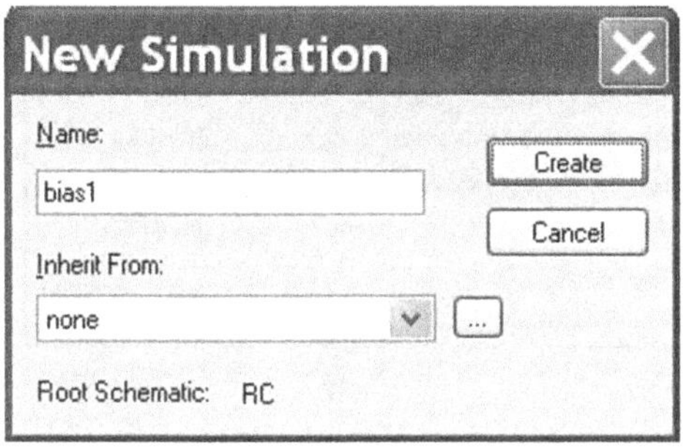

FIGURE 2.23
The simulation profile name is automatically incremented to bias1.

6. The simulation settings window will appear (Figure 2.24). If not already set by default, select the **Analysis type** to **Bias Point**. Click on **Apply** but do not exit.

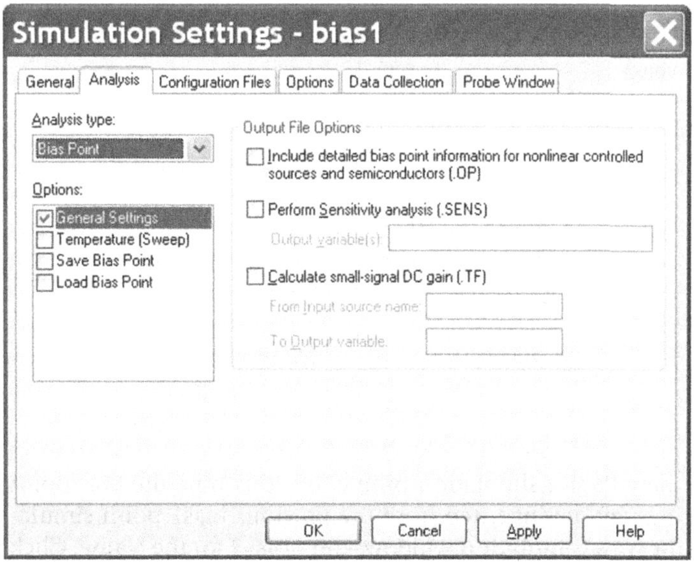

FIGURE 2.24
Setting the bias point analysis.

7. Under the **Options** tab, select **Output File** in the **Category:** box and uncheck (NOBIAS) and check (LIST) (Figure 2.25). Click on OK.

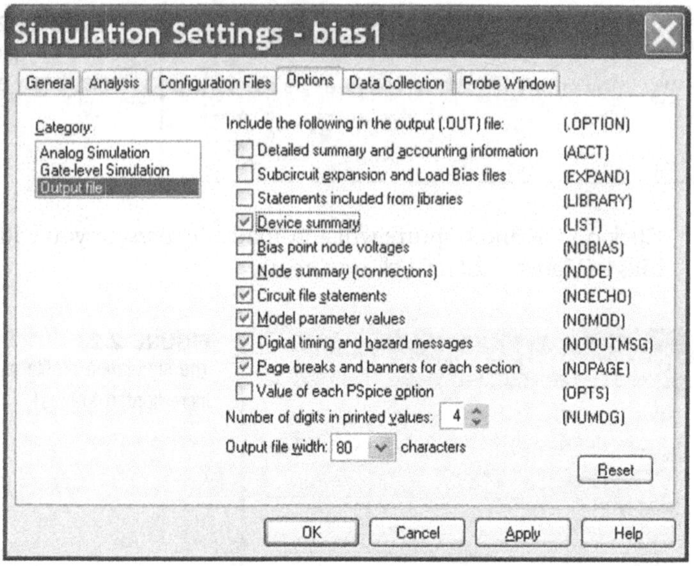

FIGURE 2.25
Output file options.

8. Run the simulation. When PSpice launches, select **View > Output** file. The output file shows a summary of the resistors, capacitor and voltage source used in the circuit, as shown in Figure 2.26. The output voltages and currents are not reported in the output file.

FIGURE 2.26
Summary list of circuit devices.

```
**** RESISTORS

NAME              NODES        MODEL      VALUE        TC1        TC2        TCE

R_R1           IN      OUT               1.00E+01
R_R2           0       OUT               1.00E+01

**** CAPACITORS

NAME              NODES        MODEL      VALUE    In. Cond.      TC1        TC2

C_C1          OUT         0              1.00E-09

**** INDEPENDENT SOURCES

NAME              NODES      DC VALUE  AC VALUE  AC PHASE

V_V1           IN         0  1.00E+01  0.00E+00  0.00E+00  degrees
```

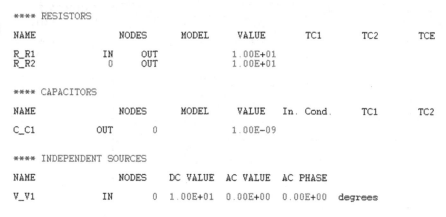

9. Create another new PSpice simulation profile, but this time the bias point simulation settings will be inherited from the previous bias1 point simulation profile. In the **New Simulation** window, enter **bias2** for the **Name**. Click

on the pull down **Inherit From** menu as shown in Figure 2.27, select bias1 and click on **Create**. In the simulation settings, you will see that Bias Point analysis is selected. Select **Options > Output file** and you will see that LIST is checked and (NOBIAS) is unchecked. Close on OK to close the new simulation profile.
You now have created three bias point simulation profiles: bias, bias1 and bias2.

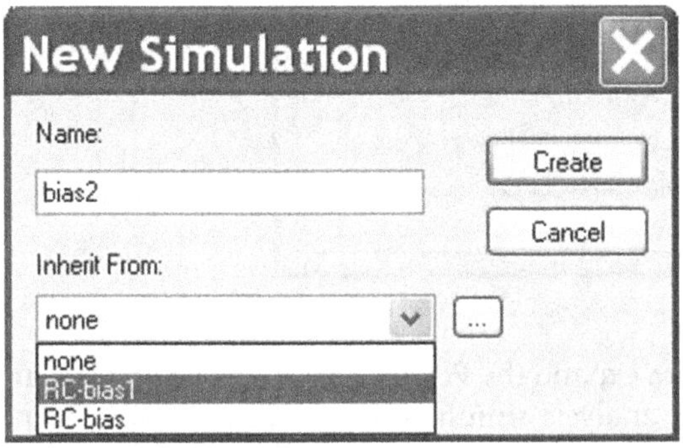

FIGURE 2.27
Inheriting an existing simulation profile settings.

10. If it is not already displayed, open the **Project Manager** window by either selecting **Window > <project path>** \RC.opj file (Figure 2.28) or clicking on the icon or .

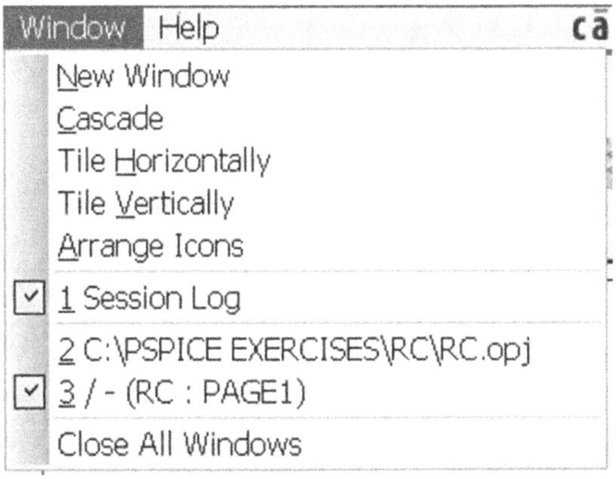

FIGURE 2.28
Selecting the Project Manager.

The Project Manager will be displayed as shown in Figure 2.29.

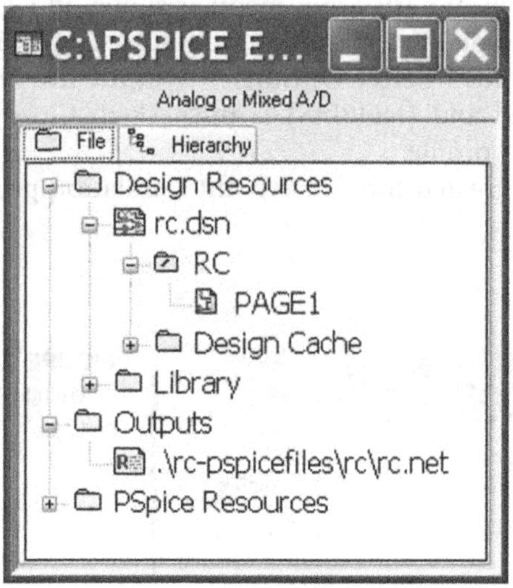

FIGURE 2.29
Project Manager window.

11. In the **Project Manager** window, expand the **PSpice Resources > Simulation Profiles**. The three bias point analysis simulation profiles that have been created are listed in the Project Manager (Figure 2.30a) and are also displayed on the left-hand side of the top toolbar (Figure 2.30b) in which a pull-down menu lists the simulation profiles, bias, bias1 and bias2.

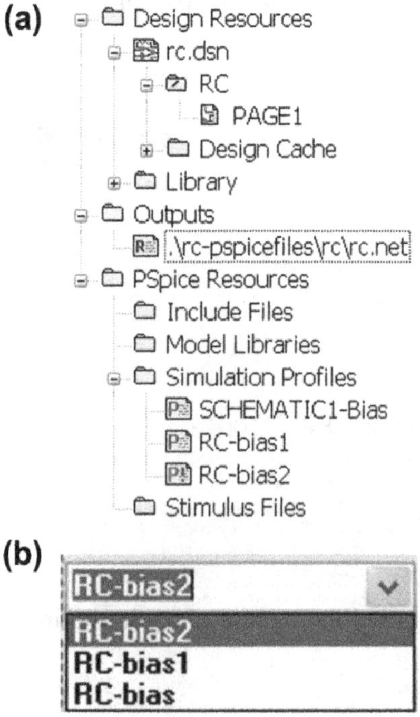

FIGURE 2.30
(a) Bias profiles listed in Project Manager; (b) list of bias profiles.

Note that bias2 is selected, which appears in the Project Manager with a red icon to its left. This indicates that this is the current or active profile. You can select another profile in the pull-down list or alternatively select the profile in the Project Manager and **rmb > Make Active**.

12. Select bias1 in the pull-down profile list in Figure 2.30b and note the change in the Project Manager Simulation Profiles section. Bias1 is now the active profile.

NOTE

This is a useful way to switch between different profiles of different simulation settings or different analysis types.

CHAPTER 3
DC Analysis

Chapter Outline

The DC analysis calculates the circuit's bias point over a range of values when sweeping a voltage or current source, temperature, a global parameter or a model parameter. The swept value can increase in a linear or a logarithmic range or can be a list of increasing values.

This is useful for example if you want to see the circuit response for a change in the supply voltage or to see how a change in a resistor value affects the circuit response. The DC sweep also allows for nested sweeps such that one of two variables is kept constant while sweeping the other variable. For example, the characteristic transistor I_C–V_{CE} curve plots the collector current against the collector–emitter voltage for fixed values of base current. The DC sweep will then contain two variables, the collector–emitter voltage V_{CE} and the base current I_B. The base current is the secondary sweep while the V_{CE} is the primary sweep. The collector current I_C is recorded by successive voltage sweeps of the collector–emitter voltage V_{CE} for stepped values of base current producing a series of curves as shown in Figure 3.1.

3.1 DC VOLTAGE SWEEP

As with other analysis types, a PSpice simulation profile needs to be created. For a DC sweep analysis, select **PSpice > New Simulation Profile** and select **DC Sweep** for the Analysis type. Make sure the **Sweep Variable** is set to **Voltage source**. The **Name** is the reference designator for the voltage source, which in this case is **V1**. The sweep type is set to **Linear**, starting from 0 V to 10 V in steps of 1 V. You can also enter a list of voltages in the **Value list**, for example 1 2 4 5 99 100, as long as the voltages are increasing in value.

Analog Design and Simulation using OrCAD Capture and PSpice. DOI: 10.1016/B978-0-08-097095-0.00003-9

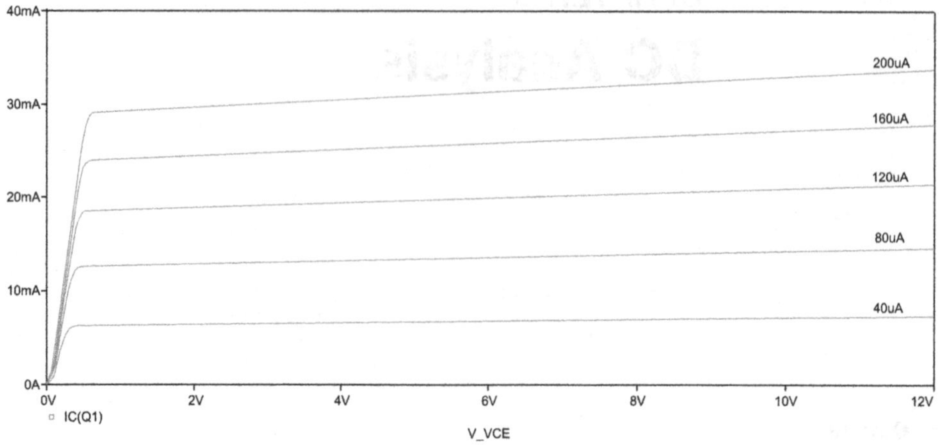

FIGURE 3.1
Nested sweep for a transistor characteristic curve.

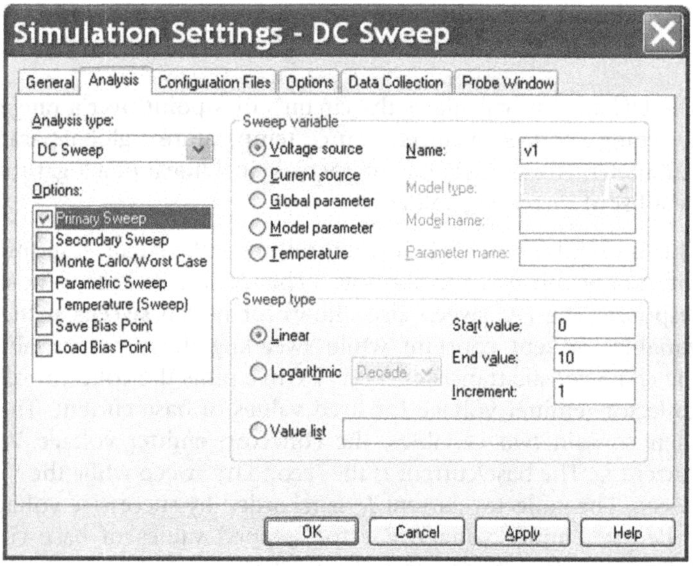

FIGURE 3.2
DC sweep simulation settings.

Figure 3.2 shows the simulation profile for a DC linear sweep of V1 from 0 to 10 V in steps of 1 V.

NOTE

You can rename the reference designator for the voltage source to anything you like as long as the first character is a V. For example, the voltage source could be named Vsupply. This also applies to other components such as resistors, especially when you want to define, for example, a load resistor, RL.

3.2 MARKERS

Markers are used to record the voltages on nodes or currents through components and are accessed from the PSpice menu. They enable data to be automatically displayed as a waveform in the PSpice waveform viewer, which is known as the **Probe** window. To add markers, select **PSpice > Markers** and then you have a choice of placing current, voltage or differential voltage markers as shown in Figure 3.3. The Advanced markers are primarily used with an AC analysis and will be covered in Chapter 4.

Figure 3.4a shows the icons for the markers in version 16.2 and Figure 3.4b the markers in version 16.3.

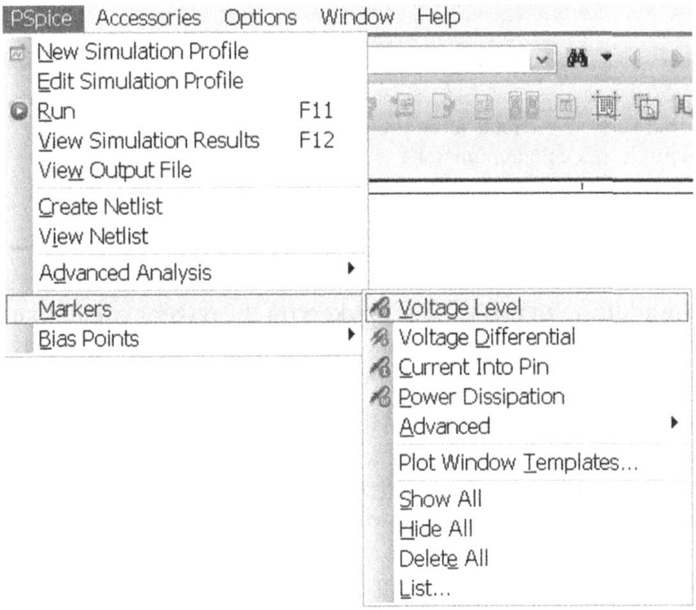

FIGURE 3.3
PSpice markers.

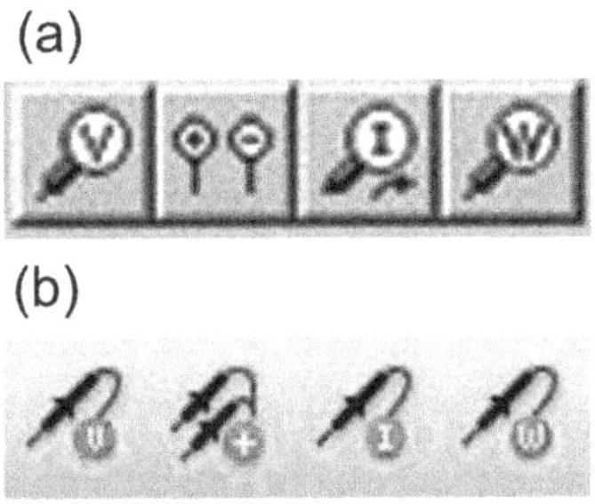

FIGURE 3.4
PSpice marker icons: (a) version 16.2; (b) version 16.3.

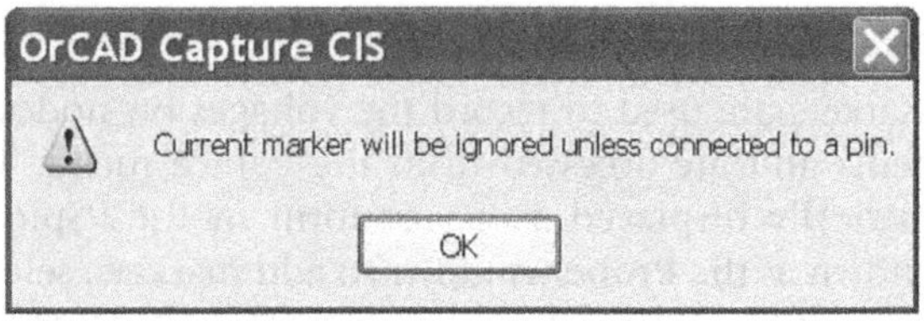

FIGURE 3.5
Warning message if you try to place a current marker on a wire.

Voltage markers are placed on wires, whereas current markers must be placed on a component pin. A message (Figure 3.5) will appear if you try to place a current marker on a wire instead of a component pin. The component pin is a different color to a wire, and in version 16.3, component pins are noticeably thinner than the connecting wires. For power markers, the markers are placed on the body of the device.

NOTE

You can only add markers to a circuit after you have set up a PSpice Simulation Profile and not before. It is a common mistake to add markers to a circuit and then set up a new simulation profile and have the markers disappear when the profile is set up.

Figure 3.6 shows the addition of two voltage markers on the resistor circuit that will record the voltage at nodes '**in**' and '**out**', respectively, and automatically display their voltage traces in the PSpice Probe waveform window.

When the simulation is run, PSpice will launch and **Probe** will plot the two voltage traces at nodes **in** and **out**. The x-axis is the swept voltage $V(\text{in})$ and the y-axis is the resultant voltage $V(\text{out})$. Note that the respective traces have the same color as the markers placed in the circuit. See Figures 3.6 and 3.7.

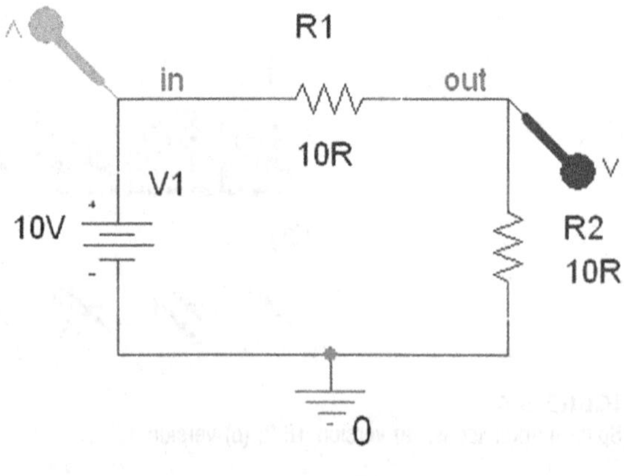

FIGURE 3.6
Addition of two voltage markers.

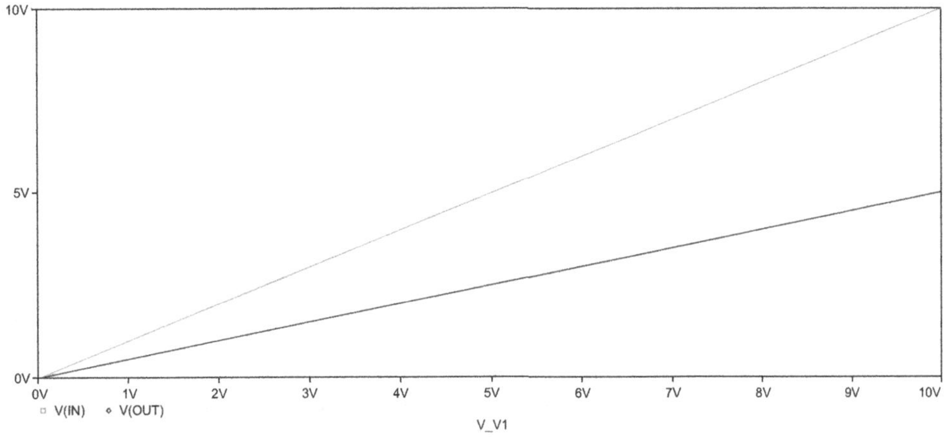

FIGURE 3.7
Probe Waveform viewer displaying the voltages at nodes, **in** and **out**.

NOTE

When you first place markers, their color is initially gray. After a simulation is run, the markers will change color and the respective waveform in **Probe** will reflect the marker's color. If you delete a trace name in Probe, the marker on the schematic will turn gray. By double clicking on the marker, the marker color will be restored and the trace name will reappear in Probe.

In release 16.3, you can change the Probe background color, which by default is black. From the top toolbar in PSpice, select **Tools > Options > Color Settings**.

The waveform trace color in Probe can be changed by highlighting the trace and **rmb > Properties**, for release 16.2 (Figure 3.8) or **rmb > Trace Property** in release 16.3. Both actions will similarly display the **Trace Properties** box shown in Figure 3.9.

In the **Trace Properties** box (Figure 3.9), the trace color, pattern, width and symbol can be changed.

FIGURE 3.8
Trace options for 16.2.

FIGURE 3.9
Changing trace properties.

NOTE

From version 16.3, you have the functionality to **rmb** on a trace to open a context-sensitive menu which displays all the related trace options as shown in Figure 3.10.

FIGURE 3.10
Trace menu options.

From version 16.3 you can change the foreground colors such as the axis and grid lines and also the background color for Probe. Select **Tools > Options > Color Settings** as shown in Figure 3.11.

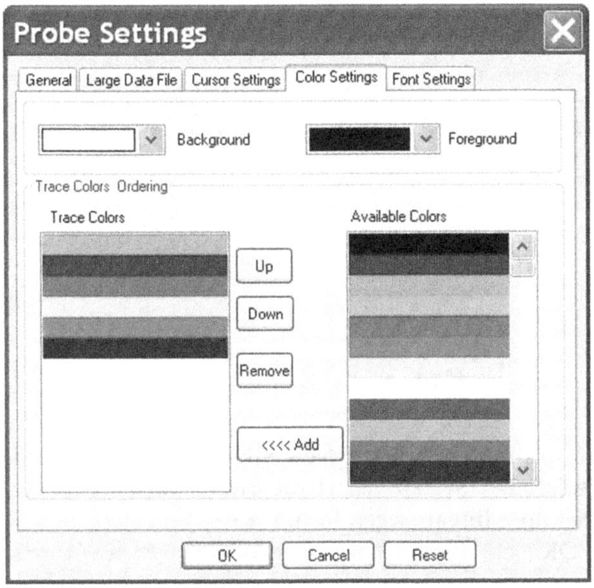

FIGURE 3.11
Changing colors in Probe.

3.3 EXERCISES

Exercise 1

The resistor network circuit in Figure 3.12 is a potential divider where the ratio of the output voltage to the input voltage is given by:

$$\frac{V_{out}}{V_{in}} = \frac{R1}{R1 + R2} \tag{3.1}$$

which can be written as:

$$V_{out} = V_{in}\frac{R1}{R1 + R2} \tag{3.2}$$

Therefore, the output voltage is determined by the ratio of the fixed values of the resistors R1 and R2. If R1 = R2, the output voltage is equal to half of the input voltage.

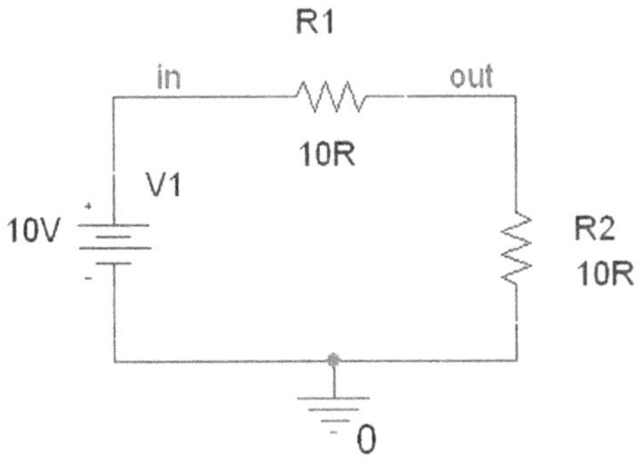

FIGURE 3.12
Resistor network.

1. Draw the resistor circuit in Figure 3.12 and name the nodes as shown by selecting **Place > Net Alias** (Figure 3.13).

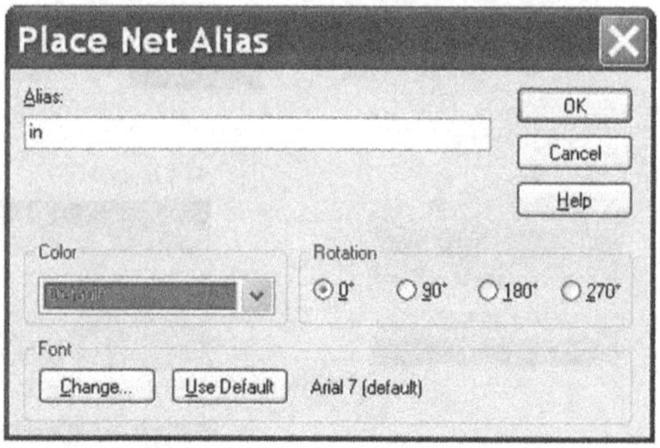

FIGURE 3.13
Placing Net Alias.

2. Create a PSpice simulation profile, **PSpice > New Simulation Profile** and for the **Analysis type**, select **DC Sweep**. Select the **Sweep variable** as the voltage source V1 and set up a **linear** sweep from 0 to 10 V in steps of 1 V (Figure 3.14). Click on OK.

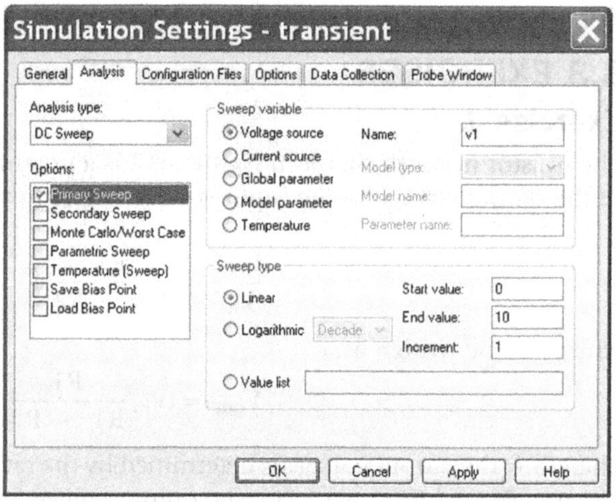

FIGURE 3.14
DC sweep simulation setting.

3. Place voltage markers on nodes **in** and **out** (Figure 3.15).

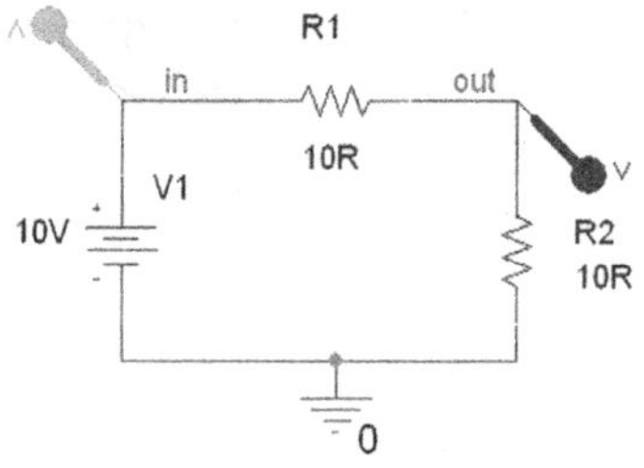

FIGURE 3.15
Placing voltage markers on the nodes.

4. Run the simulation ▶.
5. PSpice will launch and Probe will display the two traces for *V*(in) and *V*(out). Figure 3.16 shows that the trace colors are the same as the marker colors in the circuit diagram in Figure 3.15. Note that the circuit is acting as a potential divider in that the output voltage is half of the input voltage.

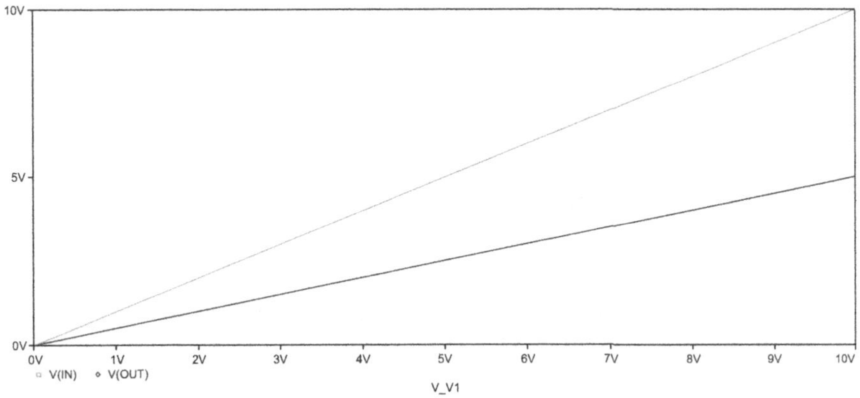

FIGURE 3.16
Output voltage response of resistor circuit for a DC sweep of V_{in}.

6. Delete the ground symbol and resimulate. You should see the warning message dialog box (Figure 3.17) and a message will be displayed asking you to check the Session Log.

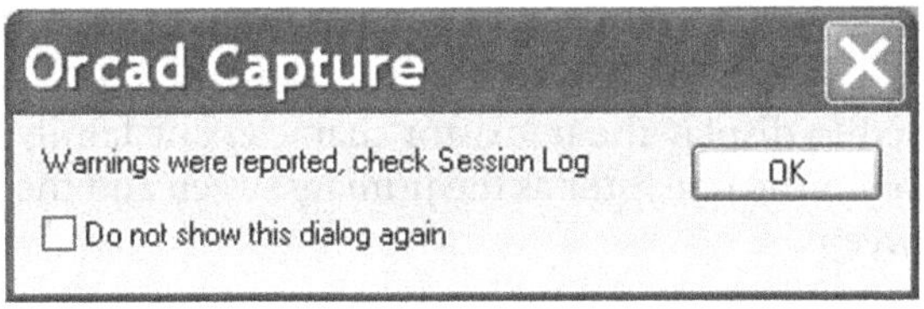

FIGURE 3.17
Warning message.

7. The Session Log is normally open at the bottom of the screen; if not, it can be found from the top toolbar, **Window > Session Log**. The warning message will read:

```
WARNING [NET0129]Your design does not contain a Ground
(0) net.
```

8. Click on OK in the Warning message (Figure 3.15) and PSpice will launch. If the Output file is not displayed, then select **View > Output File**. The output file is shown in Figure 3.18. Your circuit will have different node numbers. The removal of a 0 V symbol was covered in the exercises at the end of Chapter 2 and is mentioned here again as a reminder that in order for the analog voltages to be calculated, a 0 V node must exist in the circuit otherwise nodes will be reported as floating.

FIGURE 3.18
Output file showing floating node error messages.

```
* source RC SWEEP
V_V1           IN N00555 10V
R_R1           IN OUT   10R TC=0,0
R_R2           N00555 OUT   10R TC=0,0

**** RESUMING "DC Sweep.cir" ****
.END

ERROR -- Node IN is floating
ERROR -- Node N00555 is floating
ERROR -- Node OUT is floating
```

9. Reconnect a 0 V symbol. **Place > Ground** or press G on the keyboard and select a 0 V symbol.
10. Remove resistor R2 from the circuit and simulate.
11. PSpice will start and display the output file with an error message:

```
ERROR -- Less than 2 connections at node out
```

12. The error message relates to no DC path to ground at node **out**, i.e. the end of the resistor R2 is floating. One other requirement in PSpice is that every node in a circuit must have a DC path to ground. If you need to simulate a node as open circuit, then you can always connect a large value resistor, e.g. 100 GΩ or 1 TΩ, from a node to ground, which will provide a DC path to ground without adversely affecting the bias conditions of the circuit. Similarly, a short circuit can be implemented by a very small resistance of say 1 μΩ or less.

Exercise 2

A nested sweep will be performed to display the transistor characteristic family of curves. You will set up the VCE for the transistor as the primary sweep and the base current as the secondary sweep.

FIGURE 3.19
Transistor circuit.

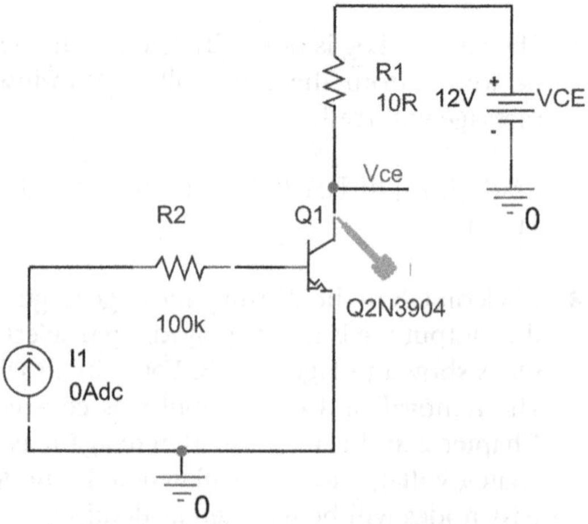

1. Draw the circuit in Figure 3.19. The transistor can be found in the **bipolar** library. In the **Place Part** menu, select the **Add Library** icon as shown in Figure 3.20 or in previous versions, click on **Add Library**. If you are using the demo version, the Q2N3904 can be found in the **eval** library. For the full version, the transistor can be found in the **bipolar** library.

FIGURE 3.20
Add Library.

This will open up the **Browse File** window shown in Figure 3.21. Scroll along or, alternatively, type in bipolar.olb in the **File name** field. Select bipolar.olb and click on **Open**.

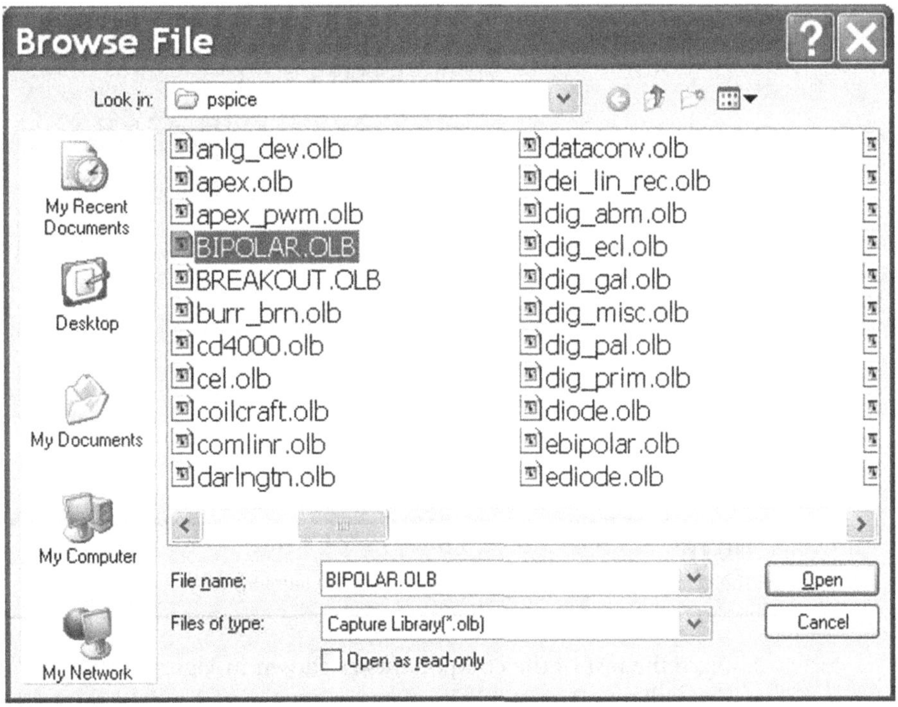

FIGURE 3.21
PSpice-Capture libraries.

2. The bipolar library will now be added to the list of libraries in the **Place Part** menu. Select the bipolar library and type in Q2N3904 (not case sensitive) in the Part box (Figure 3.22). Double click on the q2n3904 transistor and place in the schematic page.

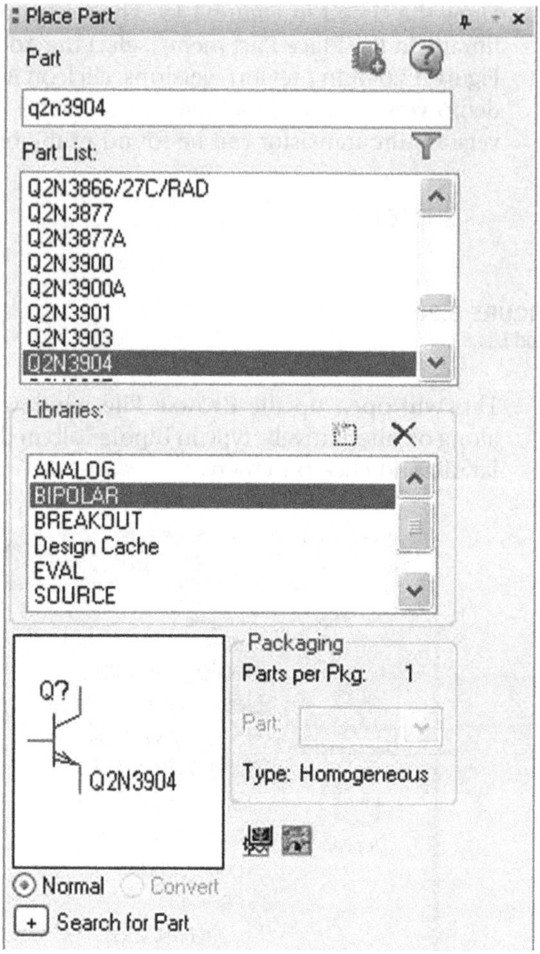

FIGURE 3.22
Q2N3904 selected from the bipolar library.

NOTE
The Q2N3904 is available in the OrCAD demo version.

3. Place the rest of the components as shown in Figure 3.19.
4. You will set up a nested DC sweep where the VCE will be the primary sweep and the base current the secondary sweep.
5. Create a simulation profile, **PSpice > New Simulation Profile**, and select the **Analysis type** to **DC Sweep**. For the **Primary Sweep**, which is shown by default, select the **Sweep variable** to be a voltage source and name the source **Vce**. The **Sweep type** is **Linear** with a **Start value** of 0 V, an **End value** of 12 V and an **Increment** of 0.1 V (Figure 3.23). Click on **Apply** but do **not** exit the Simulation Profile.

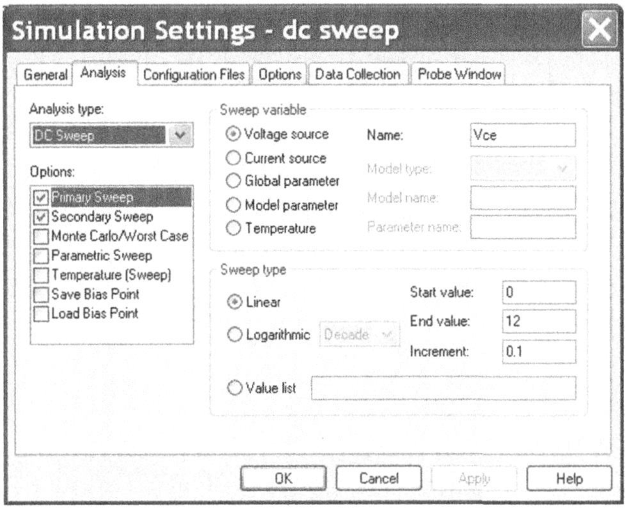

FIGURE 3.23
Primary DC sweep settings.

6. In the **Options** box, select the **Secondary Sweep**. The **Sweep variable** is the current source I1. The **Sweep type** is **Linear** with a **Start value** of 40 μA, an **End value** of 200 μA and an **Increment** of 40 μA. Make sure the **Secondary Sweep** box is checked and click on OK (Figure 3.24).

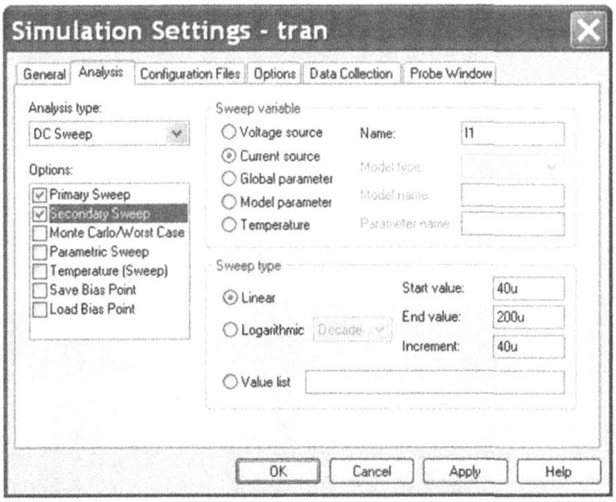

FIGURE 3.24
Secondary DC sweep settings.

7. Place a current marker on the collector pin of the transistor and simulate.
8. You should see the characteristic curves shown in Figure 3.25.

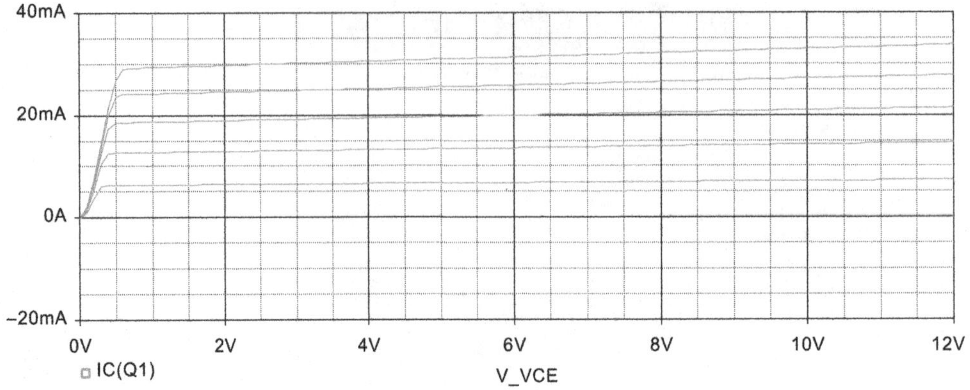

FIGURE 3.25
Nested sweep showing transistor characteristic curves.

9. Select **Plot > Axis Settings > YAxis**, change the **Data Range** to **User Defined** and enter a range from 0 mA to 40 mA. Click on OK and see the change.
10. Select **Plot > Axis Settings > YGrid** and uncheck **Automatic** and set the **Major Spacing** to 10 m. Click on OK and see the change.
11. Select **Plot > Axis Settings > XGrid** and uncheck both **Major** and **Minor Grids** to **None**. Click on OK and see the change.
12. Select **Plot > Axis Settings > YGrid** and uncheck both **Major** and **Minor Grids** to **None**. Click on OK and see the change.
13. Select **Plot > Label> Text**,change the font color to silver and add the base currents as shown in Figure 3.26.

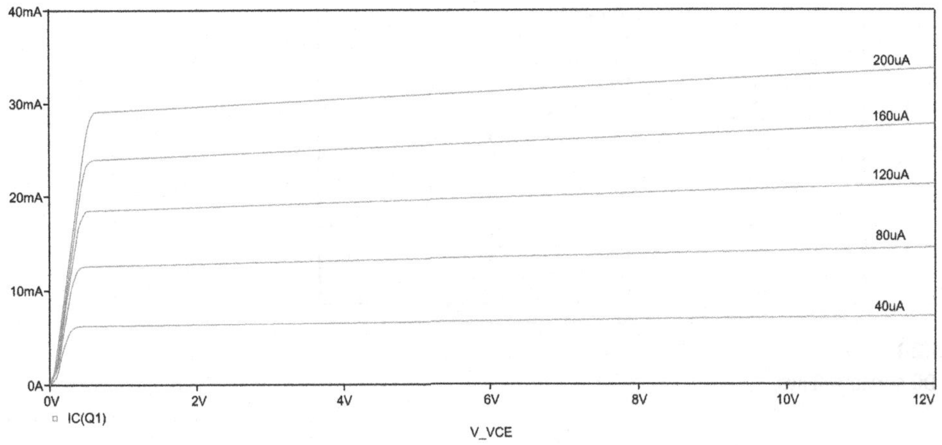

FIGURE 3.26
Nested sweep showing transistor characteristic curves for Ic against Vce for fixed values of Ib.

CHAPTER 4

AC Analysis

Chapter Outline

The AC analysis is used to calculate the frequency and phase response of a circuit by frequency sweeping an AC source connected to the circuit. The AC sweep analysis is a linear analysis and calculates what is known as the small signal response of a circuit over a range of frequencies by replacing any non-linear circuit device models with linear models. The DC bias point analysis is run prior to the AC analysis and is used to effectively linearize the circuit around the DC bias point. It must be noted that the AC analysis does not take into account effects such as clipping. You will have to run a transient analysis to determine these effects.

To perform an AC analysis, the independent voltage source V_{AC} or current source I_{AC} (Figure 4.1a) from the **source** library is used. However, any independent voltage source which has an AC property attached to the part can be used as an input to the circuit. Figure 4.1b shows the properties attached to the V_{AC} part as displayed in the Property Editor.

By default, the magnitude of the V_{AC} source is 1 V. In calculating the frequency response of a circuit, you are normally looking to calculate the gain and phase response of the circuit. Since the circuit gain is given by the ratio of V_{out} to V_{in}, setting V_{in} to 1 V, the gain or transfer function of the circuit will be equal to the output voltage, V_{out}.

4.1 SIMULATION PARAMETERS

One example in which an AC analysis is used to determine the frequency response of a circuit is the notch filter, which attenuates a narrow band of

Analog Design and Simulation using OrCAD Capture and PSpice. DOI: 10.1016/B978-0-08-097095-0.00004-0

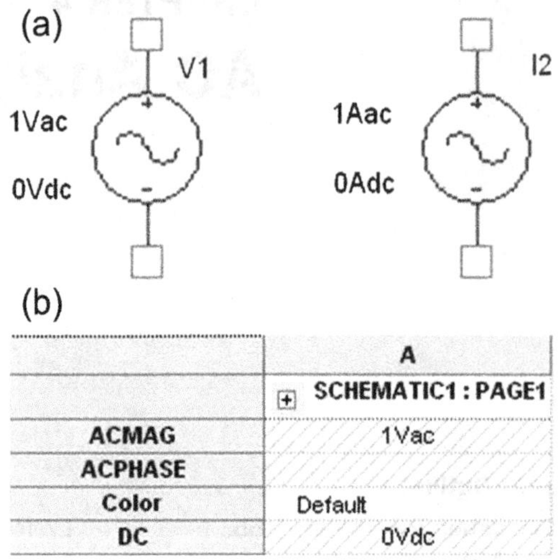

FIGURE 4.1
Independent V_{AC} and I_{AC} sources: (a) Capture parts; (b) V_{AC} properties.

unwanted frequencies – for instance, the removal of the mains frequency which can lead to unwanted 'hum' in an audio amplifier. One common implementation of a twin T notch filter is shown in Figure 4.2, where the notch frequency is given by:

$$f_o = \frac{1}{2\pi RC} \quad (4.1)$$

Figure 4.3 shows the notch filter response of the circuit where the output is attenuated by −60 dB at the notch frequency of 53 Hz.

To set up an AC analysis, a PSpice simulation profile needs to be created: **PSpice > New Simulation** Profile. In Figure 4.4, the **Analysis type** is set to **AC Sweep/ Noise** and has been set up for a logarithmic frequency sweep starting from 1 Hz to 100 kHz. You have the choice to sweep the frequency linearly over the whole

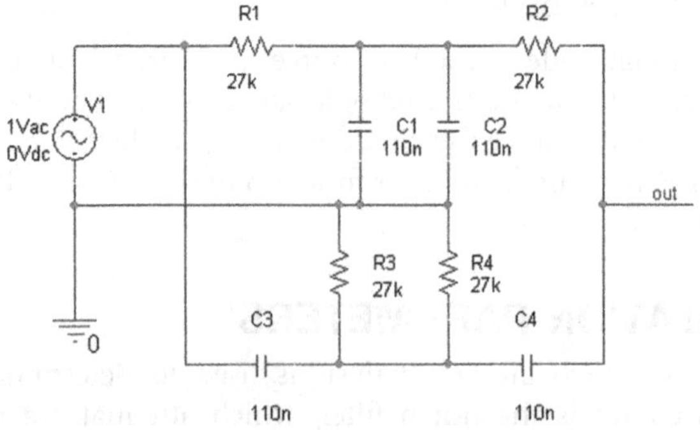

FIGURE 4.2
Twin T notch filter.

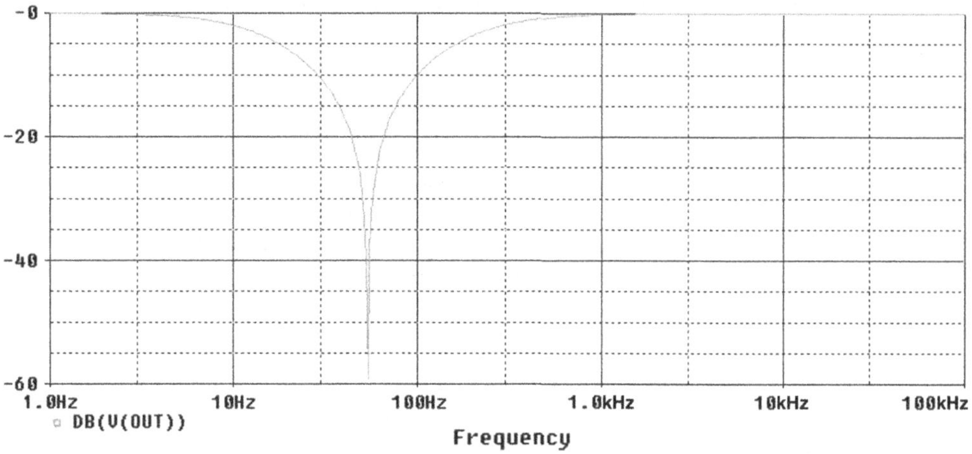

FIGURE 4.3
Notch filter response.

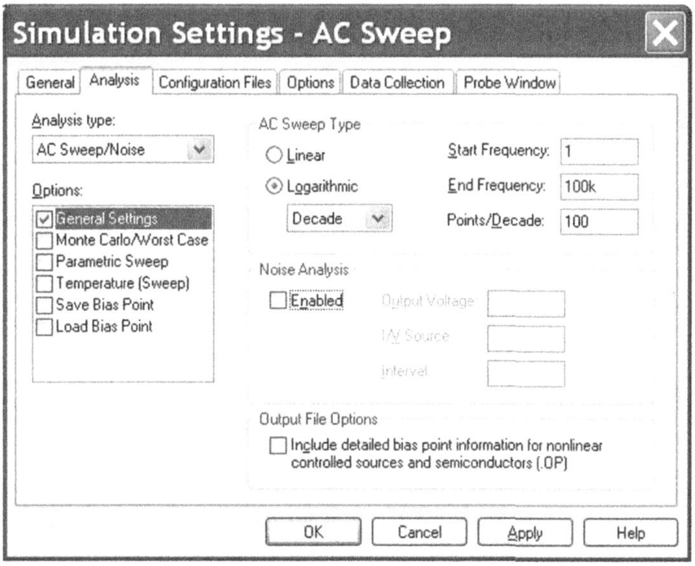

FIGURE 4.4
AC sweep simulation settings.

range or logarithmically either in decades or in octaves. If you want to use a linear range, note that the **Total Points** is applied over the whole frequency range compared to, say, a decade range where the number of points applies to each decade.

NOTE

It is a common mistake to mix up the total number of points for linear and decade (or octave). So if your AC response curve severely lacks any resolution, i.e. has been plotted showing 10 points over the complete frequency range, check your AC sweep simulation settings for the correct AC sweep type, linear of logarithmic. Another common mistake is when specifying megahertz, to use MHz, which is the same as mHz (millihertz) as PSpice is not case sensitive. For megahertz, use megHz or MEGHz or 10e6 Hz. You do not have to enter the units (Hz), for example 100meg.

4.2 AC MARKERS

AC markers can be found under the **PSpice > Markers > Advanced** menu as shown in Figure 4.5. These markers can be used to display dB magnitude, phase, group delay, and the real and imaginary parts of voltage and current. For example, a combination of these markers can be used for Bode and Nyquist plots.

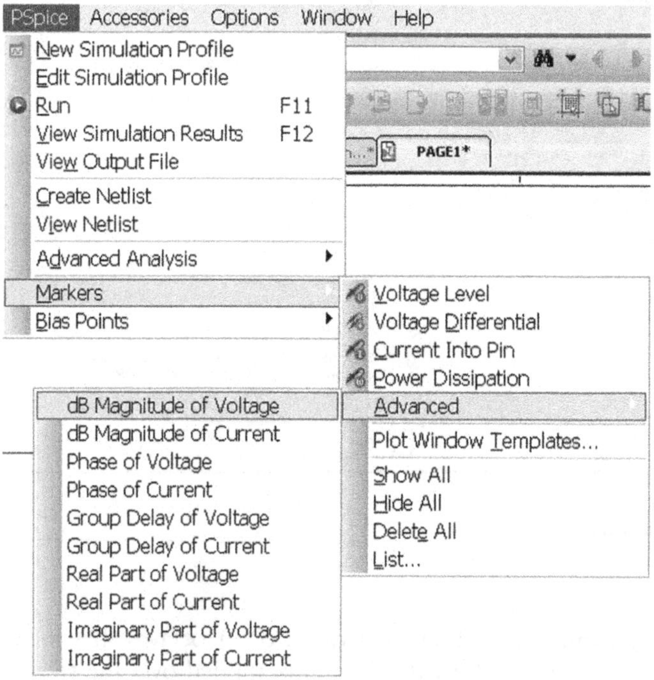

FIGURE 4.5
AC Markers menu.

4.3 EXERCISES

Exercise 1

Figure 4.6 shows a passive twin T notch filter. You will create an AC sweep and plot the frequency response of the circuit.

1. Draw the circuit of the notch filter in Figure 4.6. The V_{AC} source can be found in the **source** library.

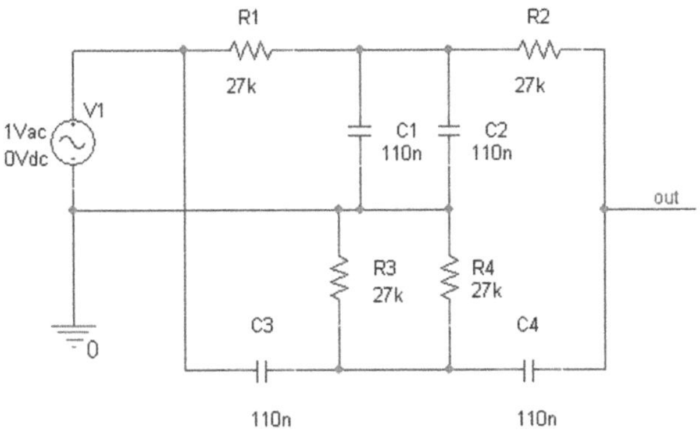

FIGURE 4.6
Twin T notch filter.

2. Create a PSpice simulation profile to perform a logarithmic sweep from 1 Hz to 100 kHz in steps of 100 points per decade (Figure 4.7).

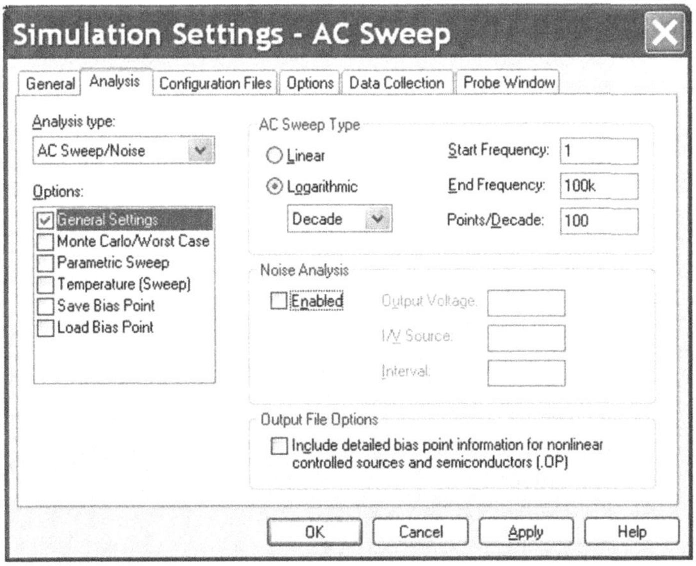

FIGURE 4.7
AC sweep simulation settings.

3. Place a V_{dB} voltage marker, which will automatically calculate the output voltage in dB: **PSpice > Markers > Advanced > dB Magnitude of Voltage** (Figure 4.8).

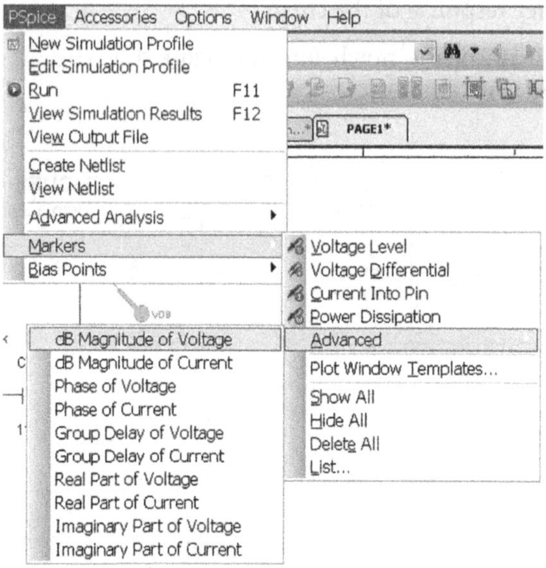

FIGURE 4.8
Adding a dB voltage marker.

4. Run the simulation and you should see the notch filter response (Figure 4.9). What we need to do now is to determine the frequency at the deepest part of the notch.

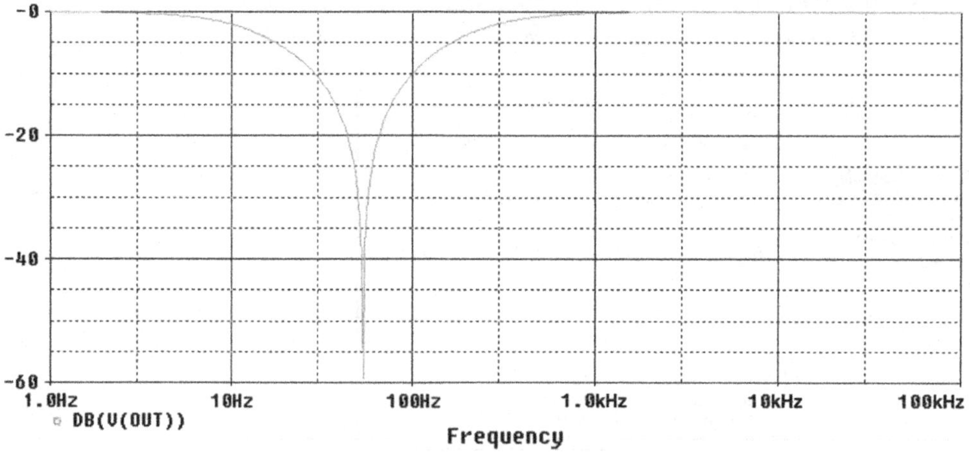

FIGURE 4.9
Notch filter response.

5. Turn on the cursor, **Trace > Cursor > Display** and place the cursor at the bottom of the notch. For a more accurate reading, zoom in towards the bottom of the trace, **View > Zoom > Area** or use the icons . Alternatively, you can use one of the cursor functions, **Trace > Cursor > Min** or .
   The **Probe Cursor** box will display a notch frequency 53.703 Hz with an attenuation of −59.348 dB as seen in Figure 4.10, depending on which software version you are using. In both software versions, the **Probe Cursor** box can be positioned anywhere in the Probe screen.

(a)

Probe Cursor		
A1 =	53.703,	-59.348
A2 =	1.0000,	-24.147m
dif=	52.703,	-59.324

(b)

Trace Color	Trace Name	Y1	Y2
	X Values	53.703	1.0000
CURSOR 1,2	DB(V(OUT))	-59.348	-24.147m

FIGURE 4.10
Cursor data: (a) version 16.2; (b) version 16.3.

6. Restore the Probe display to its original size, **View > Zoom > Fit**
   Select the trace name at the bottom of the display and press delete on the keyboard or select **Trace > Delete all Traces**. Now we are going to manually add the trace for the output voltage V(out).
7. Select **Trace > Add Trace** and the **Add Traces** window will appear as shown in Figure 4.11.

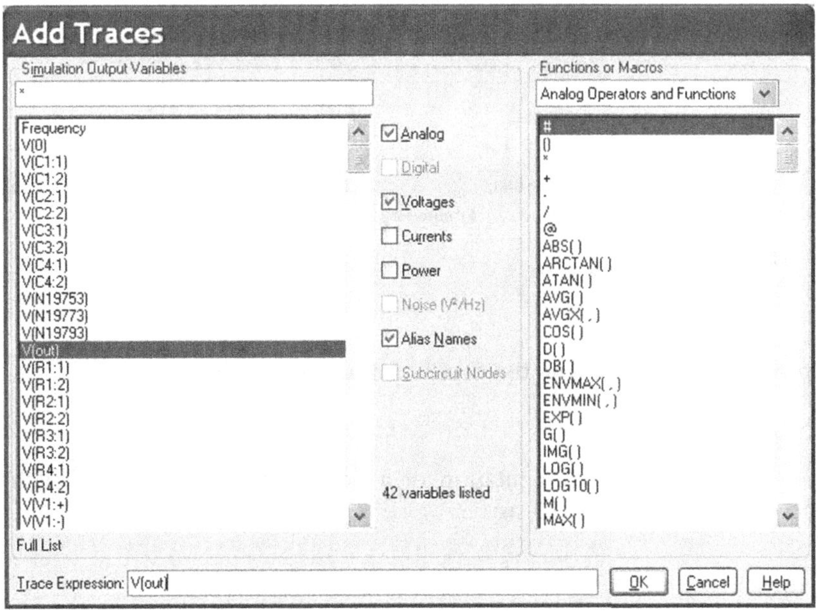

FIGURE 4.11
Add Traces window showing the list of output variables and Analog Operators and Functions.

8. The **Add Traces** window displays all the data for all the nodes and devices in the circuit. Uncheck the boxes for **Currents** and **Power** and in the list of **Simulation Output Variables**, scroll down and select the V(out) variable. Click on OK.
9. The V(out) trace will be displayed in the Probe trace window. However, what we want is the voltage in dB. Select the trace name V(out) and press the delete key to remove the trace.
10. Select **Trace > Add Trace** and in the right-hand side of **Add Traces** under **Analog Operators and Functions**, select DB().
Then, as before, select V(out) from the list of **Simulation Output Variables**. At the bottom of the window in the **Trace Expression** box, you should see DB(V(out)). The DB function automatically calculates the DB of V(out). See Figure 4.12.

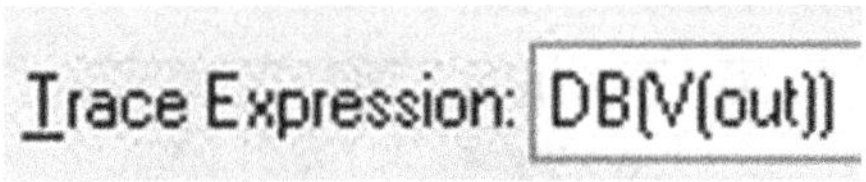

FIGURE 4.12
Conversion of V(out) to dB.

Click on OK and you should see the trace shown in Figure 4.13.

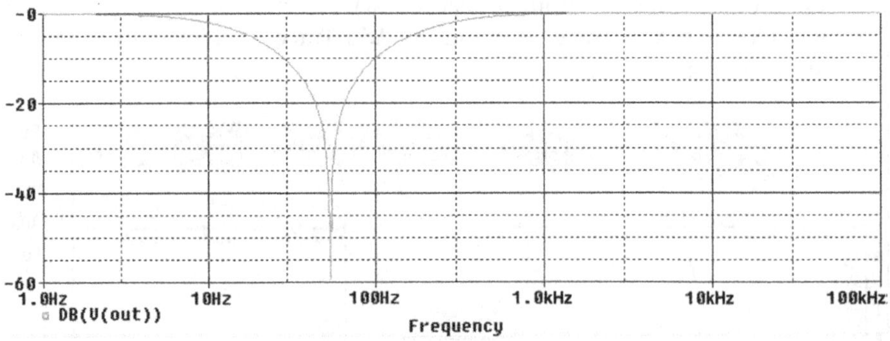

FIGURE 4.13
Notch filter response.

11. Repeat Step 5 to determine the depth of notch attenuation in dB.

Twin T notch filter

Figure 4.14 shows the implementation of a notch filter which has a notch frequency given by equation 4.1 as:

$$f_o = \frac{1}{2\pi RC}$$

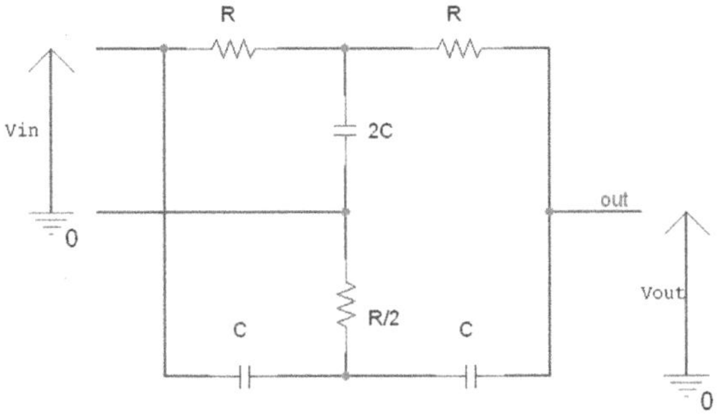

FIGURE 4.14
Twin T notch filter.

Using only one value of resistor and one value of capacitor, the circuit in Figure 4.14 can be implemented as shown in Figure 4.15.

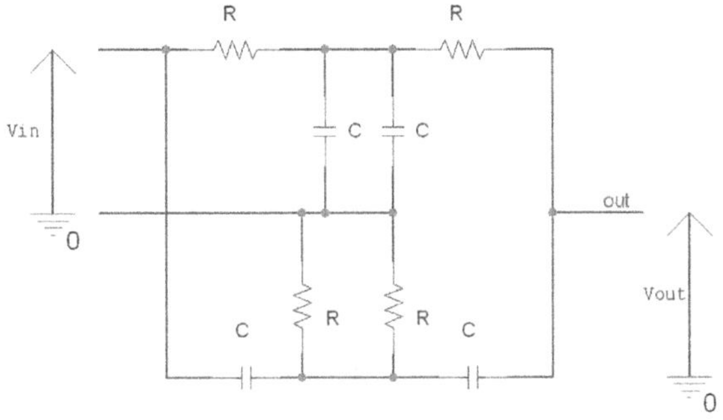

FIGURE 4.15
Twin T Notch filter implemented with one resistor and one capacitor value.

The parallel combination of the two capacitors is given by:

$$C_p = C + C = 2C$$

The parallel combination of the resistors is given by:

$$R_p = \frac{R \times R}{R + R} = \frac{R}{2}$$

which conforms to the circuit implementation shown in Figure 4.14.

For the notch filter in Figure 4.6, $R = 27\ \text{k}\Omega$ and $C = 110\ \text{n}$. Therefore, using equation 4.1, the notch frequency is given by:

$$f_o = \frac{1}{2\pi RC}$$

$$f_o = \frac{1}{2\pi \times 27 \times 10^3 \times 110 \times 10^{-9}} = 53.6\ \text{Hz}$$

CHAPTER 5

Parametric Sweep

Chapter Outline

A parametric sweep allows for a parameter to be swept through a range of values and can be performed when running a transient, AC or DC sweep analysis. Parameters that can be varied include a voltage or current source, temperature, a global parameter or a model parameter. A global parameter can represent a mathematical expression as well as a variable and is defined using the **PARAM** part from the **Special** library. To define the global variable, you have to add a new property to the **PARAM** part by editing its properties via the **Property Editor**. For example, in the resistor circuit shown in Figure 5.1, the value of resistor R2 has been replaced with a variable called {**rvariable**}; you can name the variable anything you like. The braces otherwise known as curly brackets { } are required in PSpice to define global parameters.

The **PARAM** part has a heading called **PARAMETERS:** and contains a list of defined variables and their default values. In this case, RL is defined to have a default value of 10 kΩ if no parametric sweep is performed.

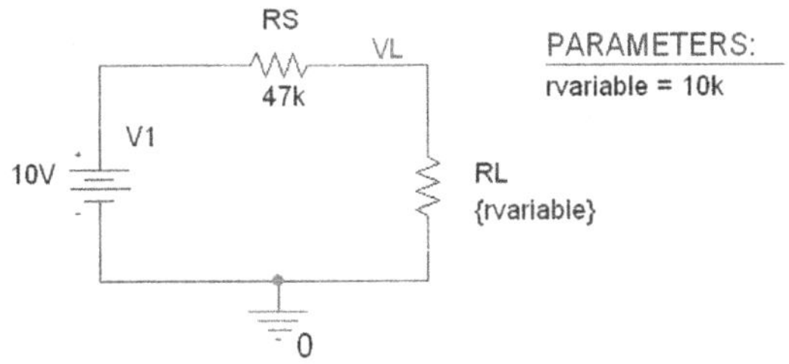

FIGURE 5.1
Defining a global parameter and a default value.

Analog Design and Simulation using OrCAD Capture and PSpice. DOI: 10.1016/B978-0-08-097095-0.00005-2

5.1 PROPERTY EDITOR

As mentioned above, the global parameter variable name and default value must be added to the **PARAM** part as a new property in the **Property Editor**. The **PARAM** part is found in the **special** library and is placed anywhere in the schematic page. By double clicking on the **PARAM** part, the **Property Editor** (Figure 5.2) will open.

The **Property Editor** is a spreadsheet that displays all the properties attached to a part. For example, a resistor will have defined properties such as a footprint, resistor value, power rating, tolerance, manufacturer's part number and PSpice model. Properties can be added to parts in the Property Editor and in the case of the **PARAM** part, we need to add a property to define the global variable that needs to be swept.

When first selected, the Property Editor will open up in one of two modes, displaying the properties in either rows (Figure 5.3a) or columns (Figure 5.3b).

It is easier to view all the properties listed in rows, as shown in Figure 5.3a. If the properties are displayed in columns (Figure 5.3b), you will have to use the scrollbar at the bottom of the Property Editor to scroll along to view all the other properties. The viewed mode can be changed, for example, from columns to rows by placing the cursor in the blank cell to the left of the **Color** property (Figure 5.3b) and when the cursor changes to an arrow, **rmb > Pivot**. The properties will now be displayed as rows.

To change the view from rows to columns, place the cursor in the blank cell above the **Color** property (Figure 5.3a). When the cursor changes to an arrow, **rmb > Pivot**.

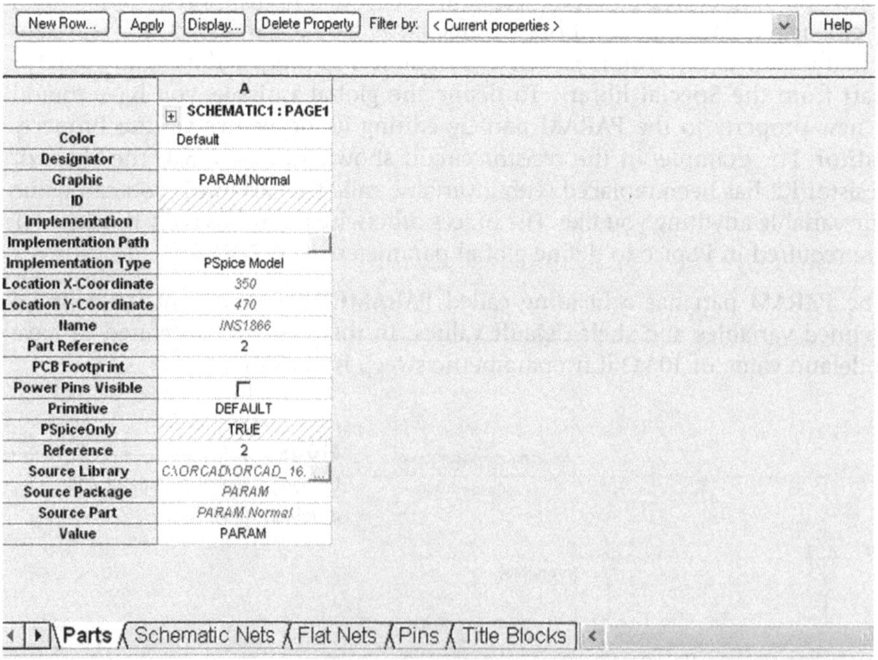

FIGURE 5.2
Property Editor.

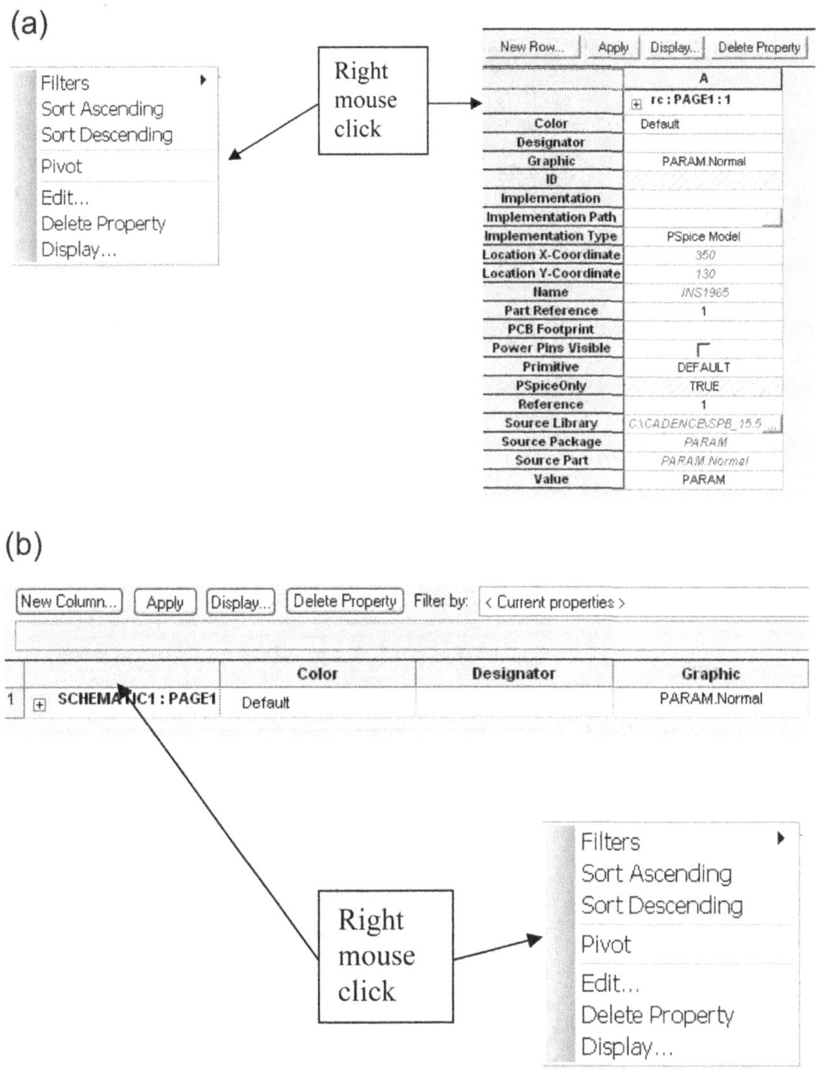

FIGURE 5.3
Property Editor displaying PARAM part properties in either (a) rows or (b) columns.

To add a property which allows a global parameter to be swept, select **New Row** as shown in Figure 5.3a and the **Add New Row** dialog box will appear (Figure 5.4). Add the **Name** of the variable and the default **Value** of the variable. Figure 5.4 shows a parameter called **rvariable** being defined with a default value of 10k and Figure 5.5 shows the new row added in the Property Editor.

The properties listed in the Property Editor have a name and a value. For example, a transistor may have a standard TO5 footprint. Footprint is the property name and its value is TO5. Another example is a resistor with a 1% tolerance. Tolerance is the property name and 1% is the value of the property. Table 5.1 gives some examples of property name and value.

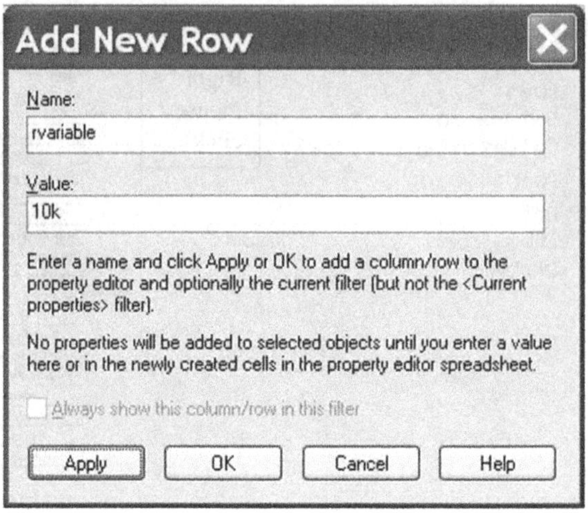

FIGURE 5.4
Adding a new property to the PARAM part.

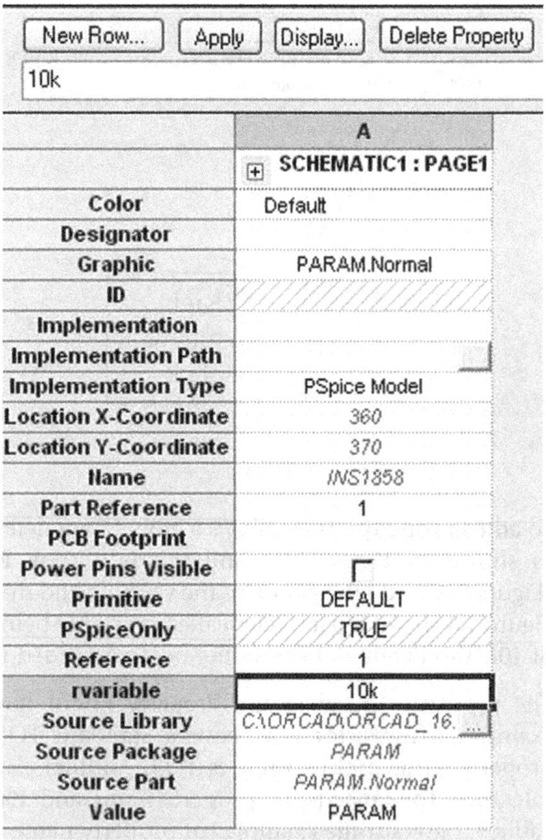

FIGURE 5.5
The new parameter variable **rvariable** with a default value of 10 k has been added to the PARAM part.

Table 5.1 Property Examples.	
Property Name	**Property Value**
Footprint	TO5
Tolerance	1%
Part reference	R1

By default, new property names and values are not displayed on the part in the schematic diagram so they have to be made visible. This is done by highlighting the property cell in the Property Editor and selecting **Display** (or **rmb > Display**). This will open up the **Display Properties** dialog box as shown in Figure 5.6.

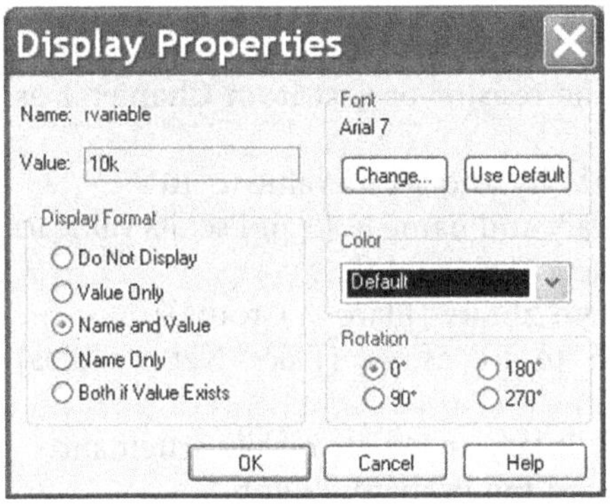

FIGURE 5.6
Display Properties controls whether property names and values are displayed on the schematic.

It is recommended for Global parameters that both property **name** and **value** are both displayed when added to the **Param** part, as seen in the resistor circuit in Figure 5.1.

NOTE
The Property Editor is closed by clicking on the lower right hand cross of the property window.

Be careful not to select the upper top cross as this will close Capture .

5.2 EXERCISES

Exercise 1

When you connect audio equipment together, for example the output of a microphone to the input of an amplifier, it is recommended that the output impedance of the microphone should match the input impedance of the amplifier. This is also true for video and radio frequency (RF) equipment. What happens is that the maximum power transfer of a signal between a source impedance and a load impedance occurs when the impedances match and the maximum power transfer that can be achieved is 50% of the source signal.

You will demonstrate simple resistance matching by plotting the load resistor power dissipation against load resistance for the circuit in Figure 5.7.

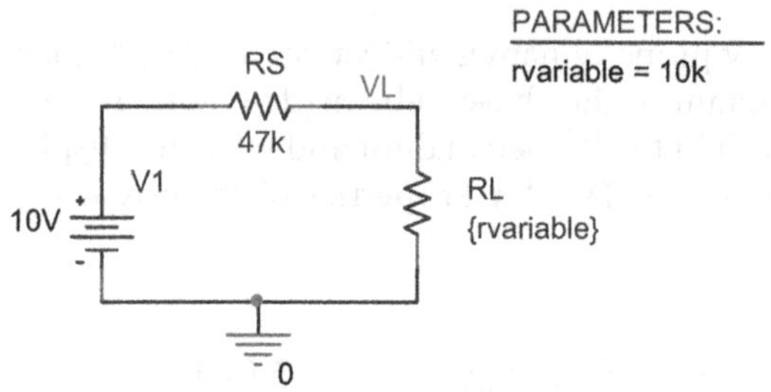
PARAMETERS:
rvariable = 10k
RS
47k
VL
V1
10V
RL
{rvariable}
0

FIGURE 5.7 Resistor network to demonstrate resistance matching.

1. Create a new PSpice project or use the resistor project from Chapter 1 as a starting point.
2. Place a V_{DC} source from the source library and set its value to 10 V.
   Place a resistor R from the analog library and name it RS and set its value to 47k. Place resistor RL and set its value to {rvariable}.
   Connect a 0 V symbol from the capsym library (Place > Ground).
   Name the net node connecting RS to RL as VL (Place Net > Alias) (Figure 5.7).
3. Place a PARAM part from the special library anywhere on the schematic.
4. Double click on the Param part to open the Property Editor.
5. Depending on how the properties are displayed in the Property Editor (rows or columns), add a new property by either clicking on **New Row...** or **New Column...** Create a new property called **rvariable** with a value of 10k as shown in Figure 5.8.

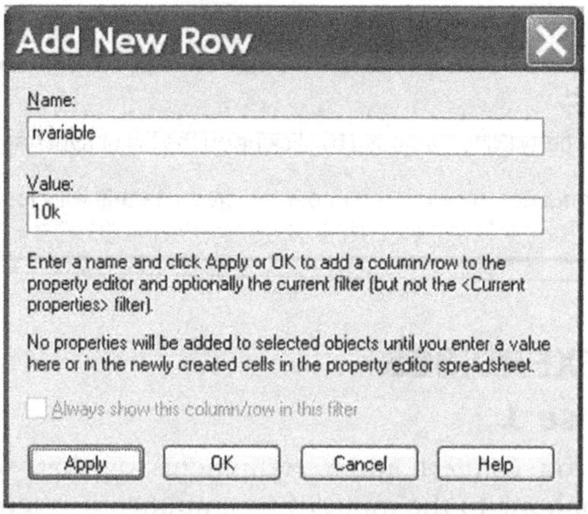
Add New Row
Name:
rvariable
Value:
10k
Enter a name and click Apply or OK to add a column/row to the property editor and optionally the current filter (but not the <Current properties> filter).
No properties will be added to selected objects until you enter a value here or in the newly created cells in the property editor spreadsheet.
Always show this column/row in this filter
Apply
OK
Cancel
Help

FIGURE 5.8 Creating a new global parameter.

The circuit is now set up with a global parameter, **rvariable**, with a default value of 10 kΩ, which will be the resistor value used for simulation if no parametric sweep is performed. Click on OK but do **not** exit the Property Editor.

6. Highlight the new property **rvariable** and select **Display**. In the **Display Properties** window (Figure 5.9) select **Name and Value**. Close the Property Editor.

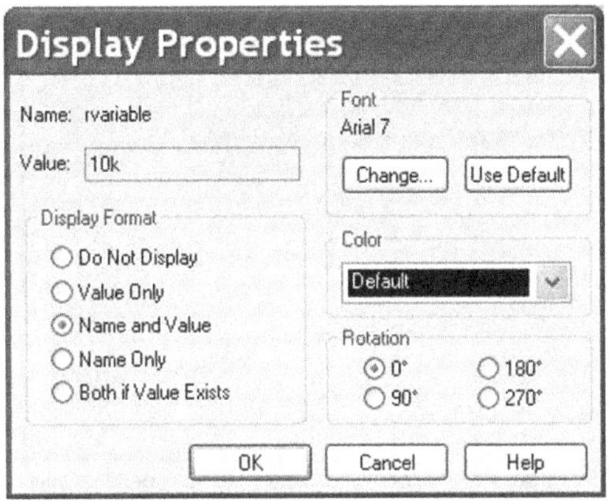

FIGURE 5.9
Display Properties.

7. You will need to set up a DC sweep with a Global parameter named **rvariable** for a linear sweep from 500 Ω to 100 kΩ in steps of 500 Ω.
Create a new simulation profile, **PSpice > New Simulation Profile**, and call it anything you like, for example, global sweep.
Select **Analysis type** to **DC Sweep** and select the **Sweep variable** as a Global parameter with a **Parameter name: rvariable.** The **Sweep type** will be linear with a start value of 500 Ω, an end value of 100 kΩ and an increment value of 500 Ω (Figure 5.10).

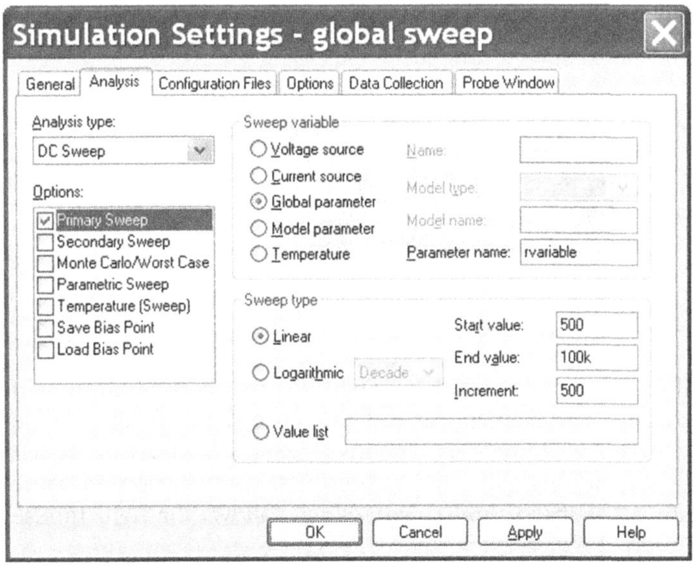

FIGURE 5.10
Simulation settings for a global parameter sweep.

8. Place a power marker on the body of RL, i.e. in the middle of RL, by selecting **PSpice > Markers > Power Dissipation**, or select the icon or .
   Your circuit should look the same as that in Figure 5.7.
9. Run the simulation (**PSpice > Run**) .
   You should see the power dissipation curve in Figure 5.11.

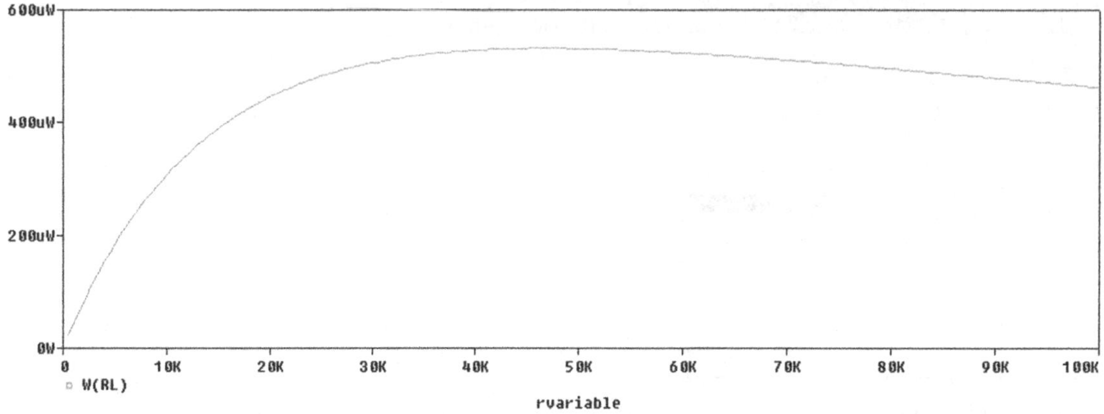

FIGURE 5.11
Resistor load power dissipation curve.

10. From the curve in Figure 5.11 the cursors can be turned on to determine the value of load resistance for maximum load power.
    Display the cursor, **Trace > Cursor > Display** or .
    In Probe, there are two cursors and they follow either the left mouse button (lmb) or the right mouse button (rmb). When you first select **Cursor > Display**, a dashed white box will appear around the symbol used for the trace name (Figure 5.12).

FIGURE 5.12
Cursor activated for W(RL) trace.

The cursor will then follow the left mouse with the left mouse button held down. The cursor box will then appear and give a coordinate reading of the cursor. In release 16.2 and previous versions (Figure 5.13a), A1 is the left mouse cursor with the first value the *x*-coordinate and the second value the *y*-coordinate. A2 is the second cursor, which follows the right mouse button. From release 16.3, the cursors are labeled as shown in Figure 5.13b, where you can also add multiple cursor measurements from other traces and plots.

(a)

Probe Cursor		
A1 =	47.000K,	5.3192u
A2 =	500.000,	221.607n
dif=	46.500K,	5.0975u

(b)

Trace Color	Trace Name	Y1	Y2
	X Values	47.000K	500.000
CURSOR 1,2	W(RL)	531.915u	22.161u

FIGURE 5.13
Cursor coordinates: (a) release 16.2; (b) release 16.3.

11. Place the cursor on the maximum point of the curve and read off the value for the load resistor.
    Alternatively, there are predefined cursor functions (Figure 5.14) which can be used to find the points on the curve such as maximum or minimum values. These can be accessed from **Trace > Cursor** or you can select the readily available icons on the top toolbar.

FIGURE 5.14
Cursor icons.

The function of each icon is shown by moving the cursor over the icons.

12. Select the **Cursor Max** function and you will see the cursor move to the maximum value on the curve. The cursor box then displays the maximum value as 47k, which is the load resistor value for maximum power transfer (Figure 5.15).

Trace Color	Trace Name	Y1	Y2
	X Values	47.000K	500.000
CURSOR 1,2	W(RL)	531.915u	43.403u

FIGURE 5.15
Cursor max.

13. With the cursor placed at the maximum value, select **Plot > Label > Mark** or click on the icon or . The coordinates of the maximum point on the curve (47k, 531.915W) will now be marked and displayed (Figure 5.16).

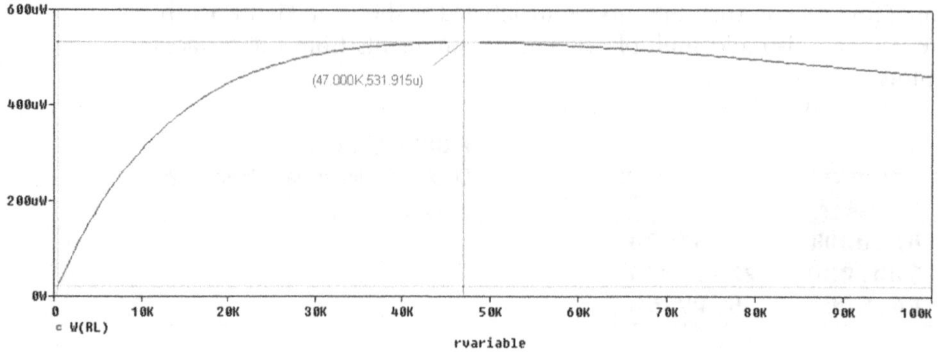

FIGURE 5.16
Load resistor power versus load resistance.

NOTE
You can add arrows and text on the curves by selecting the **Plot > Label** menu.

THEORY

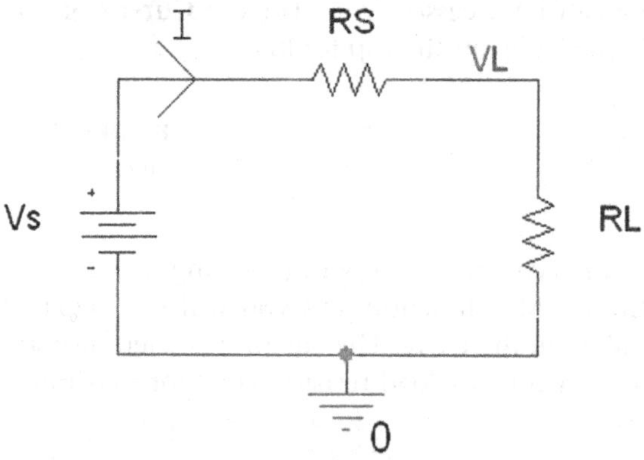

FIGURE 5.17
Resistor network.

In Figure 5.17, the current in the circuit is given by:

$$I = \frac{Vs}{Rs + RL} \tag{5.1}$$

The power dissipated in the load resistor is given by:

$$P_L = I^2 RL \tag{5.2}$$

Substitute for I from (5.1) into (5.2):

$$P_L = \left(\frac{Vs}{Rs + RL}\right)^2 RL \tag{5.3}$$

$$P_L = \frac{Vs^2}{Rs^2 + 2RsRL + RL^2} RL \tag{5.4}$$

Dividing by RL:

$$P_L = \frac{Vs^2}{\frac{Rs^2}{RL} + 2Rs + RL} \tag{5.5}$$

For maximum power in (5.5), the denominator must be a minimum. Rather than differentiating the whole equation, differentiating the denominator will give the same result.

$$\frac{dP_L}{dRL} = -\frac{Rs^2}{RL^2} + 1 \tag{5.6}$$

For a turning point,

$$\frac{dP_L}{dRL} = 0$$

Therefore,

$$0 = -\frac{Rs^2}{RL^2} + 1 \tag{5.7}$$

$$Rs = RL$$

It can be shown by differentiating (5.6) again, that the denominator is a minimum at the turning point and so when $Rs = RL$, the power is a maximum value.

Exercise 2

NOTCH FILTER

You will globally sweep the resistor values in the notch filter circuit from Chapter 4 on AC analysis to see what effect this has on the circuit response. In the circuit shown in Figure 5.18, the four resistor values have been replaced by the global parameter {Rvalue}, which has been defined by the Param symbol to have a default value of 27 k if no parametric sweep is performed.

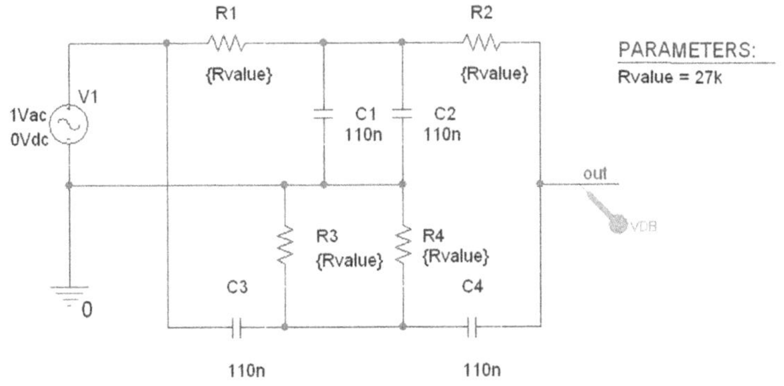

FIGURE 5.18 Notch filter resistor values set to a global default parameter value of 27 kΩ.

1. Place a **Param** part from the special library. Double click on the **Param** part and define the global variable as **Rvalue** with a default value of 27 k. Display both the property name and value for **Rvalue**. The steps are the same as in Exercise 1.
2. Create a PSpice simulation profile for an AC sweep with the same settings for the passive notch filter in Exercise 1, performing a logarithmic sweep from 10 to 10 kHz with 100 **Points/Decade** (Figure 5.19). Click on **Apply** but do exit.

FIGURE 5.19
AC sweep settings.

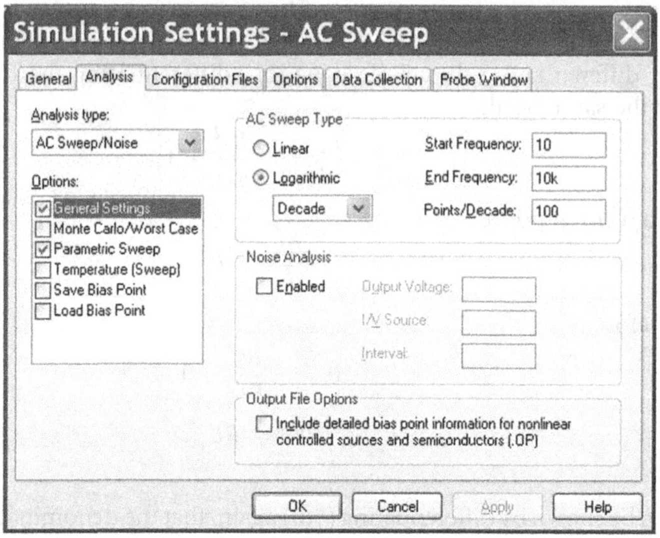

3. Select the **Parametric Sweep** in the **Options** box and set up a **Global** parametric **linear** sweep for the **Rvalue** variable starting at 24 kΩ to 30 kΩ in steps of 1 kΩ, which will perform a total of seven AC sweeps (Figure 5.20). Click on OK.

FIGURE 5.20
Global parameter settings.

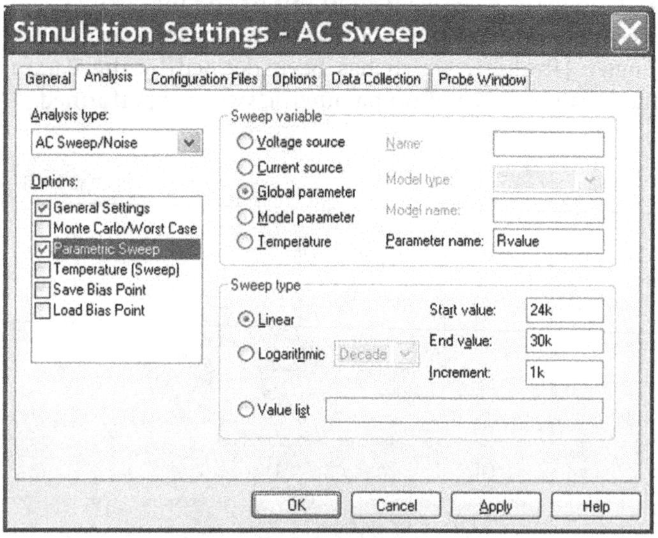

4. Place a V_{db} marker on the output node, 'out'. **PSpice > Markers > Advanced > dB Magnitude of Voltage.**

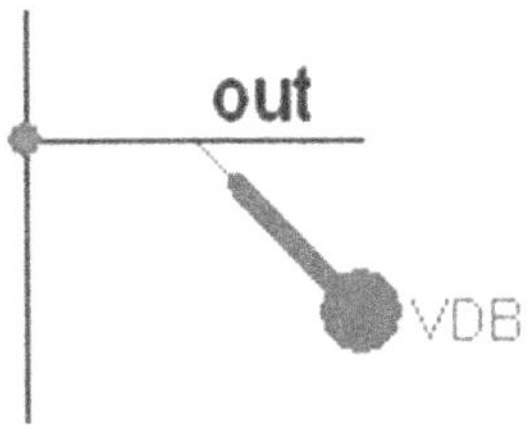

5. Run the simulation.

The notch filter response is shown in Figure 5.21. Here you can see that the notch frequency changes with a change in resistance **Rvalue**. However, the attenuation depth of the notch, also known as the *Q* of the filter, also changes with frequency.

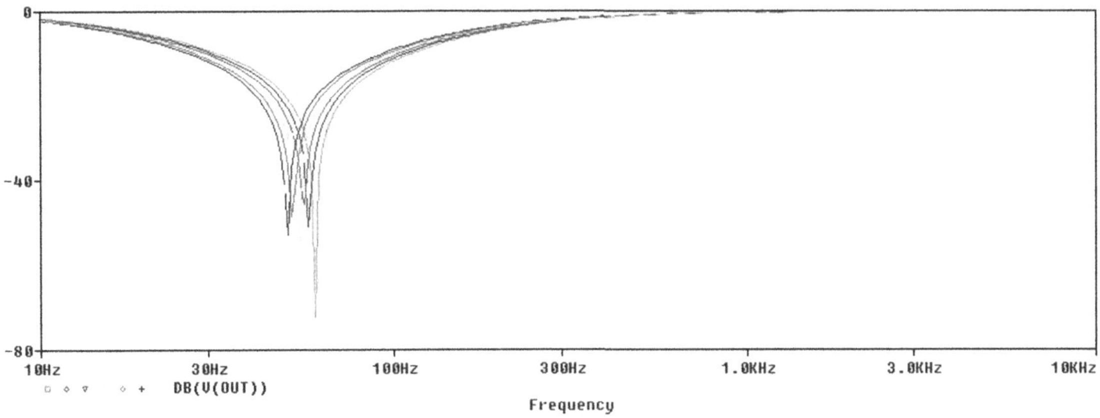

FIGURE 5.21
Passive notch filter response.

Exercise 3

ACTIVE NOTCH FILTER

Figure 5.22 shows an active notch filter built upon the previous twin T notch filter. The notch frequency is set as before by the resistor and capacitors in the passive network but the potentiometer R5 is used to set the *Q* (the sharpness of the attenuation) of the notch, which is not dependent on frequency.

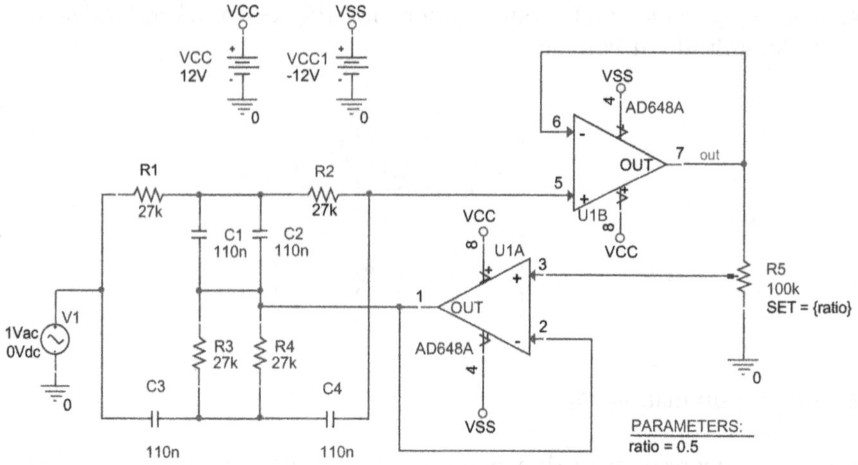

FIGURE 5.22
Active notch filter.

The potentiometer, R5, has a **SET** parameter which effectively sets up the ratio between the two resistances on either side of the potentiometer wiper (pin 2). For example, if the ratio is set to 0.4, the resistance between pins 1 and 2 takes on a value of $0.4 \times 100\,k = 40\,k$ and the resistance between pins 2 and 3 takes on a value of $(1-0.4) \times 100\,k = 60\,k$. So by varying the SET parameter between 0 and 1.0, the pot is effectively being turned through its complete resistance range.

Now, in order to automatically sweep the SET parameter through its range of 0 to 1.0, a global parameter, **ratio**, has been set up which has a default value of 0.5 corresponding to the midrange of the potentiometer (50 kΩ).

In order to draw the circuit of the active notch filter, the operational amplifiers (opamps) AD648A need to be found. Any opamps could be used, but this is a good exercise in how to search for specific parts.

NOTE

If you have the OrCAD Demo CD, search for the µA741 opamp in the eval library.

1. In the **Place Part** menu there is a **Search for Part** function which can help to locate the opamp library as shown in Figure 5.23.

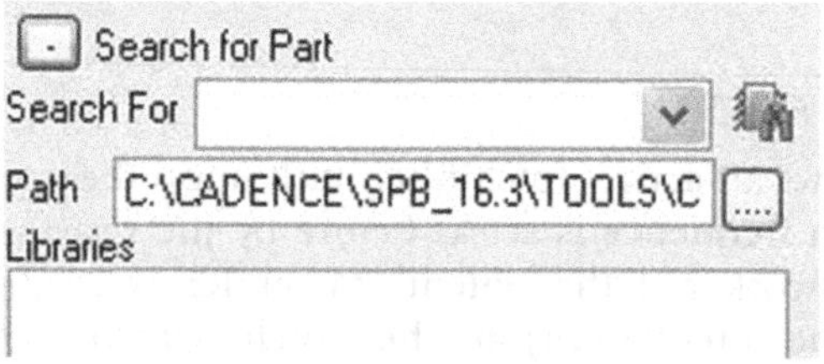

FIGURE 5.23
Search for Part.

NOTE

By default, the Search Path in Figure 5.23 does not point to the PSpice libraries but to the Capture libraries, so running a search for the opamps will not return any devices. It is a common mistake when searching for parts to forget to change the **Search for Part** path.

To change the search path, click on the Browse icon to the right of the Search Path box, which will open the **Browse File** window (Figure 5.24) showing the **[install path] > capture > library** folder selected on the right-hand side and the list of libraries in that folder shown on the left.

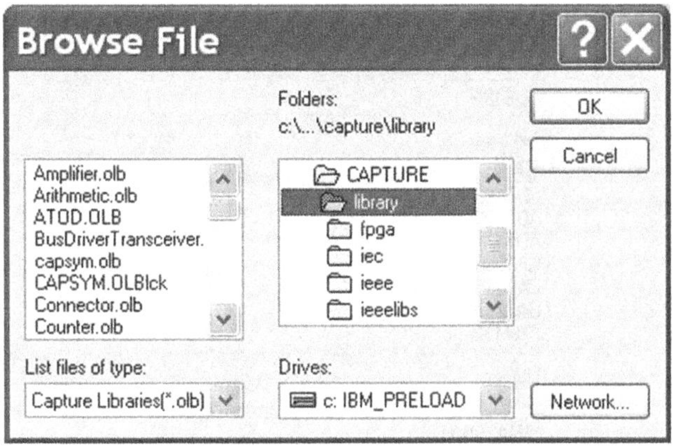

FIGURE 5.24 Browse File window displays the search part path.

In the **Folder** section, scroll down and double click on the **pspice** folder as shown in Figure 5.25. On the left-hand side, you will see a list of the available PSpice libraries. Click on OK.

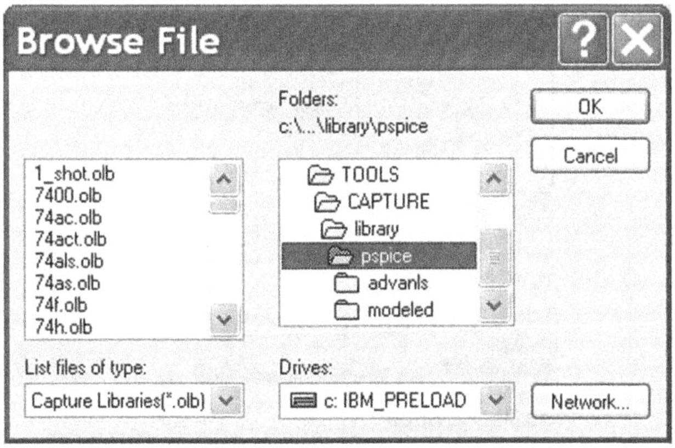

FIGURE 5.25 The PSpice library is now selected.

2. Different manufacturers append their own specific numbers and letters to standard part numbers. For example, if you are searching for the BC337 transistor and type in BC337 in the **Search For** box, you will

only see one result. However, if you type in a wildcard (*) after the transistor number, BC337*, you will see more results because the wildcard effectively ignores the manufacturer's extra characters after the transistor number when doing a search.

In the **Search for Part** enter the opamp number, AD648*, and either press return or click on the **Part Search** icon.

There should only be one instance of the AD648A from the opamp library, as shown in Figure 5.26. Double click on the AD648A and you will see the opamp library added to the list of libraries and a Capture graphical representation of the opamp.

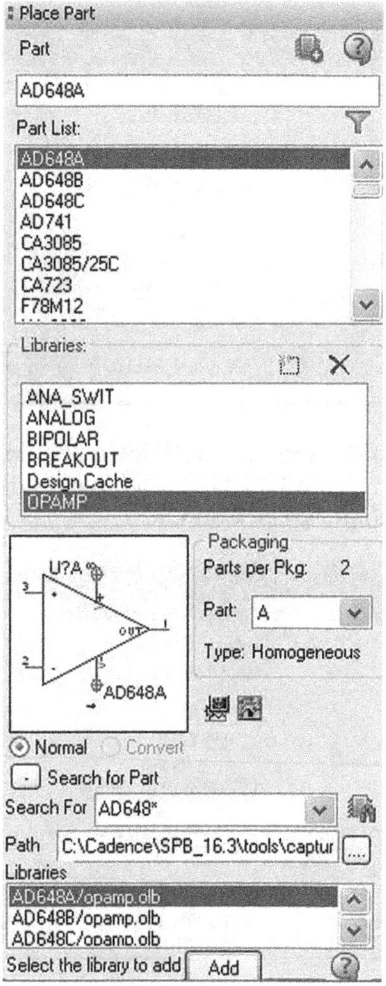

FIGURE 5.26

The opamp AD648A is found in the opamp library.

NOTE

If you have the eval software version, search for the μA741* opamp.

In the Place Part menu alongside the graphical representation of the opamp, there is a **Packaging** section which shows that there are two Parts per package. As this is a dual opamp, there are two available sections, A and B. Figure 5.26 shows Part A selected, while Figure 5.27 shows Part B selected. Note the different pin numbers and different reference designators, U?A and U?B. Type: Homogeneous indicates that both sections in a part are the same, in contrast to, for example, a relay and coil, where both sections are different and are classed as heterogeneous.

The two icons shown in Figure 5.27 indicate that the AD648A has PCB footprint and a PSpice model attached and is therefore ready for simulation. It is important that the PSpice icon is displayed when selecting parts for simulation.

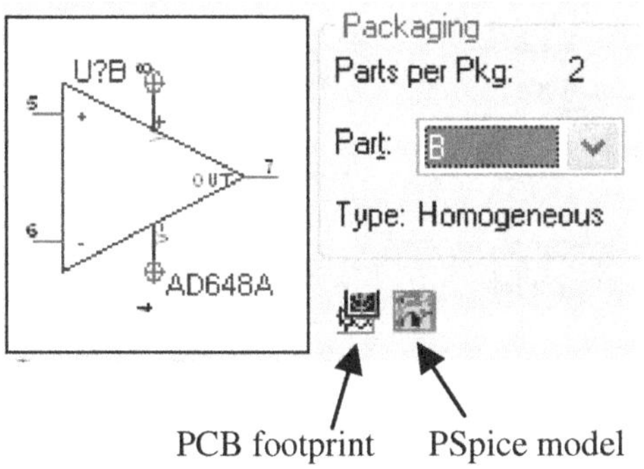

FIGURE 5.27
Part B of the opamp is selected.

3. Place the A part of the opamp in the circuit. Highlight the opamp and **rmb > Mirror Vertically** or press V. The **Mirror** and **Rotate** operations can be used even if the opamp is connected in the circuit. You do not have to delete any wires.
4. Select the B part of the opamp and place it in the circuit.
5. With all the libraries selected (left mouse click at the top of library list and drag mouse pointer down to bottom of library list), type **pot**, in the **Part** box. The **pot** is found in the **breakout** library. Place the **pot** part and change its value to 100 k. Double click on the SET property and change the default value of 0.5 to {ratio}. Do not forget the brackets.

6. We need to define **ratio** as a global parameter with a default value of 0.5. Place a Param part from the special library in the circuit.
7. Double click on the **param** part and enter a new row (or new column). Create a new property called **ratio** with a value of 0.5 and display the property name and value. The steps are the same as in Exercise 1.
8. Select **Place > Power** and scroll down to the VCC_CIRCLE symbol and in the **Name:** box, change the name to VCC and click on OK. Repeat for the VSS symbol.
9. Place and connect the remaining components as shown in Figure 5.22.
10. The parametric sweep is run in conjunction with an AC analysis. In this example we are going to run a parametric sweep on the potentiometer ratio from 0.1 to 0.9 in steps of 0.1. The first thing to do is to set up the same AC analysis as before with the passive notch filter, sweeping the frequency logarithmically from 10 to 10 kHz with 100 points/decade (Figure 5.28). Click on apply but do **not** exit.

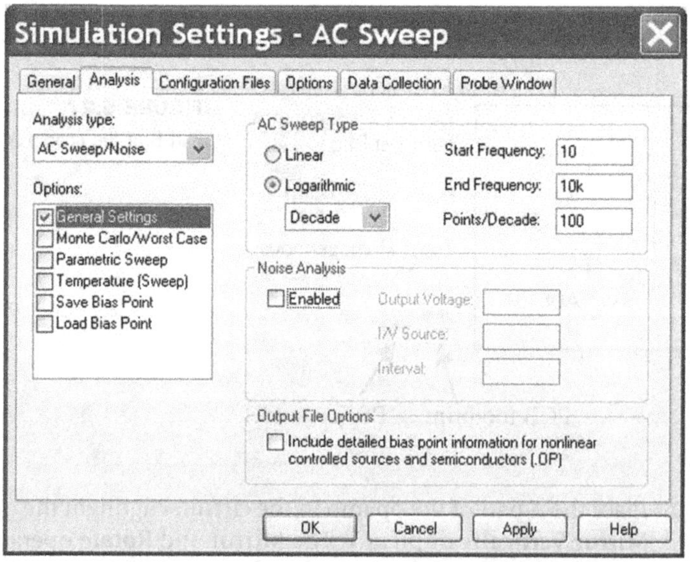

FIGURE 5.28
AC sweep settings.

11. Select the **Parametric Sweep** in the **Options** box. The Sweep variable is a **Global parameter** and the **Parameter name** is **ratio**. The **Sweep type** is linear, the **Start value** 0.1, the **End value** 0.9 and the **Increment** 0.1 (Figure 5.29). Click on OK.

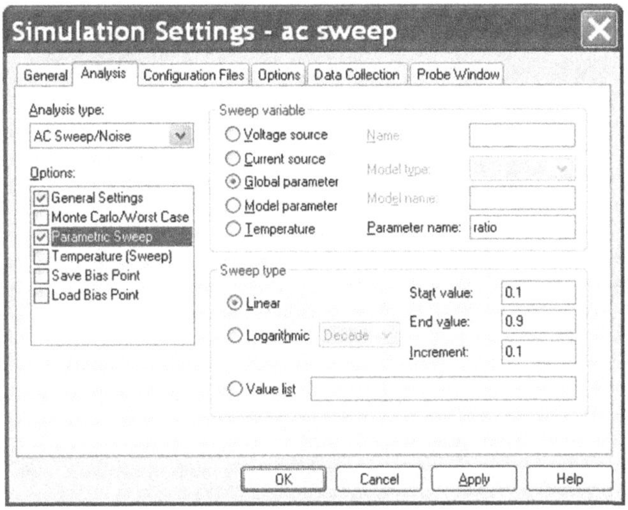

FIGURE 5.29
Global parameter settings.

12. Place a V_{db} marker on the output node 'out'. **PSpice > Markers > Advanced > dB Magnitude of Voltage.**

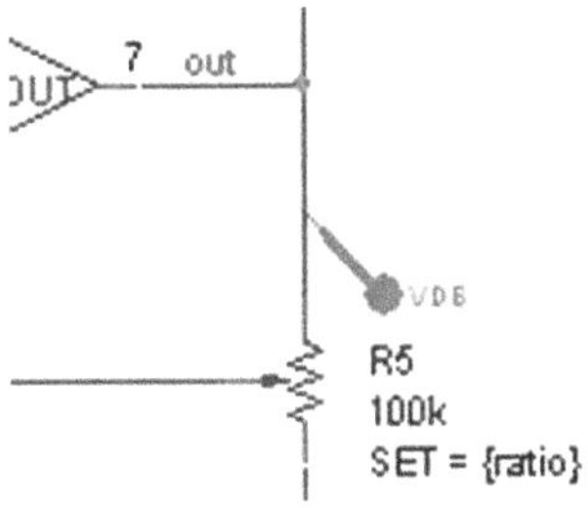

13. Run the simulation.

Figure 5.30 shows that by varying R5, the Q of the circuit (the sharpness of the notch) can be varied without a change in the notch frequency.

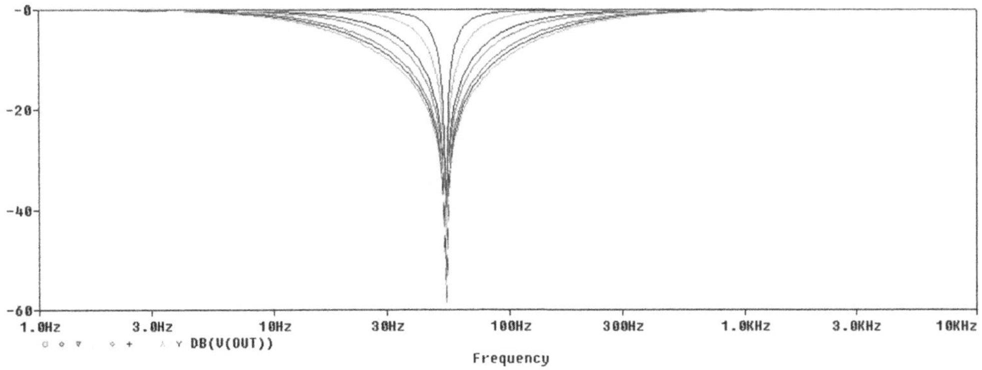

FIGURE 5.30
Response of the active notch filter.

CHAPTER 6
Stimulus Editor

Chapter Outline

The Stimulus Editor is a graphical tool to help you define transient analog and digital sources. The **sourcestm** library contains three source parts, shown in Figure 6.1, each of which provides the interface with the defined stimulus in the Stimulus Editor.

When you first place one of the sources from the **sourcestm** library, the implementation property is displayed in the schematic. This property refers to the name of the stimulus which is defined in the Stimulus Editor. Either you can enter a name of the stimulus on the schematic to start with, or you will be prompted for the stimulus name in the Stimulus Editor when started.

To start the Stimulus Editor, highlight a **sourcestm** source, **rmb > Edit PSpice Stimulus**.

FIGURE 6.1
Stimulus Editor, transient analog and digital sources.

Analog Design and Simulation using OrCAD Capture and PSpice. DOI: 10.1016/B978-0-08-097095-0.00006-4

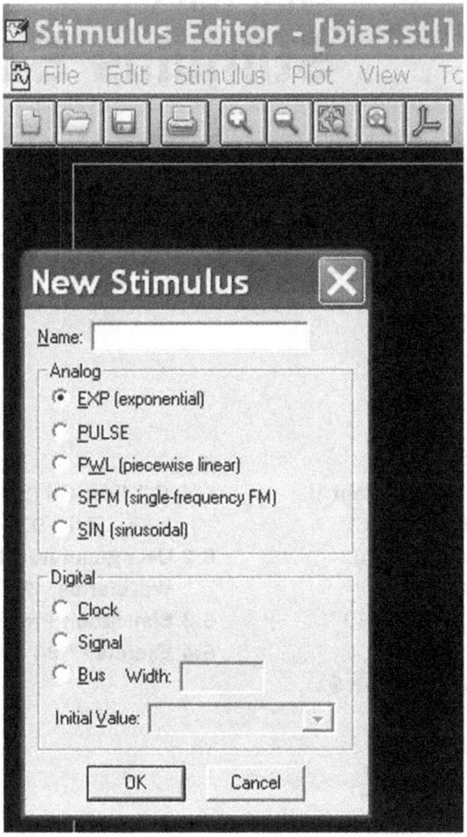

FIGURE 6.2
Stimulus Editor started.

When the Stimulus Editor starts, the **New Stimulus** window will appear as shown in Figure 6.2. Note that the stimulus file name in 16.3 has taken on the name of the PSpice simulation profile, in this case, transient.stl. In previous versions, the stimulus name takes on the name of the project name.

The New Stimulus window allows you to define analog and digital signals and prompts you to enter the stimulus name if you have not already defined the name in Capture.

6.1 STIMULUS EDITOR TRANSIENT SOURCES

6.1.1 Exponential (Exp) Source

Figures 6.3 and 6.4 show the two possible exponential waveforms which can be defined for a voltage or a current using VSTIM or ISTIM sources, respectively.

Both exponential waveforms start after a time delay (td1) and then exponentially rise or fall, using a time constant (tc1) between two voltages V1 and V2 up to a time td2. The waveform then decays or rises after td2, using a time constant (tc2).

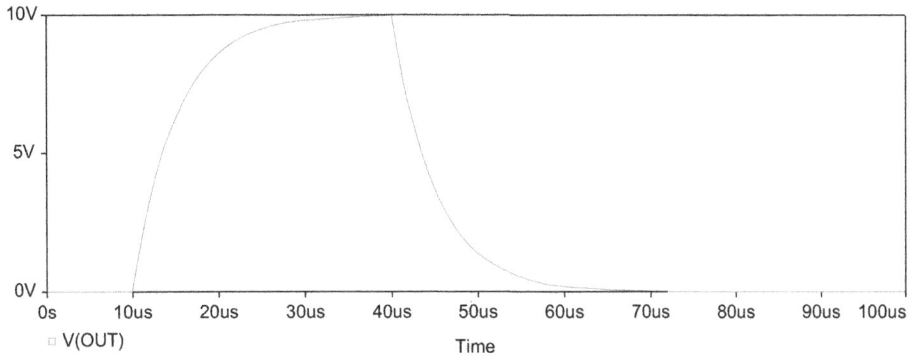

FIGURE 6.3
Exponentially rising voltage waveform.

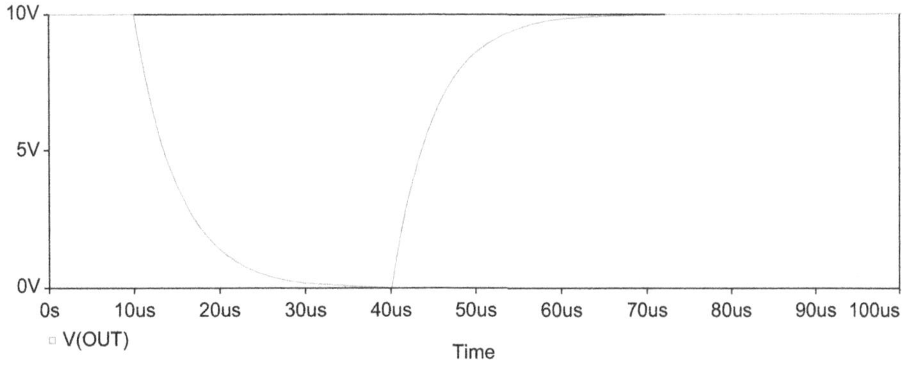

FIGURE 6.4
Exponentially decreasing voltage waveform.

For example, in Figure 6.3, the voltage is V1 (0 V) up to td1 (10 μs); then the voltage increases exponentially with a time constant given by tc1 (10 μs) towards V2 (10 V). The time for the exponential rise is defined by td2–td1 as 30 μs (40 μs– 10 μs), after which the voltage decreases exponentially with a time constant given by tc2 (5 μs) back towards V1.

Figure 6.5 shows the attribute settings for the waveform in Figure 6.3, where:

V1 – initial starting value at time 0 s,

V2 – value that voltage rises or falls to,

td1 – start time (delay) of exponential rise (or fall),

tc1 – time constant of rising (or falling) waveform,

td2 – start time (delay) of exponential fall (or rise),

tc2 – time constant of falling (or rising) waveform.

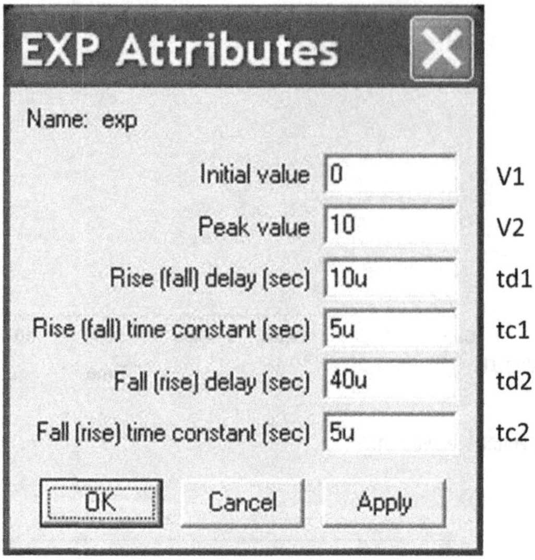

FIGURE 6.5
Exponential attributes.

Figure 6.3 was defined using V1 = 0 V, V2 = 10 V, td1 = 10 μs, tc1 = 5 μs, td2 = 40 μs and tc2 = 5 μs.

Figure 6.4 was defined using V1 = 10 V, V2 = 0 V, td1 = 10 μs, tc1 = 5 μs, td2 = 40 μs and tc2 = 5 μs.

Now the exponential voltage is defined by:

$$v(t) = (\mathrm{V2} - \mathrm{V1})\left(1 - e^{-\frac{\mathrm{time}}{\mathrm{time\ constant}}}\right)$$

So between 0 seconds and td1 the voltage is a constant:

$$v(t) = \mathrm{V1}$$

Between td1 and td2:

$$v(t) = \mathrm{V1} + (\mathrm{V2} - \mathrm{V1})\left(1 - e^{-\frac{(\mathrm{time} - \mathrm{td1})}{\mathrm{tc1}}}\right)$$

and for the time between td2 and the stop time, the voltage is given by:

$$v(t) = \mathrm{V1} + (\mathrm{V2} - \mathrm{V1})\left[\left(1 - e^{-\frac{(\mathrm{time} - \mathrm{td1})}{\mathrm{tc1}}}\right) - \left(1 - e^{-\frac{(\mathrm{time} - \mathrm{td2})}{\mathrm{tc2}}}\right)\right]$$

6.1.2 Pulse Source

Figure 6.6 shows the definition for a voltage pulse waveform, where:

V1 – low voltage,

V2 – high voltage,

TD – the time delay before the pulse starts,

TR – rise time specified in seconds, defined as the time difference between V1 and V2.

TF – fall time specified in seconds, defined as the time difference between V1 and V2.

PW – pulse width of the pulse,

PER – period of the pulse, i.e. the pulse frequency.

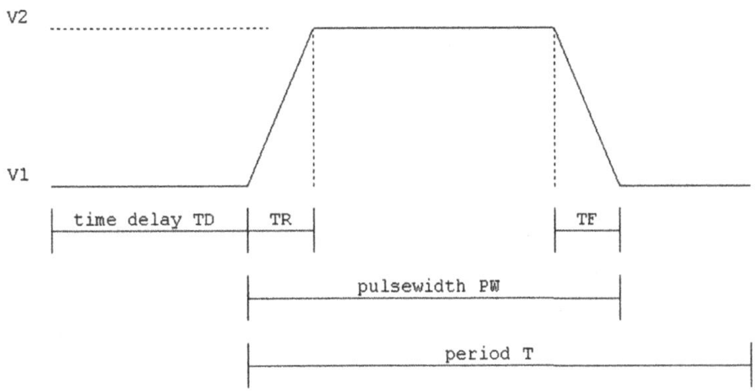

FIGURE 6.6
Pulse waveform specification.

Similarly, current pulses can be defined using the ISTIM part as shown in Figure 6.7.

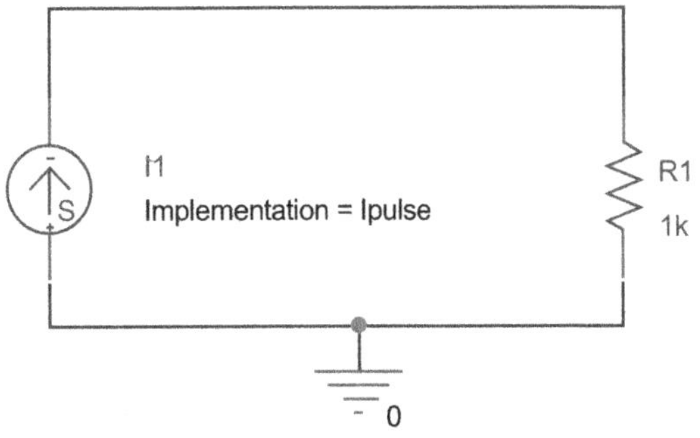

FIGURE 6.7
Using ISTIM to define a current pulse waveform.

When you first place a VSTIM, ISTIM or DigSTIM part, the implementation property name and value are shown. You only need to display the **name** of the stimulus, so double click on the **Implementation=**, which will open up the Display Properties dialog box, and select **Value Only** (Figure 6.8).

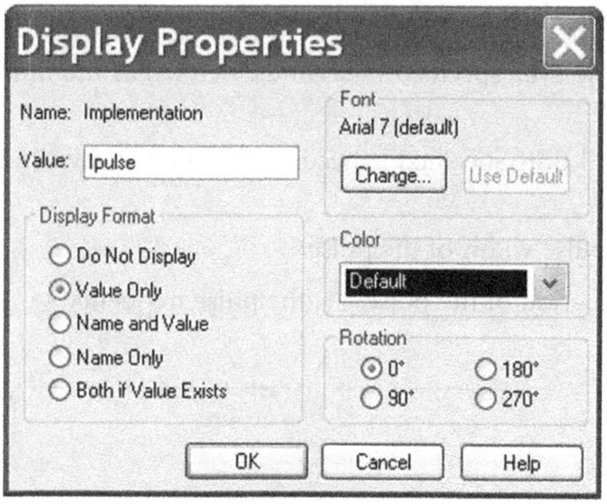

FIGURE 6.8
Making the **Implementation=** invisible.

Figure 6.9 shows the ISTIM part with the defined current stimulus, Ipulse displayed. As before, to start the Stimulus Editor, **rmb** and select **Stimulus Editor** and then select PULSE for **New Stimulus**. Figure 6.10 shows the pulse attributes defined for a current pulse using the ISTIM part. The resulting current waveform is shown in Figure 6.11.

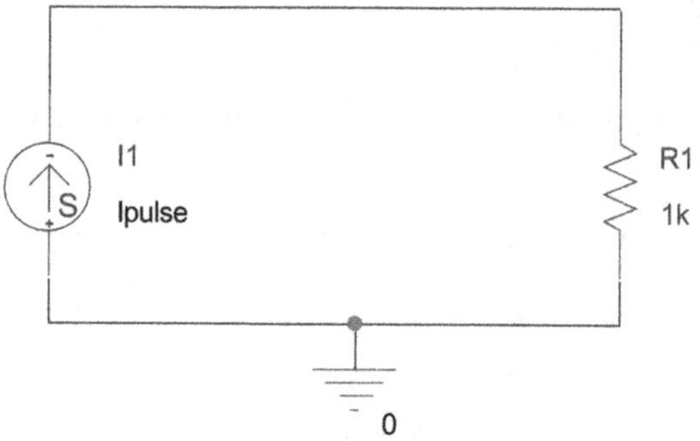

FIGURE 6.9
ISTIM part with defined current stimulus, Ipulse displayed.

6.1.3 VPWL

Piecewise linear (PWL) is where you actually draw the voltage or current waveform. You define the time and voltage (or current) axis and then use a cursor to draw the waveform. An example is given in the exercise at the end of the chapter.

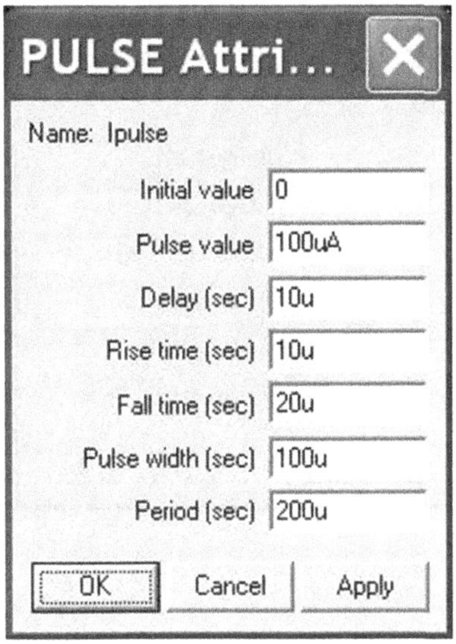

FIGURE 6.10
Current pulse attributes.

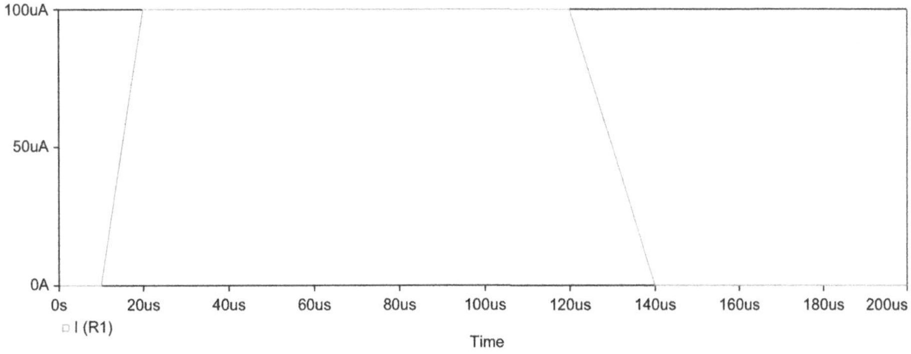

FIGURE 6.11
Current pulse waveform using the attributes in Figure 6.10.

6.1.4 SIN (Sinusoidal)

Figure 6.12 shows the attributes for a sinewave. The complete definition includes attributes for a damped sinewave, with a phase angle and an offset value. Offset value is the initial voltage or current at time 0 s, Amplitude is the maximum voltage or current, Frequency (Hz) is the number of cycles per second, Time delay (s) is the start delay, Damping factor (1/s) is the exponential decay, and Phase angle (degrees) is the phase angle.

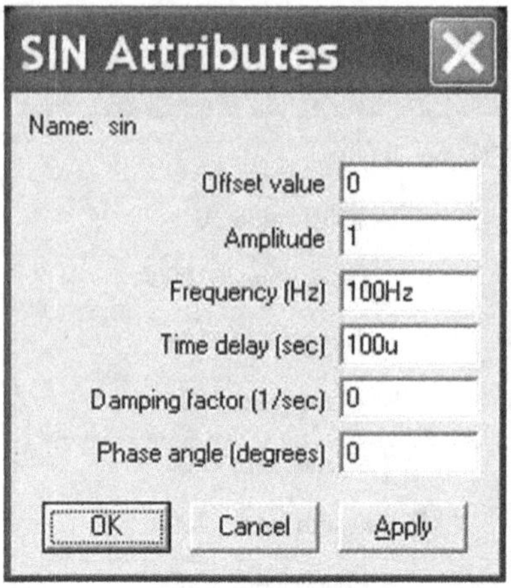

FIGURE 6.12
Sinewave attributes.

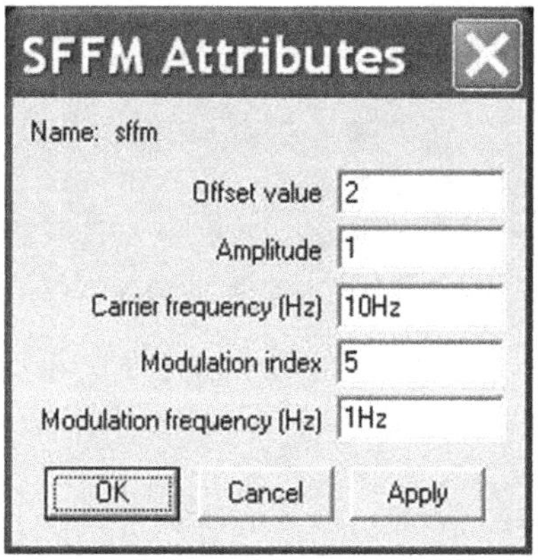

FIGURE 6.13
Frequency modulated sinewave.

6.1.5 SSFM (Single-frequency FM)

This source generates frequency-modulated sinewaves as shown in Figure 6.14, which shows the modulation of a carrier frequency. The sinewave is given by:

$$v(t) = V_{off} + V_{ampl} \times \sin[(2\pi f_c t + (\text{mod} \times \sin(2\pi f_m \text{time}))]$$

where V_{off} is the offset voltage, V_{ampl} is the maximum value of voltage, mod is the modulation index, f_c is the carrier frequency, and f_m is the modulation frequency.

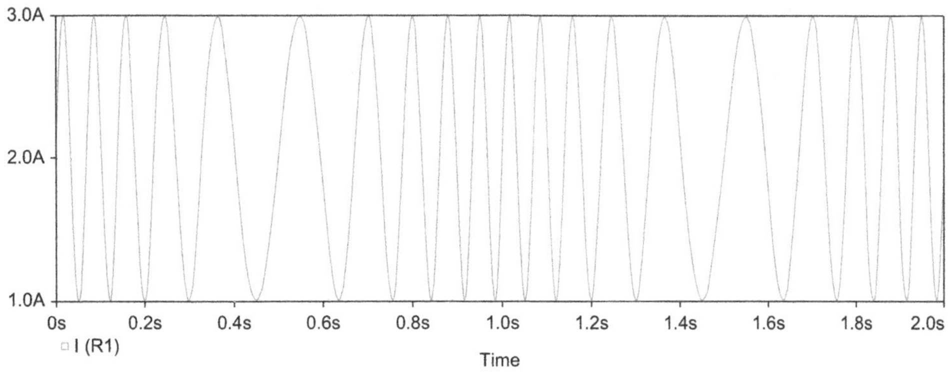

FIGURE 6.14
Single-frequency FM attributes.

6.2 USER-GENERATED TIME–VOLTAGE WAVEFORMS

You can also use the waveforms generated by a transient analysis in Probe as a time–voltage source. In Probe, select **File > Export**, which gives the options shown in Figure 6.15.

An alternative method to create a time–voltage text file is to select the trace name in Probe, select **copy** and **paste** the data into a text file.

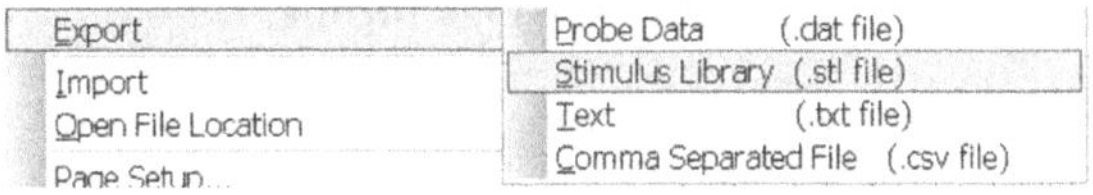

FIGURE 6.15
Exporting time–voltage data.

6.3 SIMULATION PROFILES

Prior to version 16.3, when you launch the Stimulus Editor, a stimulus file with the name of the project is created; for example a project named stimulus will create a stimulus.stl. All the stimuli you create are saved in the stimulus.stl file and so in order to select a different stimulus, all you need to do in the schematic is to change the name shown on the VSTIM, ISTIM or DigSTIM source.

From version 16.3 onwards, the stimulus file is associated with the current active simulation profile and can be accessed via the simulation profile under the Configuration Files tab. In previous versions, there were separate tabs for Stimulus, Library and Include options.

Under configured files you will see the stimulus.stl file (see Figure 6.23). If you do not see the stimulus file then you can browse for the Filename. You can then add the stimulus file to the profile (Add to Profile). However, there are other options:

- Add as Global: all designs will have access to the stimulus file
- Add to Design: only the current design will have access.

Adding the stimulus file as Global is useful if you have created a standard set of stimuli to test all your circuits. You can add several stimulus files and arrange the order by clicking on the up and down arrows. The red cross deletes the selected file.

6.4 EXERCISE

NOTE

From release 16.3 onwards, there are some differences compared to previous versions. When you first create a new project in any release, a PSpice bias simulation profile is created by default. This can be seen in the Project Manager under **PSpice Resource** > **Simulation Profiles**. In order to keep compatibility between releases, delete the bias simulation profile in the Project Manager.

1. Create a project called stimulus.
2. In the Project Manager, expand **PSpice Resources > Simulation Profiles** as shown in Figure 6.16 and delete the SCHEMATIC1-Bias profile.

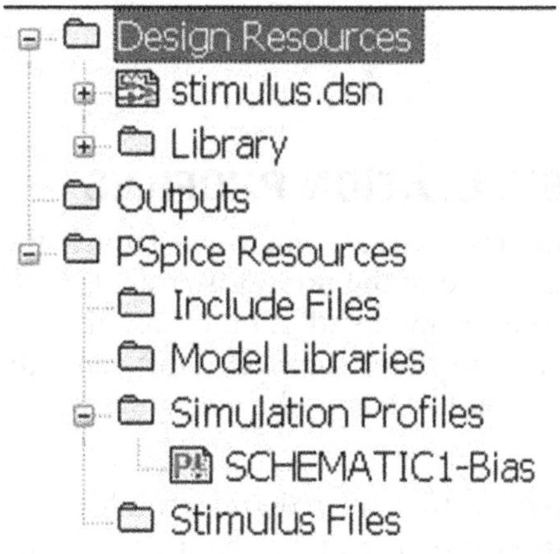

FIGURE 6.16
Bias simulation profile in Project Manager.

3. Draw the circuit diagram in Figure 6.17. The VSTIM source (V1) is from the Sourcestm library.

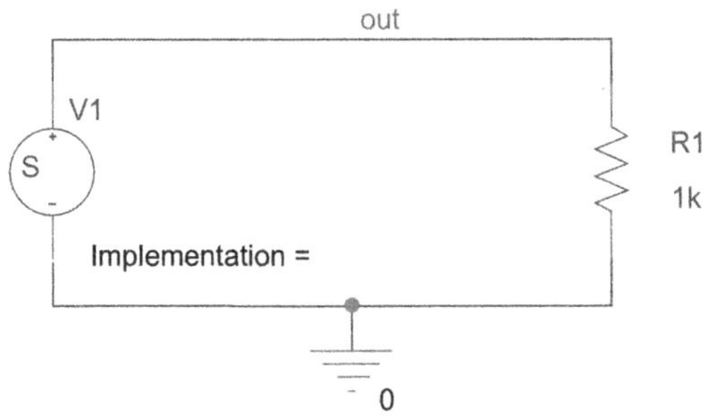

FIGURE 6.17
VSTIM 100 Hz sinewave generation.

4. Highlight VSTIM and **rmb > Edit PSpice Stimulus**. In the **New Stimulus** window, name the source as sin100Hz and select a **SIN** source (Figure 6.18).

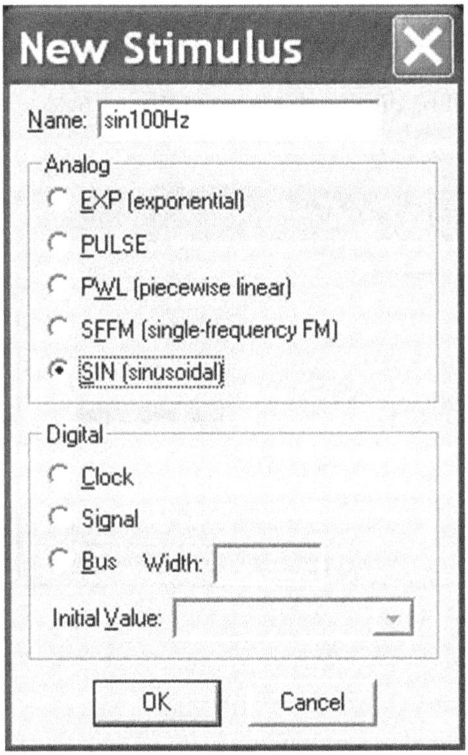

FIGURE 6.18
New sinewave source.

5. Create a 100 Hz sinewave with no offset and amplitude of 1 V. Leave all the other values at their default value of 0 (Figure 6.19). Click on OK and save the stimulus source and **Update Schematic** when prompted and exit the Stimulus Editor.

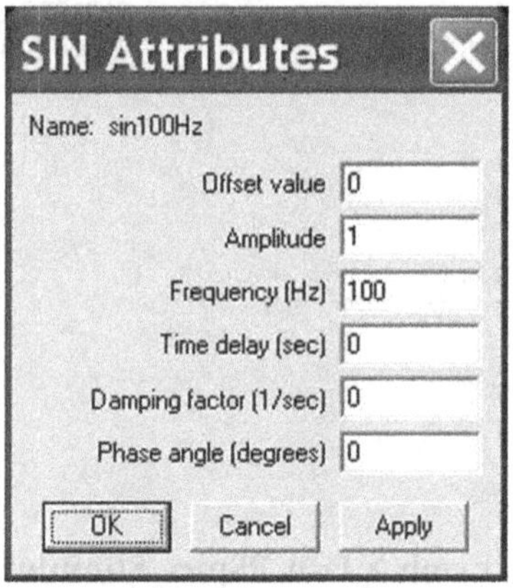

FIGURE 6.19
Creation of a 100 Hz sinewave source.

6. In Capture, the name of the stimulus is now shown on V1. Double click on **Implementation= sin100Hz** and select **Value Only** (Figure 6.20).
   Only the name of the stimulus will be displayed, as seen in Figure 6.21.

FIGURE 6.20
Making the implementation name invisible.

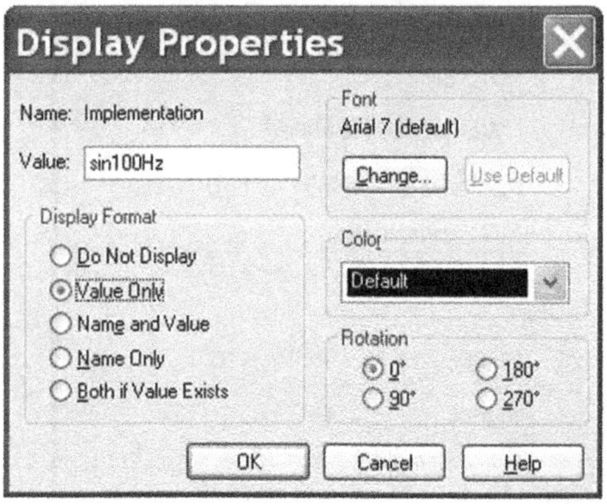

FIGURE 6.21
Displaying only the stimulus name.

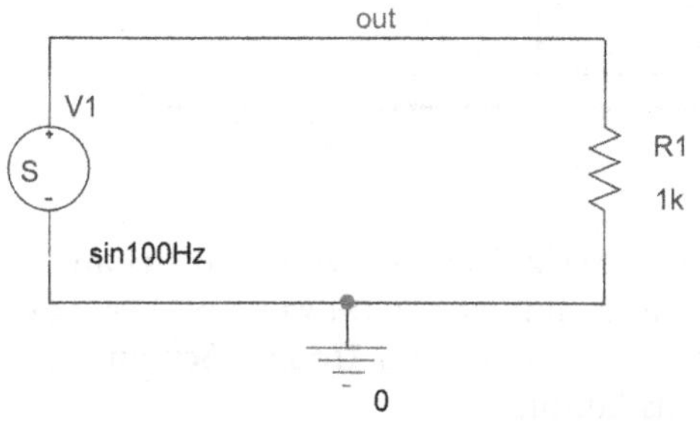

7. Create a PSpice simulation profile (PSpice > New Simulation Profile) and call it **transient**. In the **Analysis type:** pull-down menu, select **Time Domain (Transient)** and set the **Run to time:** to 20 ms (Figure 6.22). Click on **Apply** but do **not** exit the profile.

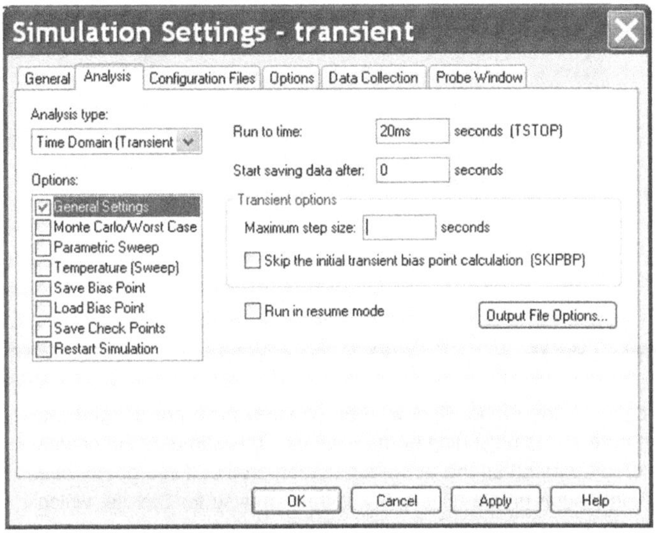

FIGURE 6.22
Simulation settings.

8. Select **Configuration Files > Category > Stimulus** and you should see the stimulus.stl file listed (Figure 6.23).
Highlight the stimulus.stl file and click on **Edit**. The Stimulus Editor will launch showing the sin100Hz stimulus. This is a quick way to see what stimuli are available in stimulus files. Close the simulation profile.

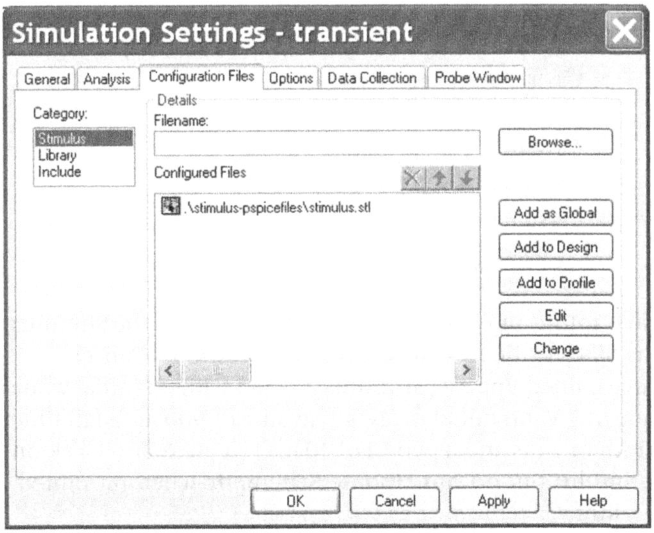

FIGURE 6.23
Stimulus files listed in Simulation profile.

9. Place a voltage marker on node **out** (Figure 6.24), run the simulation (**PSpice > Run**) and confirm that a sinewave voltage appears across the resistor.

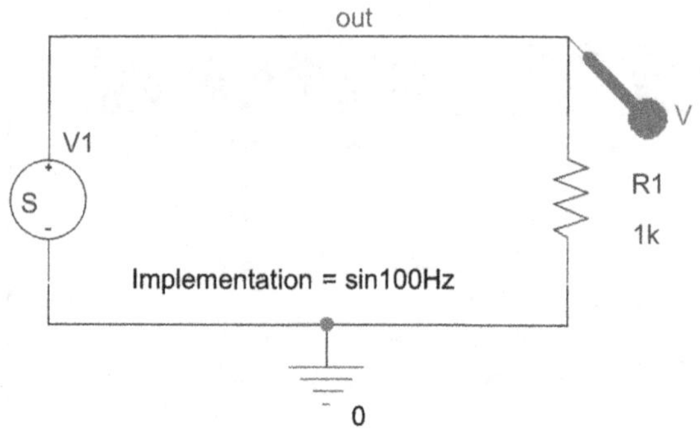

FIGURE 6.24
Placing a voltage marker.

NOTE

If you see a flat voltage line in the Probe window, then you may have not deleted the default bias.stl file. See note at the beginning of the exercise. The Stimulus Editor was then invoked with the default bias.stl as the active simulation profile. If you do not see the stimulus file in the simulation profile (Figure 6.23) then Browse for the file, which can be found in the bias folder, and then select Add to Design.

---

10. In the Project Manager, expand **PSpice > Resources > Stimulus Files** and you should see the **transient.stl** stimulus file as shown in Figure 6.25. You can double click on the file here to open the Stimulus Editor to check the stimulus.

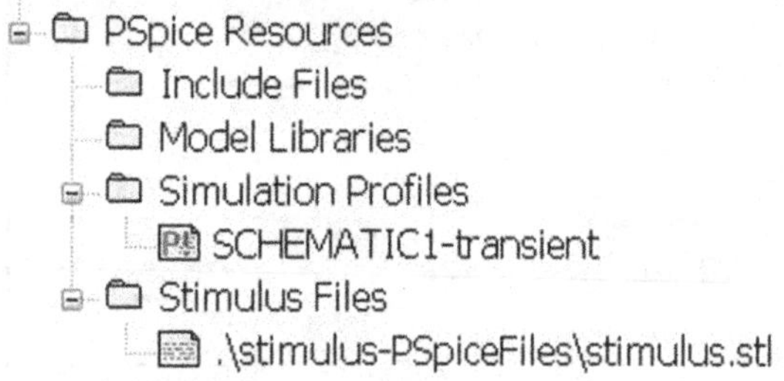

FIGURE 6.25
Location of stimulus file, transient.stl.

11. Highlight VSTIM in Capture and launch the Stimulus Editor. In the Stimulus Editor the previous SIN Attributes will be displayed. Click on Cancel.
12. Create a pulse source named **Vpulse (Stimulus > New)** with an initial value of 0 V, an amplitude of 1 V, no initial delay, a rise time of 500 μs, a fall time of 1 ms, a pulse width of 2 ms and a period of 10 ms (Figure 6.26). Click on OK and save the stimulus but **do not Update Schematic** when prompted. Exit the Stimulus Editor.

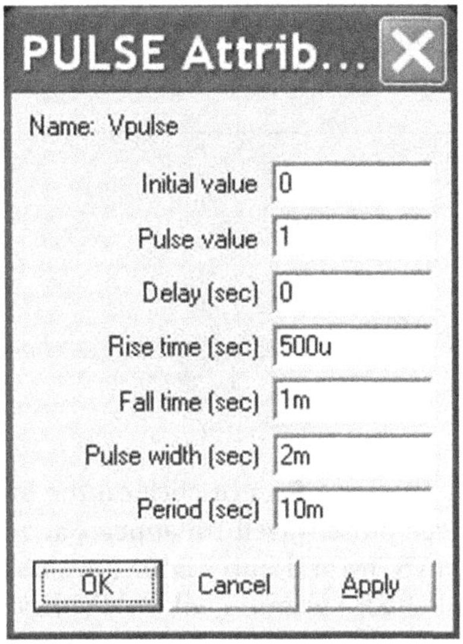

FIGURE 6.26
Vpulse attributes.

NOTE

When entering attribute values, press the TAB button on the keyboard to move down to the next attribute box.

NOTE

The stimulus is already named as sin100Hz in Capture and cannot be updated from the Stimulus Editor. In previous software releases, if you say yes to Update Schematic, the cursor will change to an hourglass and just sit there. You will need to switch to Capture, where you will see the dialog box in Figure 6.27. Just click on OK.

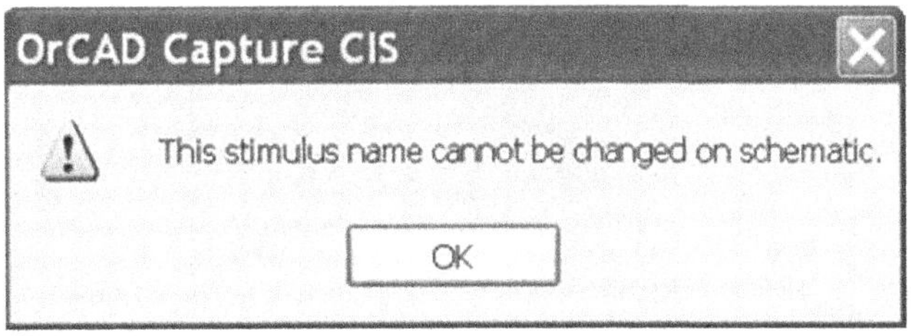

FIGURE 6.27
Stimulus name change warning.

---

13. In Capture, double click on the stimulus name, sin100Hz, and change it to Vpulse as shown in Figure 6.28.

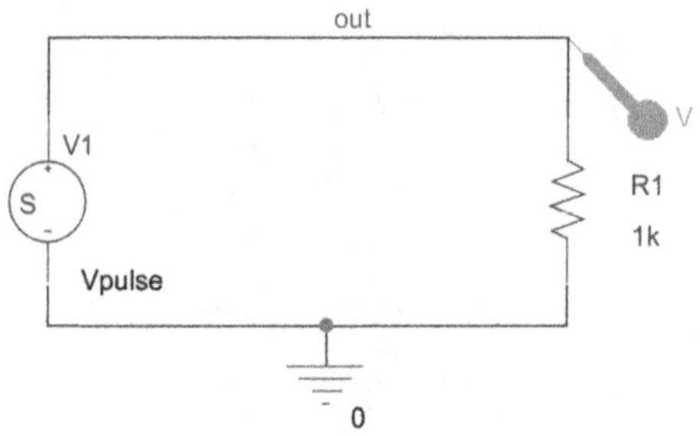

FIGURE 6.28
Using the Vpulse source.

14. Run the simulation. **PSpice > Run** or click on the blue play button.
15. Confirm that a voltage pulse waveform appears across the resistor.
16. In Capture, highlight VSTIM and **rmb > EditStimulus Editor**. In the Stimulus Editor the previous PULSE Attributes will be displayed. Click on Cancel.
17. Create a new stimulus, **Stimulus > New**. Name the stimulus Vin and select PWL (piecewise-linear).

NOTE

You may be asked if you want to change the axis settings.

---

18. From the top toolbar, select **Plot > Axis Setting**. Set the waveform drawing resolution to that shown in Figure 6.29.

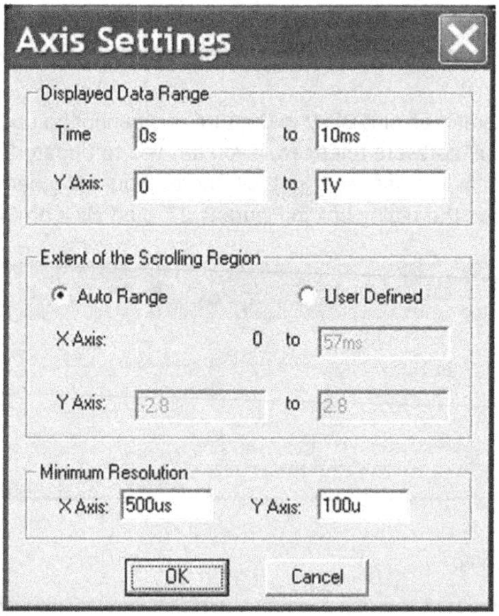

FIGURE 6.29
Axis settings.

19. A pen cursor will appear. Draw a corresponding piecewise linear graph approximating the PWL voltage shown in Figure 6.30. The first point at (0,0) has already been selected. The accuracy does not matter as long as there are three peaks defined. This stimulus will be used in Chapter 7 on transient analysis. Press escape to exit draw mode.

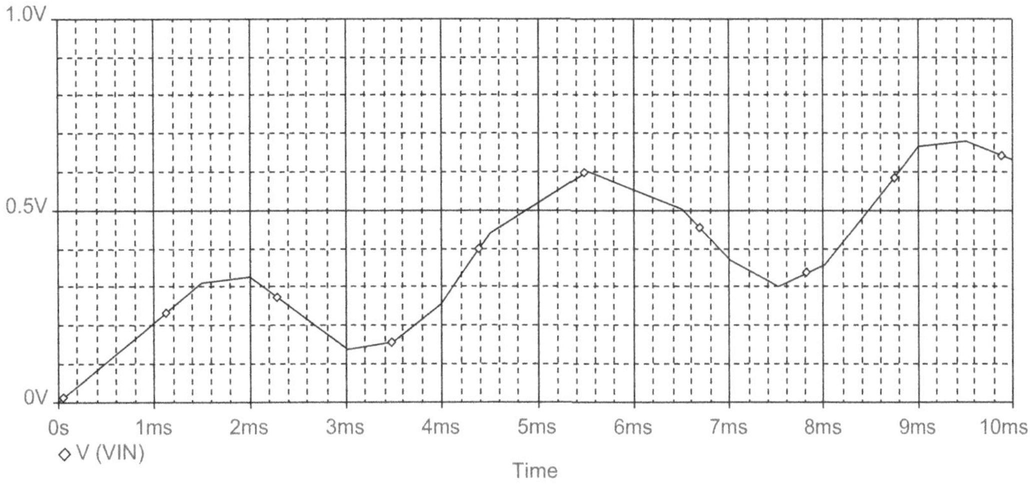

FIGURE 6.30
Piecewise linear waveform.

20. If you want to delete or move a point, press escape out of draw mode and place the cursor on a point, which will turn red, and then delete or move the point. To return to draw mode, select **Edit > Add** or select the icon  or .

NOTE

Prior to release 16.3, you can only place points forward in time; you cannot go backwards. If you want to delete or move a point, press escape out of draw mode and place the cursor on a point, which will turn red, and then delete or move the point.

---

21. Save the stimulus file and exit the Stimulus Editor but **do not Update Schematic**.

22. Change the name of the stimulus from Vpulse to Vin and simulate, and confirm that the piecewise voltage waveform appears across the resistor.

NOTE

The Vin source will be used in Chapter 7.

CHAPTER 7

Transient Analysis

Chapter Outline

Transient analysis calculates a circuit's response over a period of time defined by the user. The accuracy of the transient analysis is dependent on the size of internal time steps, which together make up the complete simulation time known as the **Run to time** or **Stop time**. However, as mentioned in Chapter 2, a DC bias point analysis is performed first to establish the starting DC operating point for the circuit at time $t = 0$ s. The time is then incremented by one pre-determined time step at which node voltages and current are calculated based on the initial calculated values at time $t = 0$. For every time step, the node voltages and currents are calculated and compared to the previous time step DC solution. Only when the difference between two DC solutions falls within a specified tolerance (accuracy) will the analysis move on to the next internal time step. The time step is dynamically adjusted until a solution within tolerance is found.

For example, for slowly changing signals, the time step will increase without a significant reduction in the accuracy of the calculation, whereas for quickly changing signals, as in the case of a pulse waveform with a fast leading edge rise time, the time step will decrease to provide the required accuracy. The value for the maximum internal time step can be defined by the user.

If no solution is found, the analysis has failed to converge to a solution and will be reported as such. These convergence problems and solutions will be discussed in more detail in Chapter 8.

Analog Design and Simulation using OrCAD Capture and PSpice. DOI: 10.1016/B978-0-08-097095-0.00007-6

There are some circuits where a DC solution cannot be found, as in the case of oscillators. For these circuits, there is an option in the simulation profile to skip over the initial DC bias point analysis. If you add an initial condition to the circuit, the transient analysis will use the initial condition as its starting DC bias point.

7.1 SIMULATION SETTINGS

Figure 7.1 shows the PSpice simulation profile for a transient (time domain) analysis. In this example, the simulation time has been set to 5 μs. The **Start saving data after:** specifies the time after which data are collected to plot the resulting waveform in Probe in order to reduce the size of the data file.

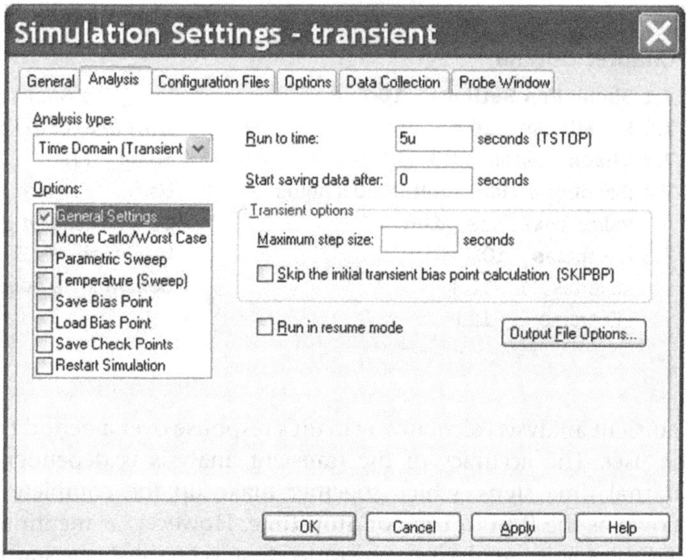

FIGURE 7.1 Transient analysis simulation profile.

Maximum step size: defines the maximum internal step size, which is dependent on the specified **run to time** but is nominally set at the **run to time** divided by 50.

Skip the initial transient bias point calculation will disable the bias point calculation for a transient analysis.

7.2 SCHEDULING

Scheduling allows you to dynamically alter a simulation setting for a transient analysis; for example, you may want to use a smaller step size during periods that require greater accuracy and relax the accuracy for periods of less activity. Scheduling can also be applied to the simulation settings runtime parameters, RELTOL, ABSTOL, VNTOL, GMIN and ITL, which can be found in **PSpice > Simulation Profile > Options**. You replace the parameter value with the scheduling command, which is defined by:

{SCHEDULE(t1,v1,t2,v2...tn,tn)}

Note that t1 always starts from 0.

For example, it may be more efficient to reduce the relative accuracy of simulation from 0.001% to 0.1%, RELTOL, during periods of less activity by specifying a change in accuracy every millisecond. The format will be defined as:

{schedule(0,0, 1m,0.1, 2m,0.001, 3m,0.1, 4m,0.001)}

The simulation settings will be discussed in more detail in Chapter 8.

7.3 CHECK POINTS

Check points were introduced in version 16.2 to allow you to effectively mark and save the state of a transient simulation at a check point and to restart transient simulations from defined check points. This allows you to run simulations over selective periods. This is useful if you have convergence problems in that you can run the simulation from a defined check point marked in time before the simulation error, rather than having to run the whole simulation from the beginning.

Check points are only available for a transient simulation and are selected in the simulation profile in **Analysis > Options** box (Figure 7.2) as **Save Check Points** and **Restart Simulation**. Check points are defined by specifying the time interval between check points. The simulation time interval is measured in seconds and the real time interval is measured in minutes (default) or hours. The time points are the specific points when the check points were created.

Before you restart a simulation from a saved check point, you can change component values, parameter values, simulation setting options, check point restart and data save options. Figure 7.3 shows the **Restart Simulation** option selected.

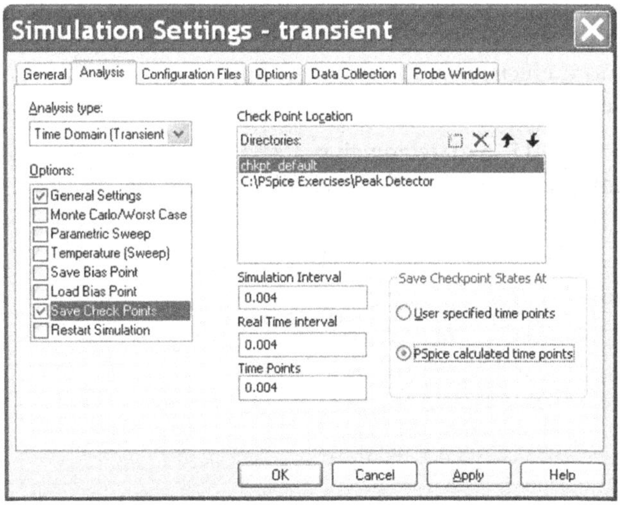

FIGURE 7.2
Saving a check point.

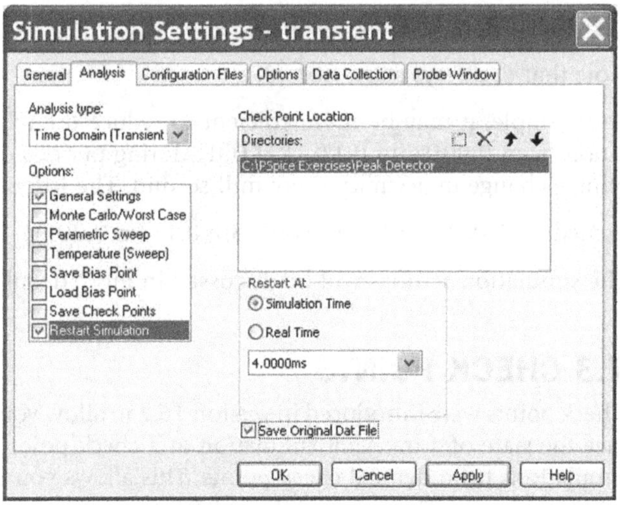

FIGURE 7.3
Restarting a simulation using a saved check point.

The saved check point data are set to simulation time in seconds such that **Restart At** shows 4 ms, which was specified in the saved check point data file. The simulation will then start at 4 ms using the saved state of the transient simulation.

7.4 DEFINING A TIME–VOLTAGE STIMULUS USING TEXT FILES

The piecewise linear stimulus was introduced in Chapter 6, where a graphically drawn voltage waveform was used as an input waveform to a circuit. Input waveforms can also be defined using pairs of time–voltage coordinates, which can be entered in the Property Editor or read from an external text file.

Figure 7.4 shows the voltage VPWL and current IPWL sources and the corresponding time and voltage properties (Figure 7.5) in the Property Editor. By default, eight time–voltage pairs are displayed in the Property Editor for the VPWL and IPWL parts, but, as seen in Figure 7.5, more time–voltage pairs have been added. It is more efficient and easier to define a large number of time–voltage pairs in a text file.

Figure 7.6 shows the VPWL_FILE part referencing a text file which contains time–voltage pairs as shown in Figure 7.7. For example, at 1 ms the voltage is

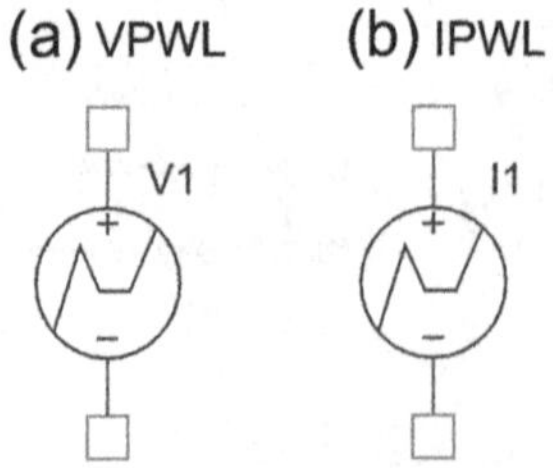

FIGURE 7.4
Piecewise linear sources for (a) voltage and (b) current.

T1	0
T2	1ms
T3	2ms
T4	3ms
T5	4ms
T6	5ms
T7	6ms
T8	7ms
T9	8ms
T10	9ms
T11	10ms
V1	0
V2	0.2055
V3	0.3273
V4	0.1382
V5	0.2852
V6	0.5182
V7	0.5527
V8	0.3727
V9	0.3584
V10	0.6673
V11	0.6291
Value	VPWL

FIGURE 7.5
VPWL and IPWL time–voltage properties displayed in the Property Editor.

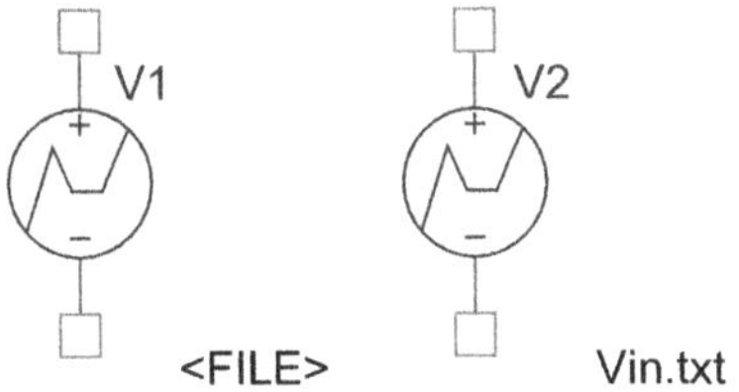

FIGURE 7.6
Piecewise linear part VPWL_FILE referencing a file.

```
* Stimulus Vin
0, 0
0.001, 0.2055
0.0015, 0.3109
0.002, 0.3273
0.0025, 0.2345
0.003, 0.1382
0.0035, 0.1564
0.004, 0.2582
```

FIGURE 7.7
Time–voltage data points describing the input voltage waveform, V_{in}.

0.2055 V, at 2 ms 0.3273 V, and so on. It is always a good idea to make the first line a comment as PSpice normally ignores the first line.

When you reference a text file such as Vin.txt, you need to specify the location of the text file. You can use absolute addressing specifying the direct path to the file or relative addressing specifying the path location relative to the project location. Figure 7.8 shows the hierarchy of a project showing the different folders in which the Vin.txt file can be placed and the corresponding <FILE> name for the referenced Vin.txt on the VPWL_FILE part.

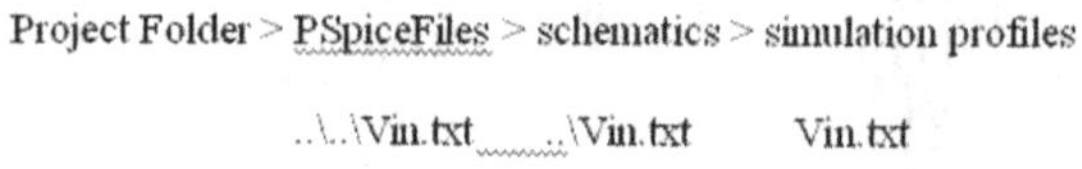

FIGURE 7.8
Referencing the Vin.txt time–voltage text file for VPWL_FILE.

For example, if you place the Vin.txt file in the same folder which contains the schematics, then you enter ..\Vin.txt in the <FILE> property of the VPWL_FILE.

Project Folder > PSpiceFiles > schematics > simulation profiles
..\..\Vin.txt ..\Vin.txt Vin.txt

You can also provide an absolute path to a text file. For example, if you had a folder named stimulus, then you enter C:\stimulus\Vin.txt.

In the **source** library there are other VPWL and IPWL parts which allow you to make a VPWL periodic for a number of cycles or repeat forever. These are given as:

VPWL_F_RE_FOREVER
VPWL_F_RE_N_TIMES
VPWL_RE_FOREVER
VPWL_RE_N_TIMES
IPWL_F_RE_FOREVER
IPWL_F_RE_N_TIMES
IPWL_RE_FOREVER
IPWL_RE_N_TIMES

The above source will be introduced in the exercises.

7.5 EXERCISES

Exercise 1

This exercise will demonstrate the effect that the maximum time step has on the resolution of a simulation and introduce the use of the scheduling command.

1. Draw the circuit in Figure 7.9, which consists of a VSIN source from the **source** library, connected to a load resistor R1.

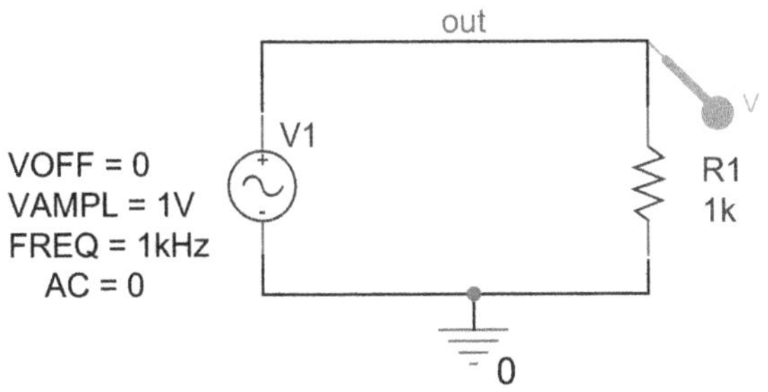

FIGURE 7.9
Sinewave voltage applied to a load resistor.

2. Create a PSpice simulation profile called **transient** and select **Analysis type:** to **Time Domain (Transient)** and enter a **Run to time** of 10 ms, which will display 10 cycles of the sinewave (Figure 7.10).

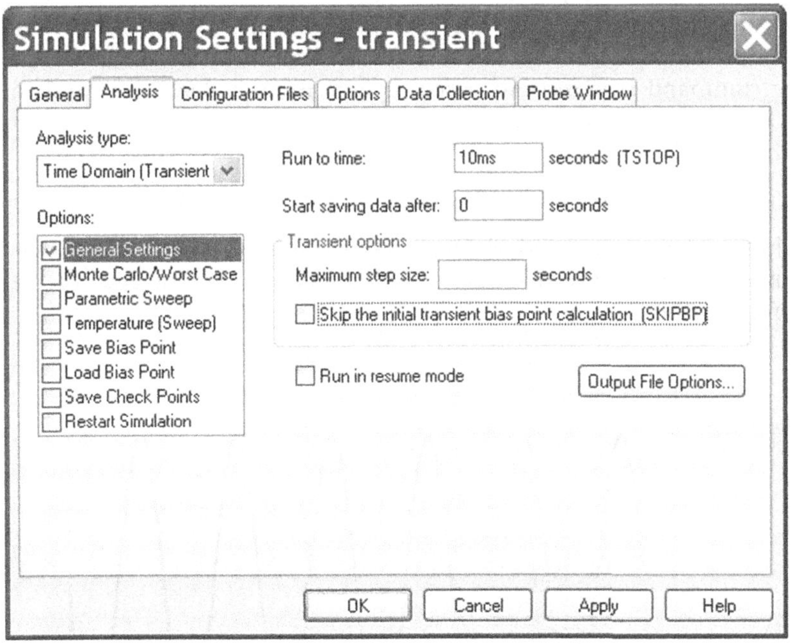

FIGURE 7.10
Simulation settings for a transient analysis.

Place a voltage marker on node '**out**' and run the simulation. You should see the resultant waveform as shown in Figure 7.11, which is lacking in resolution.

3. In Probe select **Tools > Options** and check the box for **Mark Data Points** or click on the icon . You will see the data points that make up the sinewave.

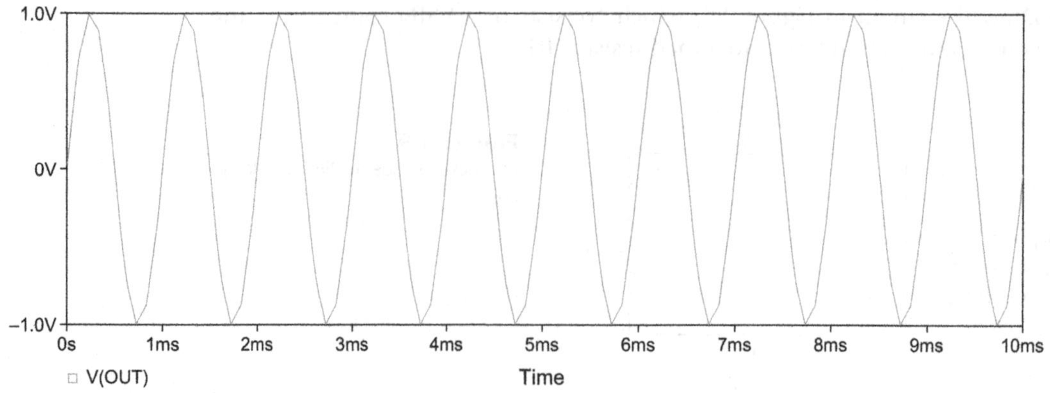

FIGURE 7.11
Distorted resultant sinewave lacking resolution.

4. In the simulation profile, set up a schedule command to decrease the time step at set time points. You can enter the schedule command in the **Maximum step size** box, but because of the small field in which to type in the command it is recommended to type the schedule command in a text editor such as Notepad and cut and paste the following command into the box:

   {schedule(0,0, 2m,0.05m, 4m,0.01m, 6m,0.005m, 8m,0.001m)}

5. Run the simulation. As the **Mark Data Points** is still on, you should see the resolution of the waveform improve with a decrease in the limit of the maximum step size (Figure 7.12).

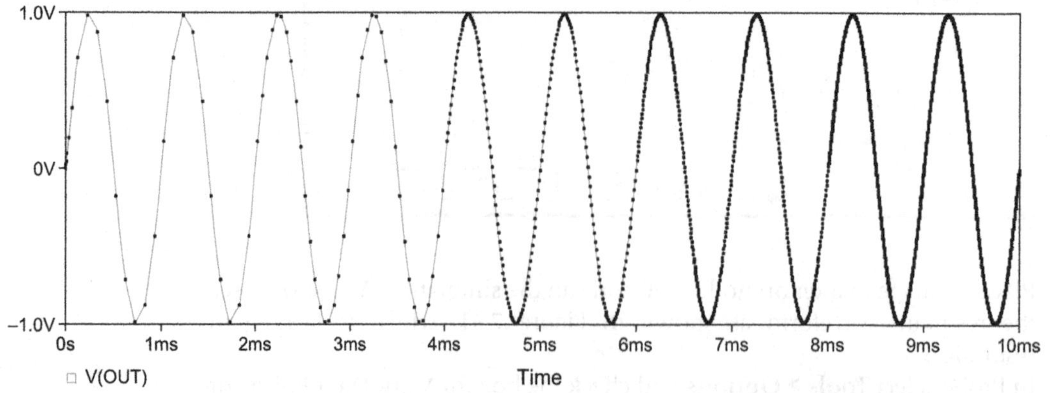

FIGURE 7.12
Improved sinewave resolution with a successive decrease of the maximum time step using the schedule command.

Exercise 2

Figure 7.13 shows a peak detector circuit, where the input stimulus can be either the Vin Sourcstm source created in the Stimulus Editor exercise or a file containing a time–voltage definition of the input waveform. Both implementations will be described.

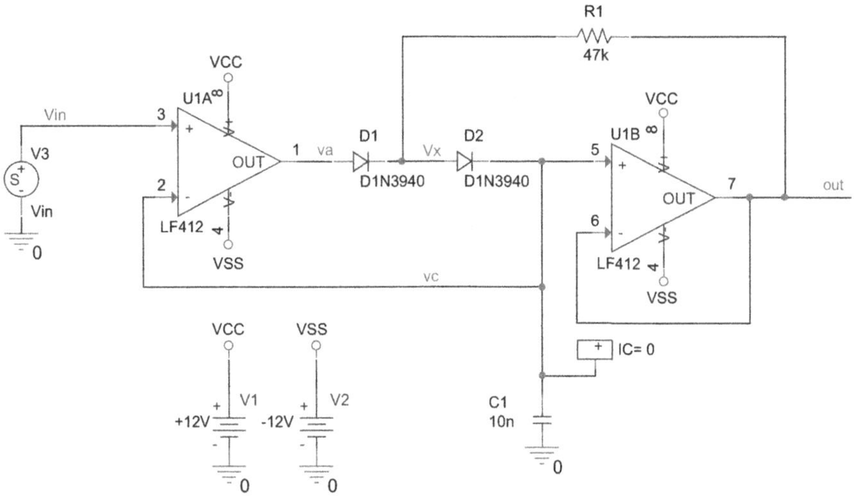

FIGURE 7.13
Peak detector circuit.

1. Create a project called Peak Detector and draw the circuit in Figure 7.13. If you are using the demo CD, use the uA741 opamps.
2. Rename the SCHEMATIC1 folder to Peak Detector.
3. You need to set up an initial condition (IC) on the capacitor, C1, by using an IC1 part from the special library. This ensures that at time $t=0$, the voltage on the capacitor is 0 V 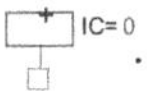.

   Alternatively, you can double click on the capacitor, C1, and in the **Property Editor** enter a value of 0 for the IC property value (Figure 7.14). This ensures that at time $t=0$, the voltage on the capacitor is 0 V. If you change the capacitor, then you have to remember to set the initial condition, whereas an IC1 part will always be visible on the schematic.

FIGURE 7.14
Setting an initial value of 0 V on the capacitor.

4. Create a simulation profile, make sure you name it **transient** and set the run to time to 10 ms. Close the simulation profile.

   Two methods are described to define the input waveform V_{in} for the Peak Detector.

USING THE GRAPHICALLY CREATED WAVEFORM IN THE STIMULUS EDITOR

5. For the input stimulus, using the predefined Vin sourcestm in Chapter 6, edit the simulation profile and select the **Configuration Files** tab, select **Category** to **Stimulus** and **Browse** to the location of the **stimulus.stl** file. Click on **Add to Design** as shown in Figure 7.15. An explanation of stimulus files added to the simulation profile was given in Chapter 6.

FIGURE 7.15 Adding the stimulus.stl file to the simulation profile.

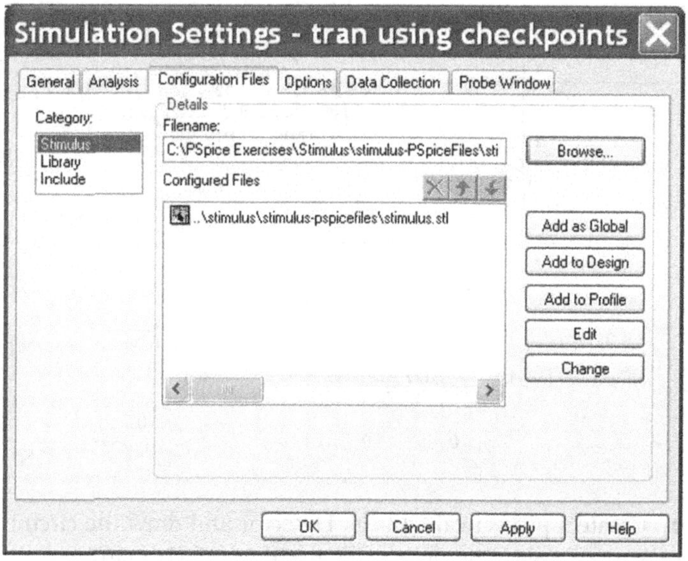

6. Check the stimulus by highlighting the stimulus name and click on **Edit**. This will launch the Stimulus Editor and display the V_{in} waveform. Close the simulation profile.
7. Go to Step 12.

USING A FILE WITH TIME–VOLTAGE DATA DESCRIBING THE INPUT WAVEFORM

8. Enter the time–voltage data points in Figure 7.16 in a text editor such as Notepad. By default, the simulator ignores the first line, so do **not** enter data on the first line. However, it is always a good idea to add a description or a comment to the data file using an asterisk * character to describe, for example, what the data is. The simulator will ignore any lines beginning with a * character.
   Name the file Vin and save the file as a text file in the PSpice folder for the Project, **Peak Detector > peak detector-PSpiceFiles** (Figure 7.17).
   Make sure the file has been saved with a .txt extension as Vin.txt.

```
* Stimulus Vin
0, 0
0.001, 0.2055
0.0015, 0.3109
0.002, 0.3273
0.0025, 0.2345
0.003, 0.1382
0.0035, 0.1564
0.004, 0.2582
0.0045, 0.44
0.005, 0.5182
0.0055, 0.6018
0.006, 0.5527
0.0065, 0.5018
0.007, 0.3727
0.0075, 0.3
0.008, 0.3564
0.0085, 0.5109
0.009, 0.6673
0.0095, 0.6782
0.01, 0.6291
```

FIGURE 7.16
Time–voltage data points describing the input voltage waveform, V_{in}.

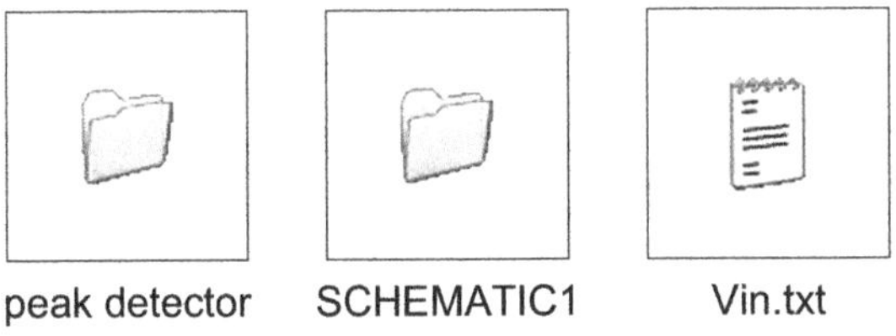

FIGURE 7.17
Place the V_{in} text file in the PSpiceFiles folder.

9. Place a VPWL_FILE from the **source** library and rename the <FILE> shown in Figure 7.18 as ..\..\Vin.txt

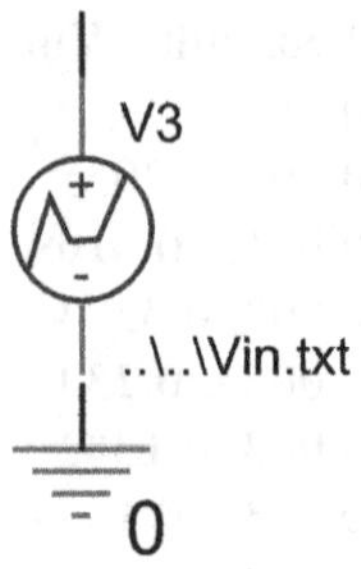

FIGURE 7.18
Adding a VPWL_FILE.

10. Your Peak Detector circuit will be as shown in Figure 7.19.

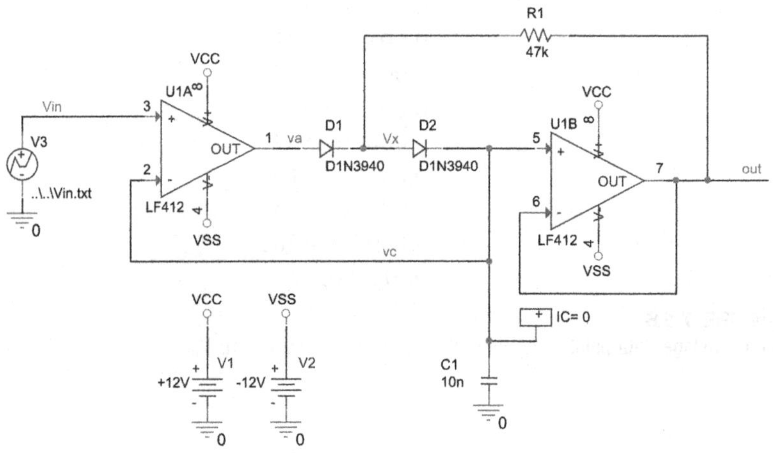

FIGURE 7.19
Peak detector circuit using a text file.

11. Go to Step 12.
12. Place voltage markers on nodes **in** and **out** and run the simulation. Figure 7.20 shows the simulation response of the peak detector to the input voltage V_{in}.

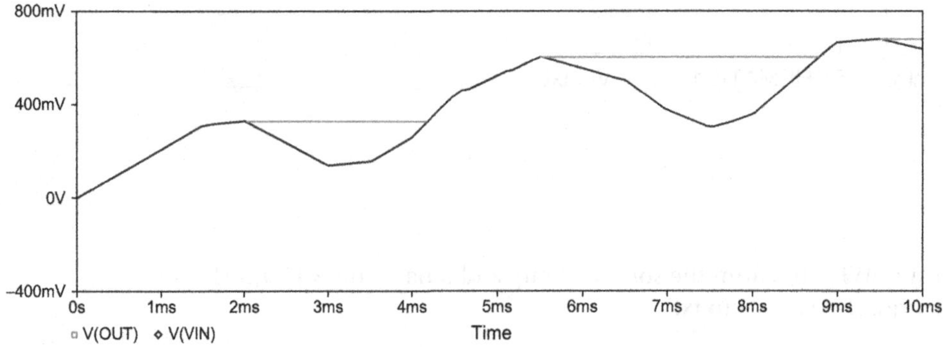

FIGURE 7.20
Peak detector transient response.

GENERATING A PERIODIC V_{IN}

13. Delete the VPWL_FILE source and replace it with a VPWL_F_RE_FOREVER from the source library. Double click on <FILE> and, as in Step 9, enter ..\..\Vin.txt

14. Edit the PSpice Simulation Profile, increase the simulation run to time to 50 ms and run the simulation. You should see the response as shown in Figure 7.21, where V_{in} is now periodic (repeats forever).

15. Investigate the VPWL_F_RE_N_TIMES source.

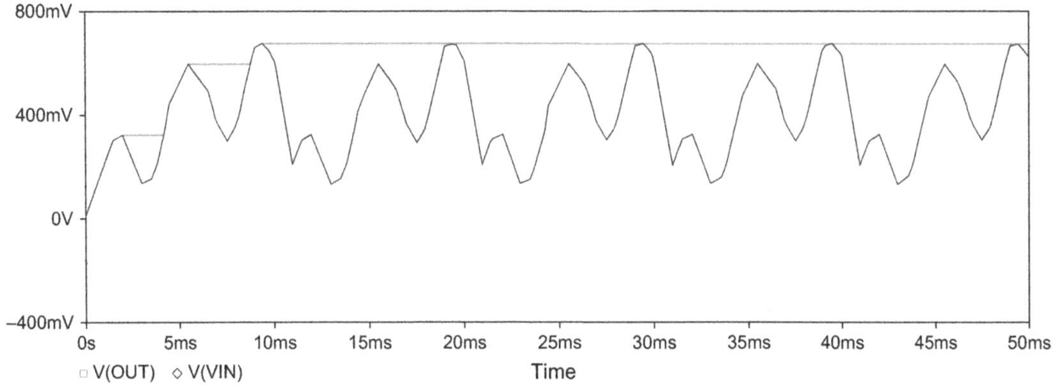

FIGURE 7.21
The V_{in} signal is periodic.

CHAPTER 8

Convergence Problems and Error Messages

Chapter Outline

PSpice uses the Newton–Raphson iteration method to calculate the nodal voltages and currents for non-linear circuit equations. The algorithm will start off with an initial 'guess' to the solution and perform an iterative process until the voltages and currents converge to a consistent solution. As discussed in Chapter 7, at every time step, the node voltages and currents are calculated and compared to the previous time step DC solution. Only when the difference between two DC solutions falls within a specified tolerance (accuracy) will the analysis move on to the next internal time step. The time step is dynamically adjusted until a solution within tolerance is found. However, if a solution cannot be found, PSpice will report that the simulation has failed owing to a convergence problem. There are also occasions where the time step becomes too small for the iteration process to continue. This can occur if there is a fast moving signal in the circuit such as a pulse with a very short unrealistic rise time.

Simulations can also fail because of circuit errors and missing or incorrect parameters specified. Some of these common errors will be discussed in this chapter.

8.1 COMMON ERROR MESSAGES

```
Error - Node <name> is floating
```

There is no zero volt node '0' in the circuit. See Chapter 2, Exercise 1, and Chapter 3, Exercise 1.

Analog Design and Simulation using OrCAD Capture and PSpice. DOI: 10.1016/B978-0-08-097095-0.00008-8

```
Error - Missing DC path to ground
```

There is no direct DC path to ground at a node. Add a large value resistor from the node either to the 0 node or to a DC ground path. See Chapter 3, Exercise 1.

```
Error - Less than two connections at node <name>
```

There is no PSpice model attached to the Capture part. The Capture part is missing the required properties for PSpice simulation. This error can also happen if nets are left floating in that a wire is left 'dangling'.

```
Error - Voltage source or inductor loop
```

Voltage sources are modeled as ideal in that they have no internal resistance; therefore, connecting two voltage sources in parallel will result in an infinite current, which will exceed the maximum current limit. Voltages and currents are limited to ± 1e10 V and ± 1e10 A.

An inductor is essentially a time-varying voltage source and if connected in parallel with another inductor or voltage source, the same error message will result. Inductors are modeled as ideal in that they have zero series winding resistance.

8.2 ESTABLISHING A BIAS POINT

The bias point analysis is the starting point for a transient analysis and a DC sweep. However, if PSpice cannot calculate the bias point for a circuit, the power supplies will be reduced from 100% towards zero, where the non-linearities of the circuit will effectively be linearized and hence improve the chances of a bias point solution being found. The power supplies are then stepped back up to 100% to establish a bias point upon which a DC sweep or a transient analysis can be started.

When you run a transient analysis, PSpice launches and the simulation progress is shown in the **Simulation Status Window** and **Output Window**. Figure 8.1 shows an example of the Output window reporting that the bias point was calculated, the transient analysis was started and finished and the simulation is complete.

FIGURE 8.1
Output Window in PSpice.

Reading and checking circuit
Circuit read in and checked, no errors
Calculating bias point for Transient Analysis
Bias point calculated
Transient Analysis
Transient Analysis finished
Simulation complete

The information displayed helps us to determine whether the convergence failed during the bias point analysis or the subsequent transient, DC or AC analysis.

8.3 CONVERGENCE ISSUES

If there is a convergence issue then the simulation pauses and the PSpice Runtime Settings window appears as shown in Figure 8.2. You can then change a simulation parameter and resume the simulation.

The output file will also be displayed and will report the last node voltages tried and which devices failed to converge, which may relate to the nodes where the problem is.

There are no clearly defined rules for solving convergence problems. What you need to do is to try to localize the problem. For large circuits, methodically remove and simulate smaller portions of the circuit. A hierarchical design, made up of blocks of circuitry, can help in solving convergence problems. Each block can be simulated separately and successively in the hierarchy, building up to the complete design. Hierarchical designs are covered in Chapter 20.

Another approach for large circuits is to replace parts of circuits with Analog Behavioral Models (ABMs), which use mathematical expressions or tables to model components or circuit behavior. These devices simulate more quickly and can help to localize which circuits are not converging by a process of replacement and elimination. However, ABMs, if not used properly, can cause convergence problems in their own right, especially if a mathematical expression contains a denominator variable, which can under certain circuit conditions be set to equal zero, resulting in large numbers exceeding the PSpice limits of $\pm$ 1e10 V and $\pm$ 1e10 A.

Small circuits consisting of a few components can also cause convergence problems; even the humble diode, if not modeled with a series resistance, can cause currents and voltages to exceed the PSpice limits of $\pm$ 1e10 V and $\pm$ 1e10 A.

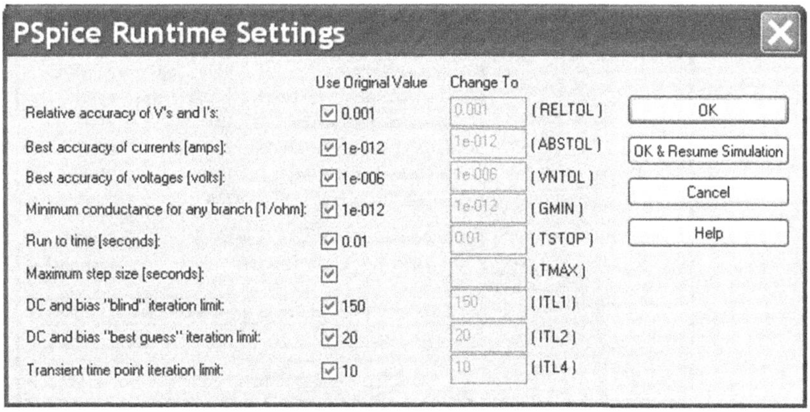

FIGURE 8.2
PSpice Runtime Settings.

Ideally, models from semiconductor vendors are complete, have been tested for simulation and should not be the main cause of convergence problems. The only issue is that some semiconductor models are represented by a subcircuit, especially power MOSFETs, which can lead to convergence problems.

8.4 SIMULATION SETTINGS OPTIONS

The simulation settings can be accessed via the simulation profile and selecting the Options tab as shown in Figure 8.3.

The calculated voltages and currents are based on previously calculated values, such that the condition for convergence, for a node voltage, is given by:

$$|v(n-1) - v(n)| > RELTOL^{*}v(n) + VNTOL \qquad (8.1)$$

Similarly, for a branch current:

$$|i(n-1) - i(n)| > RELTOL^{*}i(n) + ABSTOL \qquad (8.2)$$

where RELTOL is the relative tolerance and VNTOL and ABSTOL are the absolute tolerances for voltage and currents, respectively. RELTOL has a default value of 0.001, which is equivalent to a 0.1% accuracy. Convergence will only occur when the relative differences between consecutive voltages and consecutive currents are calculated within the specified accuracy of simulation.

If you have a high-voltage circuit in which the output voltage is 100 V, then VNTOL which by default is 1 μV, can be increased to 10 mV or 100 mV without affecting the resolution of accuracy and may help with subsequent convergence problems. The same applies to high-current circuits, where the default value of 1 pA for ABSTOL can be increased.

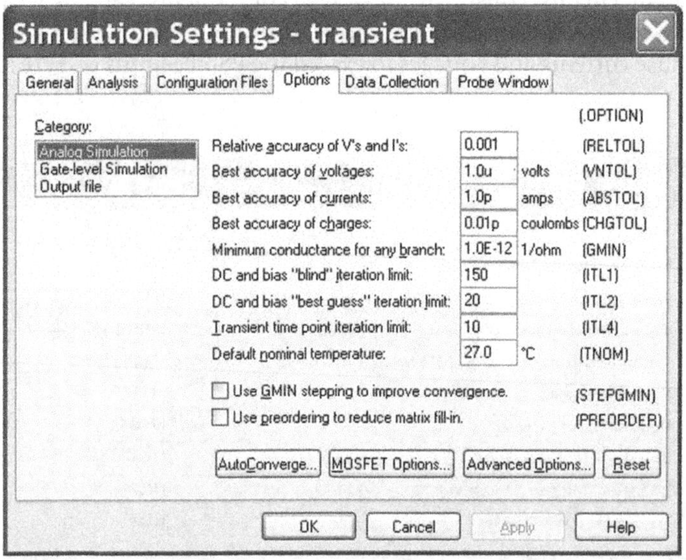

FIGURE 8.3 Simulation Settings options.

The number of times the simulator will try to reach convergence is set by the iteration limits, ITL1, ITL2 and ITL4. In some cases, just increasing the number of iteration limits will help a circuit to converge without having to reduce the accuracy of any of the simulation parameters. In Figure 8.3, you can see that ITL1 and ITL2 are used in the calculation of the DC bias point and ITL4 is used in a transient analysis. When running a transient analysis it is always a good idea to see whether the convergence problem occurs during the DC bias point calculation or the transient analysis so that you can increase the appropriate iteration limit.

There is a new feature in the simulation settings **option** window, labeled **AutoConverge**, which allows modified simulation settings to be used in a subsequent simulation run if there is a convergence problem. You enter the maximum **Relaxed limit** such that the simulator will automatically relax the simulation limit in an optimal manner subject to the maximum values set (Figure 8.4). The simulator will automatically start again at time $t = 0$ using the modified values.

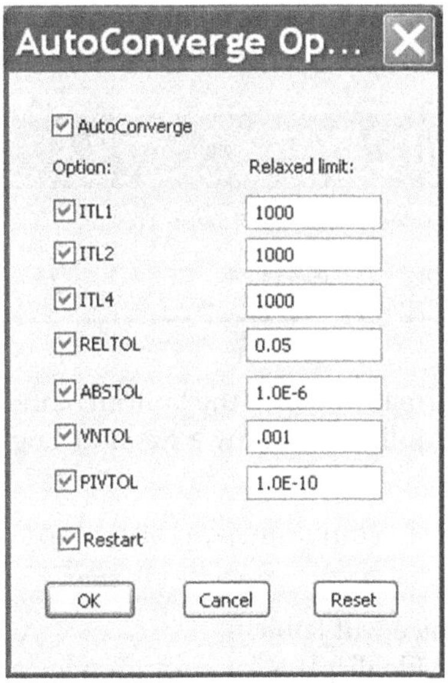

FIGURE 8.4
AutoConverge settings.

Across each semiconductor there is a small conductance which provides a small conducting path for currents such that initial currents and voltages can be calculated for the initial DC bias point solution. This conductance is called GMIN and is globally available as one of the simulation setting options shown in Figure 8.3. This is particularly useful when you have power MOSFETs or diodes with a large off resistance. By default, GMIN is 1.0E-12 siemens, but this can be increased by a factor of 10 or 100. There is an option to automatically step GMIN, **Use GMIN stepping to improve convergence**, which is very useful.

8.5 EXERCISES

Exercise 1

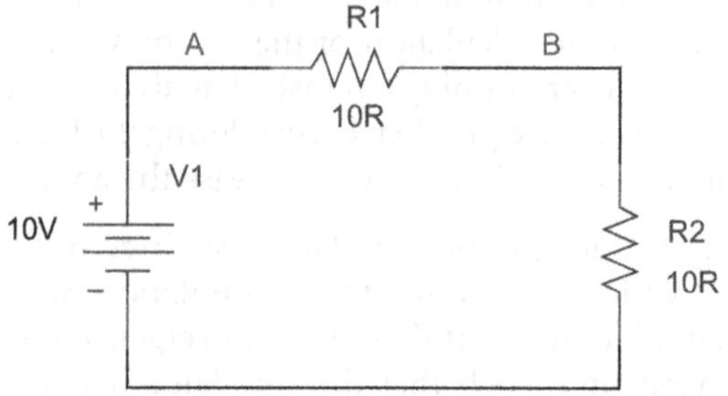

FIGURE 8.5
Missing zero volt node.

1. Draw the circuit in Figure 8.5 and create a bias point analysis, **PSpice > New Simulation Profile**. In the **Analysis** type select **Bias Point** and click on OK.
2. Run the Simulation, **PSpice > Run** or click on the run button.
   You should see the warning message dialog box (Figure 8.6) and a message will be displayed asking you to check the Session Log.

FIGURE 8.6
Warning message.

   The Session Log is normally open at the bottom of the screen; if not, it can be found from the top toolbar, **Window > Session Log**. The warning message will read:

   ```
   WARNING [NET0129] Your design does not contain a Ground
   (0) net.
   ```

   Click on OK and PSpice will launch.
3. In PSpice, the output file displays an error message as shown in Figure 8.7. All nodes are reported as floating as there is no reference to 0 V. Connecting a ground 0 V symbol will allow the circuit to simulate.

FIGURE 8.7
Floating node error message.

```
V_V1          A N00514 10V
R_R1          A B  10R TC=0,0
R_R2          N00514 B  10R TC=0,0

**** RESUMING bias.cir ****
.END

ERROR -- Node A is floating
ERROR -- Node N00514 is floating
ERROR -- Node B is floating
```

Exercise 2

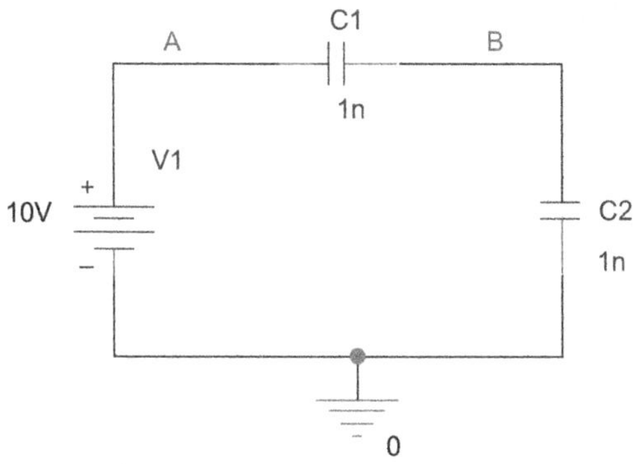

FIGURE 8.8
Floating node.

1. Draw the circuit in Figure 8.8 and create a bias point simulation, **PSpice > New Simulation Profile**. In the **Analysis** type select **Bias Point** and click on OK.
2. Run the Simulation, **PSpice > Run** or click on the run button .
3. In PSpice, the output file displays an error message as shown in Figure 8.9. This is because there is no DC path to ground at node B, which will always be the case for two capacitors connected in series. Connecting a large value resistor across C1 or C2 will allow the circuit to simulate.

```
* source CAPACITORS
C_C1            A B  1n   TC=0,0
V_V1            A 0 10V
C_C2            0 B  1n   TC=0,0

**** RESUMING bias.cir ****
.END

ERROR -- Node B is floating
```

FIGURE 8.9
Floating node error message.

Exercise 3

1. Draw the circuit in Figure 8.10 and create a bias point simulation profile.

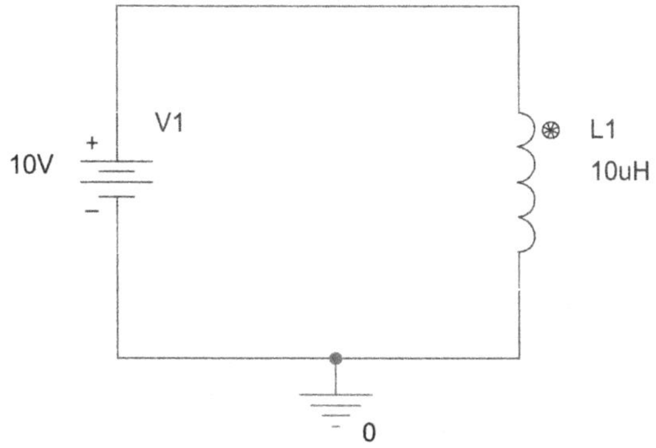

FIGURE 8.10
Inductor loop.

2. Run the simulation ▶.
3. The PSpice output file will display an error message as shown in Figure 8.11. Inductors do not contain a series resistance, so adding a series resistor will allow the circuit to simulate.

```
* source SINEWAVE
V_V1          N00502 0 10V
L_L1          N00502 0  10uH

**** RESUMING transient.cir ****
.END

ERROR -- Voltage source and/or inductor loop involving V_V1
You may break the loop by adding a series resistance
```

FIGURE 8.11
Inductor loop error message.

Exercise 4

1. Draw the circuit in Figure 8.12 using the Q2N3904 transistor from the Capture **transistor** library. Label the nodes as shown (Place > Net Alias).

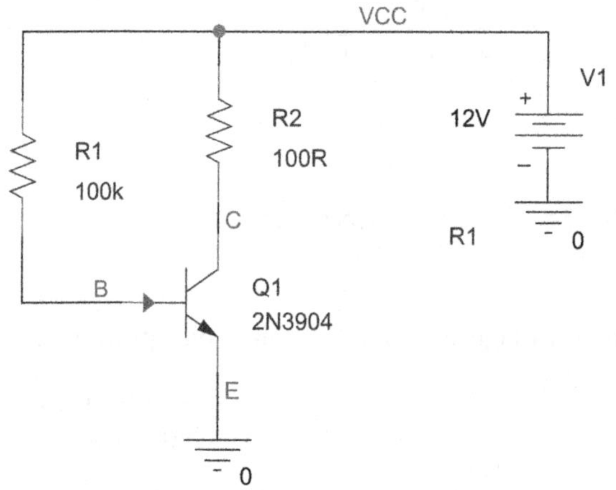

FIGURE 8.12
No PSpice model.

2. Create a bias point simulation profile.
3. Run the simulation ▶.
4. The PSpice output file will display an error message as shown in Figure 8.13. Note that the transistor does not appear in the netlist.

```
* source SINEWAVE
R_R2          C VCC  100R TC=0,0
V_V1          VCC 0 12V
R_R1          B VCC  100k TC=0,0

**** RESUMING transient.cir ****
.END

ERROR -- Less than 2 connections at node C
ERROR -- Less than 2 connections at node B
```

FIGURE 8.13
No PSpice model.

5. In Capture a green circle will appear next to the transistor. Click on the circle and a warning message will appear as shown in Figure 8.14, stating that there is no PSpice Template for Q1. The PSpice Template is a required property for a Capture part to be simulated in PSpice. This will be covered in more detail in Chapter 16.

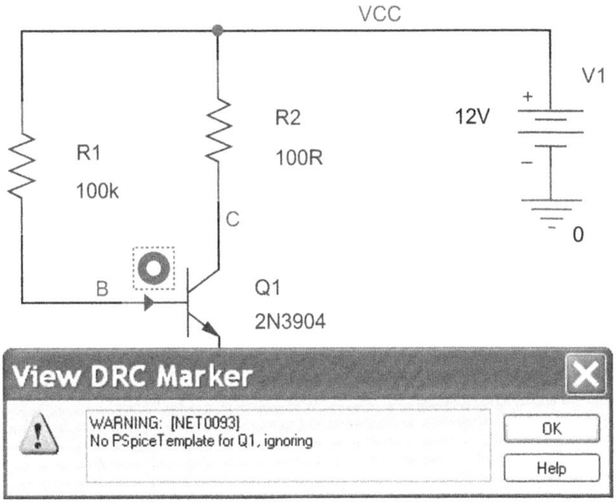

FIGURE 8.14
Missing PSpice template.

CHAPTER 9
Transformers

Chapter Outline

A transformer is implemented by magnetically coupling two or more coils (inductors) together. For air core transformers, a K_Linear coupling device from the **analog** library is used, whereas for non-linear transformers, a magnetic core model is referenced by the K coupling device. When creating a linear transformer, the coils are specified in units of henry (H), whereas for non-linear transformers, you specify the number of turns for the inductors.

The PSpice magnetic cores model hysteresis effects and include a coupling coefficient which is used to define the proportion of flux linkage between the coils and has a value between 0 and 1. For coils wound on the same magnetic core, the coupling coefficient has a value almost equal to 1. For air cored coils, the flux linkage is smaller.

9.1 LINEAR TRANSFORMER

Figure 9.1 shows a step-down linear transformer using inductors for the primary and secondary windings which are specified in Henry's. The two coils are magnetically coupled together using the K_Linear part, K1, which shows that L1 and L2 are coupled together. Ideally, the primary and secondary circuits are electrically isolated. However, for PSpice simulation, as mentioned in Chapter 2, there must be a DC path to ground for every node. This is achieved by using a large value resistor R4 connected from the secondary to 0 V, which will not have a significant effect on the accuracy of simulation.

Analog Design and Simulation using OrCAD Capture and PSpice. DOI: 10.1016/B978-0-08-097095-0.00009-X

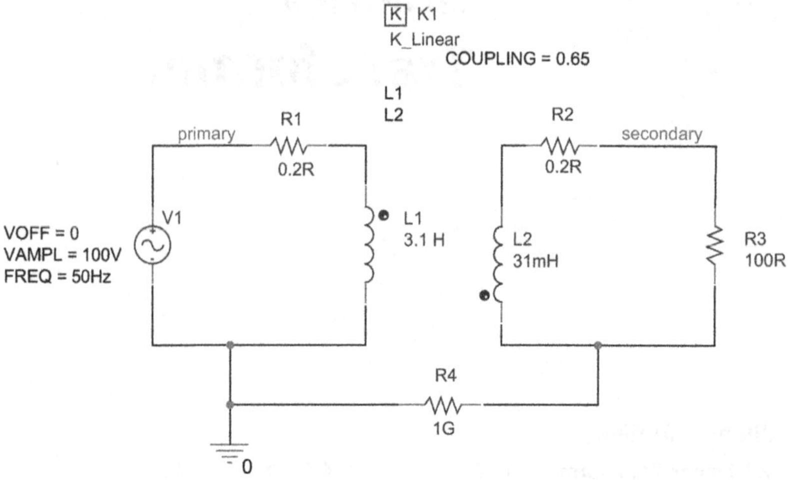

FIGURE 9.1
Linear air core transformer.

NOTE

The dot convention has now been added to inductors to indicate the direction of current flow and subsequent voltage polarity, which is related to how the coils are wound relative to each other. In previous OrCAD versions, you have to display the pin numbers for the inductors, which can then be aligned relative to each other.

9.2 NON-LINEAR TRANSFORMER

Figure 9.2 shows a circuit for a non-linear transformer created using three inductors for the coils, L1, L2 and L3. The reference designators, L1, L2 and L3, are added to the K device in the Property Editor and displayed on the schematic to indicate which coils make up the transformer. The standard K coupling

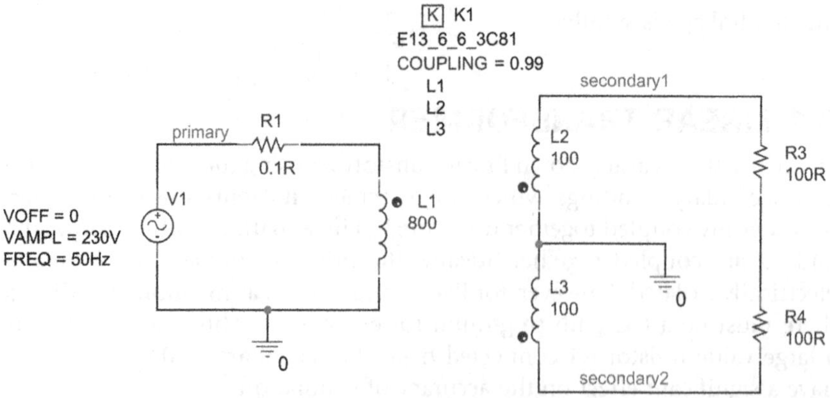

FIGURE 9.2
Non-linear center-tapped transformer.

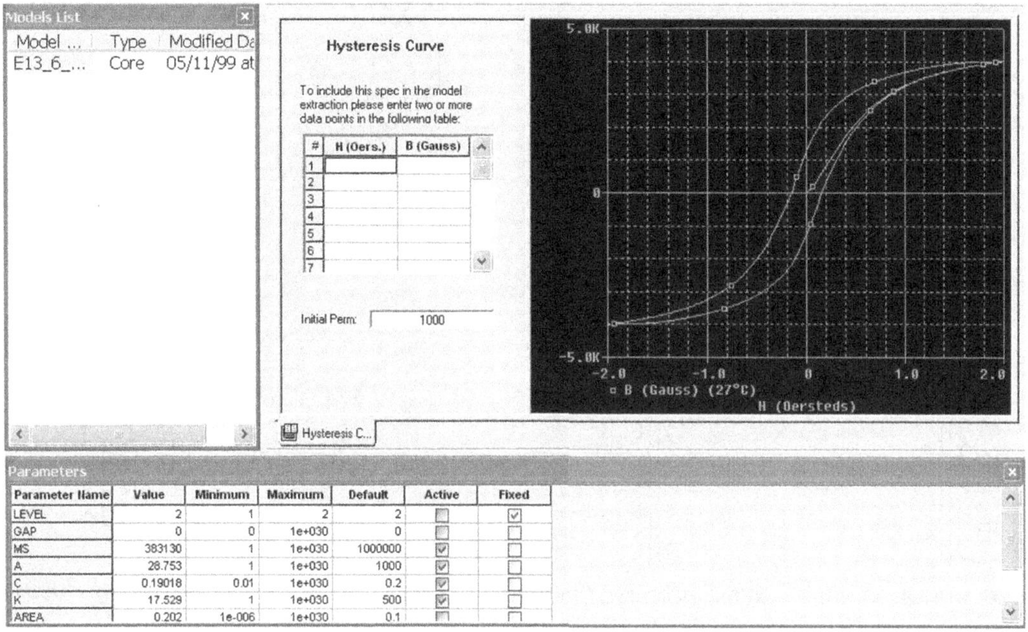

FIGURE 9.3
Model parameters and hysteresis curve for the E13_6_6_3C81 magnetic core.

devices allow for up to six inductors to be coupled together. Manufacturer's magnetic core models are found in the **magnetic** library. In this example, the E13_6_6_3C81 magnetic core is used.

The hysteresis curve for the magnetic cores can be displayed by selecting a K device and **rmb > Edit PSpice Model**, which will open the PSpice Model Editor. In the Model Editor the text description for the model will be displayed. By selecting **View > Extract Model** and clicking on **Yes** to the message window, the characteristic hysteresis curve will be plotted. At the bottom of the Model Editor, is a table of the magnetic core parameters. For each core, there is a gap parameter which can be specified. There is also an entry table whereby you enter B-H curve data and extract your own model parameters. Figure 9.3 shows the model parameters and hysteresis curve for the E13_6_6_3C81 magnetic core. The Model Editor will be covered in more detail in Chapter 16.

9.3 PREDEFINED TRANSFORMERS

A linear transformer, XFRM_LINEAR (Figure 9.4), is available in the **analog** library. Non-linear transformers, which include center-tapped primary and secondary windings, can be found in the **breakout** library (Figure 9.5). These transformers have properties that enable you to enter the inductance, coil resistances and number of turns. Double click on the transformers to access the properties in the Property Editor.

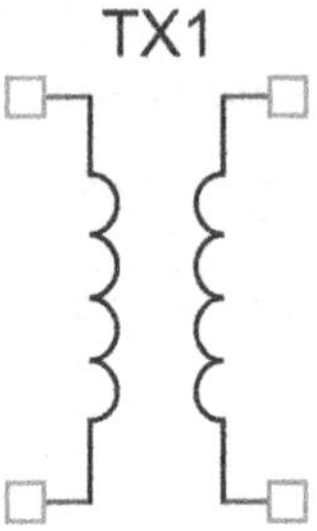

FIGURE 9.4
Linear transformer XFRM_LINEAR.

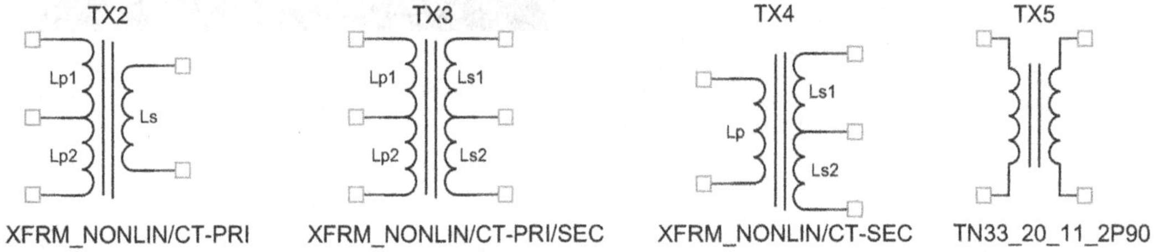

FIGURE 9.5
Non-linear transformers.

9.4 EXERCISES

Exercise 1

FIGURE 9.6
Linear transformer.

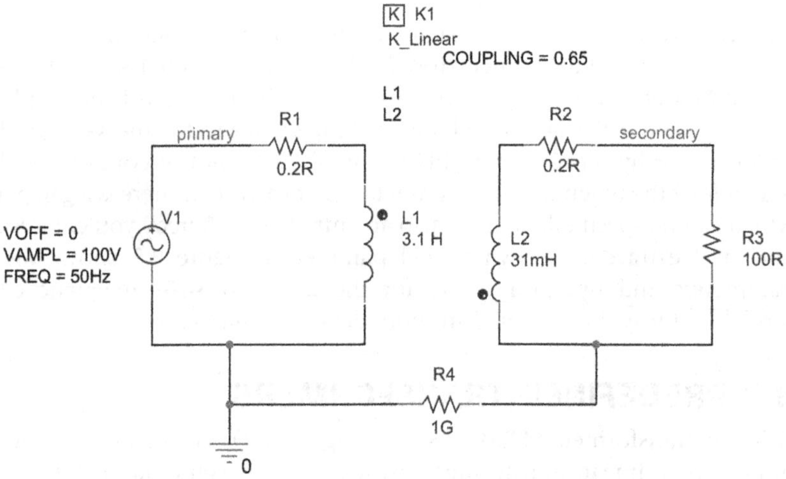

You will create a step-down transformer circuit with a primary inductance winding inductance of 3.1 H and a winding resistance of 0.2 Ω. The secondary winding has an inductance of 31 mH and a winding resistance of 0.2 Ω. The transformer provides a step-down ratio of 10 and is connected to a 100 Ω load resistance.

1. Create a new project called Linear Transformer and draw the circuit in Figure 9.6. The inductors, resistors and K_Linear are all from the **analog** library. V1 is a VSIN source from the **source** library. Set the coupling coefficient to 0.65.
2. If you have a previous version of OrCAD which does not have the dot convention shown on the inductors, then display pin 1 of the inductors and orientate the inductors accordingly.
3. Double click on the K_Linear device to open the Property Editor and add the inductor reference designators as shown in Figure 9.7. Highlight L1 and L2, then **rmb >Display** and select **Value Only** (Figure 9.8).

L1	L1
L2	L2
L3	
L4	
L5	
L6	

FIGURE 9.7
Defining which coils are magnetically coupled together.

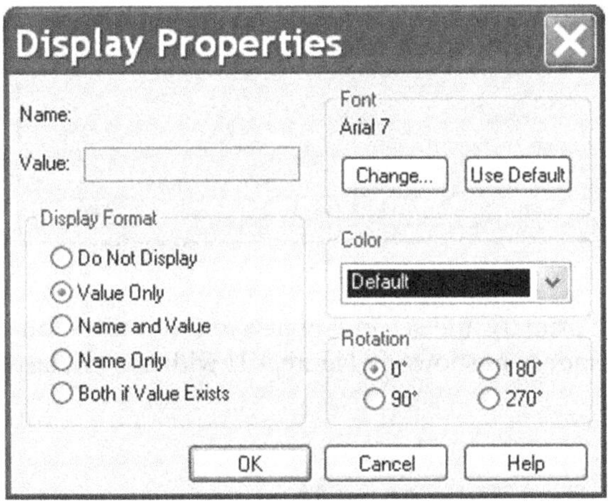

FIGURE 9.8
Displaying the reference designators for the inductors.

NOTE

A common misunderstanding is that the coil reference designators must be entered as L1 to L6 in the Property Editor. If you have coils with reference designators L3, L4 and L5 in a circuit, then you enter the designators as shown in Figure 9.9. It is recommended that you show the reference designators on the schematic for a K core device, especially if you annotate and change the coil numbering.

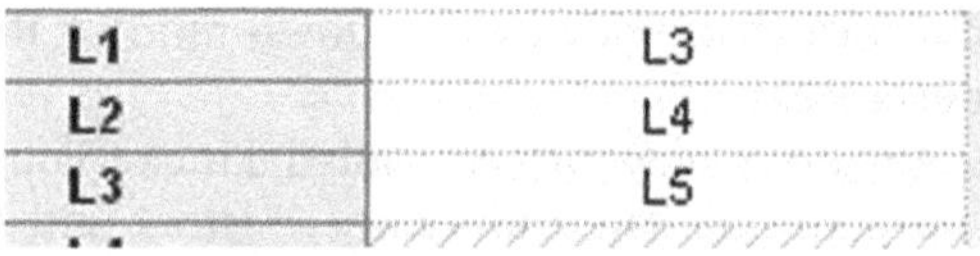

FIGURE 9.9
Entering coil reference designators.

4. Set up a transient simulation profile with a **Run to time** of 50 ms (Figure 9.10).

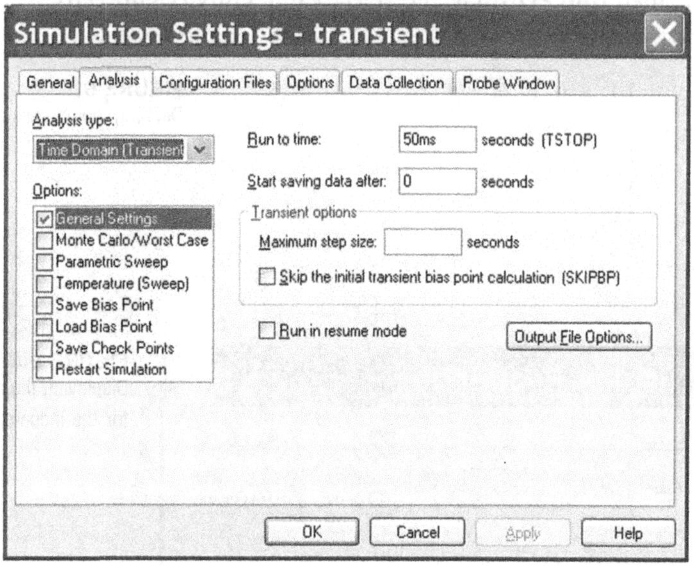

FIGURE 9.10
Transient simulation profile.

5. Place voltage markers on the primary and secondary nets and run the simulation. You should see the response as shown in Figure 9.11 with a secondary voltage of 6.4 V.

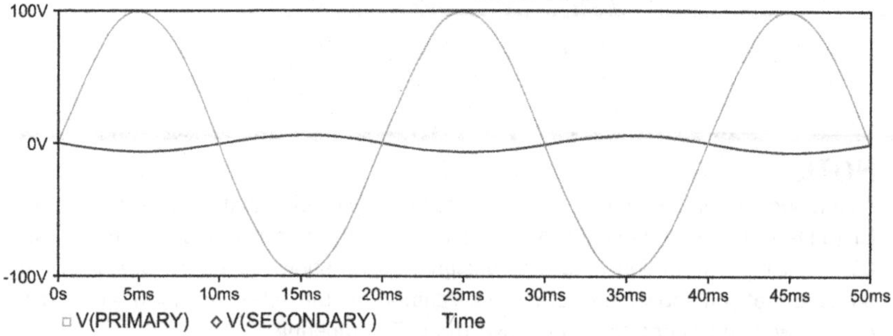

FIGURE 9.11
Primary and secondary voltage waveforms of the step-down transformer.

If you have not placed markers on the circuit then when the PSpice window appears, select **Trace > Add Trace** and select from the left-hand side under **Simulation Output Variables, V(primary)** and **V(secondary)**.

6. The relationship between the primary and secondary voltages and the primary and secondary inductances for air cored transformers is given by:

$$\frac{V_S}{V_P} = \sqrt{\frac{L_S}{L_P}}$$

So, if you increased *k* to 1 you would see a secondary voltage of approximately 10 V. In practice, air cored transformers have a low coupling coefficient.

TIP

When running simulations with transformers, you may need to limit the maximum step size if you see distorted sinewave outputs.

Exercise 2

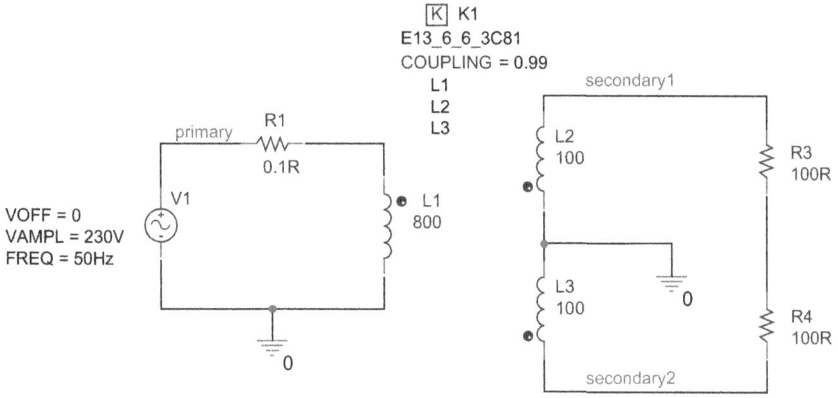

FIGURE 9.12
Non-linear, center-tapped transformer circuit.

You will create a non-linear, center-tapped transformer circuit.

1. Create a new project called Non Linear Transformer and draw the circuit in Figure 9.12. Enter the inductor values as the number of turns as shown. The E13_6_6_3C81 core model is found in the **magnetic** library. Enter a value of 0.99 for the coupling coefficient.
2. As in Exercise 1, double click on the K core device and in the Property Editor, add and display the reference designators L1, L2 and L3.
3. Create a transient simulation profile with a run to time of 50 ms.
4. Place voltage markers on the nets, **primary, secondary1** and **secondary2** and run the simulation.

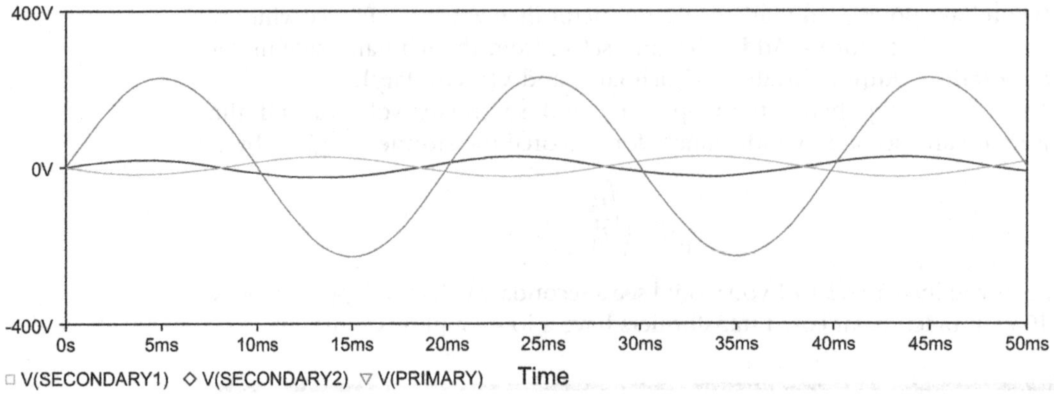

FIGURE 9.13
Primary and secondary waveforms.

5. Figure 9.13 shows the voltage waveforms for the non-linear transformer. You should see 25.1 V on each secondary winding.
6. Investigate the other transformers in the analog and breakout libraries.

CHAPTER 10

Monte Carlo Analysis

Chapter Outline

Monte Carlo analysis is essentially a statistical analysis that calculates the response of a circuit when device model parameters are randomly varied between specified tolerance limits according to a specified statistical distribution. For example, all the circuits encountered so far have been simulated using fixed component values. However, discrete real components such as resistors, inductors and capacitors all have a specified tolerance, so that when you select, for example, a 10 kΩ ± 1% resistor, you can expect the actual measured resistor value to be somewhere between 9900 and 10 100 Ω. Other discrete components and semiconductors in a circuit will also have tolerances and so the combined effect of all the component tolerances may result in a significant deviation from the expected circuit response. This is especially the case in filter designs where applied component tolerances may result in a deviation from the required filter response.

What the Monte Carlo analysis does it to provide statistical data predicting the effect of randomly varying model parameters or component values (variance) within specified tolerance limits. The generated values follow a statistically defined distribution. The circuit analysis (DC, AC or transient) is repeated a number of specified times with each Monte Carlo run generating a new set of randomly derived component or model parameter values. The greater the number of runs, the greater the chances that every component value within its tolerance range will be used for simulation. It is not uncommon to perform hundreds or even thousands of Monte Carlo runs in order to cover as many possible component values within their tolerance limits. Monte Carlo, in effect,

Analog Design and Simulation using OrCAD Capture and PSpice. DOI: 10.1016/B978-0-08-097095-0.00010-6

predicts the robustness or yield of a circuit by varying component or model parameter values up to their specified tolerance limits.

Although the results of a Monte Carlo analysis can be seen as a spread of waveforms in the PSpice waveform viewer (Probe), a **Performance Analysis** can be used to generate and display histograms for the statistical data together with a summary of the statistical data. This provides a more visual representation of the statistical results of a Monte Carlo analysis.

10.1 SIMULATION SETTINGS

A Monte Carlo analysis is run in conjunction with another analysis, AC, DC or transient analysis. Tolerances are applied to parts in the schematic via the **Property Editor** and the required analysis is created in the simulation profile. For the band pass filter in Figure 10.1, component tolerances have been added to the resistors and capacitors and displayed on the circuit. The circuit response is analyzed by performing an AC sweep from 10 Hz to 100 kHz. Monte Carlo will run an initial analysis with all nominal values being used and then run subsequent analysis using randomly generated component values up to the number of Monte Carlo runs specified.

Figure 10.2 shows the simulation profile for an AC sweep running a Monte Carlo analysis.

OUTPUT VARIABLE

You need to specify a node voltage or an independent current or voltage source. In this example, the output variable is V(out).

FIGURE 10.1
1500 Hz band pass filter.

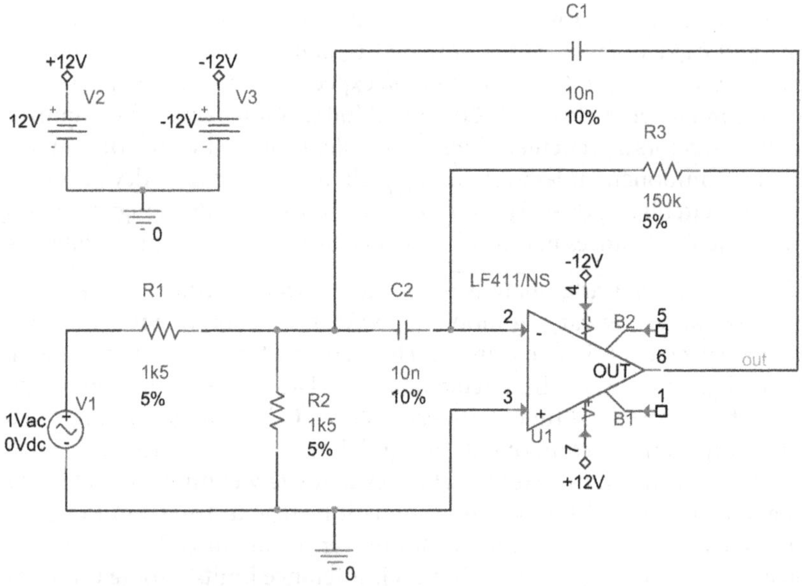

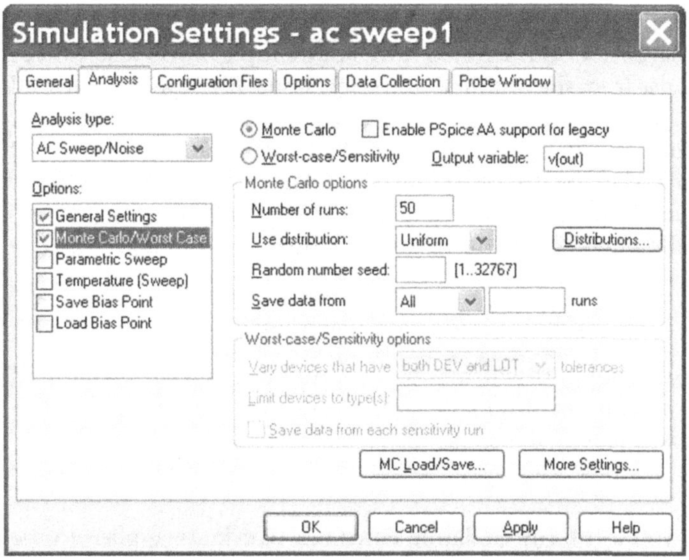

FIGURE 10.2 Monte Carlo simulation settings.

NUMBER OF RUNS

This is the number of times the AC, DC or transient analysis is run. The maximum number of waveforms that can be displayed in Probe is 400. However, for printed results, the maximum number of runs has now been increased from 2000 to 10 000. The first run is the nominal run where no tolerances are applied.

USE DISTRIBUTION

The model parameter deviations from the nominal values up to the tolerance limits are determined by a probability distribution curve. By default, the distribution curve is uniform; that is, each value has an equal chance of being used. The other option is the Gaussian distribution, which is the familiar bell-shaped curve commonly used in manufacturing. Component values are more likely to take on values found near the center of the distribution compared to the outer edges of the tolerance limits.

You can also define your own probability distribution curves using coordinate pairs specifying the deviation and associated probability, where the deviation is in the range −1 to +1 and the probability is in the range from 0 to 1. More information can be found in the PSpice A/D Reference Guide in <install dir> doc\pspcref\pspcref.pdf.

RANDOM NUMBER SEED

As with most random number generators, an initial seed value is required to generate a set of random numbers. This value must be an odd integer number from 1 to 32 767. If no seed number is specified, the default value of 17 533 is used.

SAVE DATA FROM

This allows you to save data from selected runs. For example, if you just want to see the circuit response from the nominal run, select none. If you want to save all the runs, select **All.** If you want to select every third run from the nominal run, i.e. the fourth, seventh, tenth, etc., select **Every,** then enter 3 in the **runs** box. If you want to save data from the first three runs, select **First** and then enter 3 in the **runs** box. If you want to save data from the third, fifth, seventh and tenth run, select **Runs (list)** and enter 3, 5, 7, 10 in the **runs** box. Saved runs will be displayed in Probe.

MC LOAD/SAVE

PSpice now has history support which allows you to save randomly generated model parameters or component values from a Monte Carlo run in a file, which can be used for subsequent analysis.

MORE SETTINGS

This option allows you to specify Collating Functions which are applied to the output waveforms returning a single resulting value. For example, the MAX function searches for and returns the maximum value of a waveform. The YMAX function returns the value corresponding to the greatest difference between the nominal run waveform and the current waveform. The RISE and FALL functions search for the first occurrence of a waveform crossing above or below a set threshold value. For each function, you can specify the range over which you want to apply these functions. Figure 10.3 shows the available collating functions.

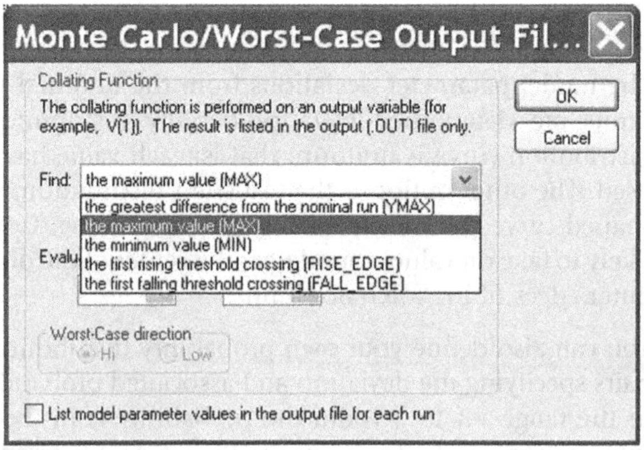

FIGURE 10.3
Collating functions.

The collating functions are summarized as:

YMAX: Find the greatest difference from the nominal run
MAX: Find the maximum value
MIN: Find the minimum value
RISE_EDGE: Find the first rising threshold crossing
FALL_EDGE: Find the first falling threshold crossing

10.2 ADDING TOLERANCE VALUES

Tolerances can now be added to the discrete R, L and C parts in the Property Editor as these parts now include a tolerance property. You must make sure that you include the % symbol when entering the tolerance value i.e. 10%.

In previous OrCAD versions the only way to add tolerances to discrete parts was to use generic **Breakout** parts, which allowed you to edit the PSpice model. For example, to add a tolerance to a resistor, you had to use an Rbreak part from the **breakout** library, edit the model (rmb > Edit PSpice Model) and add tolerance statements to the PSpice model. The default model definition for an Rbreak is given by:

```
.model Rbreak RES R=1
```

where Rbreak is the model name that can be changed and appears on the schematic, RES is the PSpice model type and R is a resistance multiplier. There are two types of tolerance you can add, **dev** and **lot**, as shown below:

```
.model Rmc1 RES R=1 lot = 2% dev=5%
```

Here, the model name has been changed to Rmc1 and two tolerance types have been added to the model statement. The dev tolerance causes the tolerance values of devices that share the same model name (Rmc1 in this example) to vary independently from each other, while the lot tolerance will cause the tolerance values of devices, with the same model name, to track together.

Dev is the same as applying a standard 5% tolerance to all the resistors, where each resistor with the same model name will be assigned its own random resistance value independent from all the other resistors with the same model name.

Lot is where all the resistors with the same model name will be treated as one group and will track together by as much as ± 2%. This was mainly used in integrated circuit (IC) design where a change in temperature was equally applied to groups of components. The combined total tolerance for Rmc1 can therefore be as much as ± 7%.

One example in which you would use both lot and dev is for a single in-line (SIL) or dual in-line (DIL) resistor pack; each resistor will have a dev tolerance defined, and therefore each randomly generated resistance value will have a different value from the others. If there is a rise in temperature, the lot tolerance will ensure that all resistance values increase together by the same percentage.

As mentioned above, only discrete parts have an attached tolerance property which can be edited in the Property Editor. If you want to add for example a specified manufacturer's tolerance to the B_f of a transistor to see its effect on a circuit's performance, you will have to edit the transistor model and add, for example, the dev tolerance to the B_f model parameter as:

```
.model Q2N3906 PNP (Is=1.41f Xti=3 Eg=1.11 Vaf=18.7
                 Bf=180.7 dev=50% Ne=1.5 Ise=0
```

In this example, a dev tolerance of 50% with a default uniform distribution has been added to the B_f of the transistor. However, the distribution for the B_f is more likely to be Gaussian, so you would add dev/gauss=12.5% as PSpice limits a Gaussian distribution to $\pm 4\sigma$:

```
.model Q2N3906 PNP (Is=1.41f Xti=3 Eg=1.11 Vaf=18.7
                 Bf=180.7 dev/gauss=12.5% Ne=1.5 Ise=0
```

10.3 EXERCISES

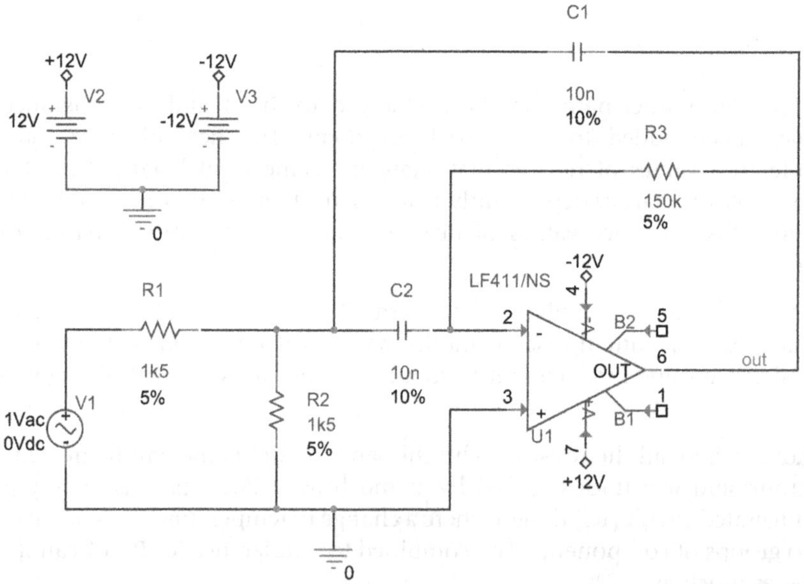

FIGURE 10.4
1500 Hz band pass filter.

Exercise 1

Figure 10.4 shows a Sallen and Key 1500 Hz band pass filter circuit. Tolerances will be added to the resistors and capacitors and a Monte Carlo analysis will be performed to predict the statistical variation of the band pass frequency.

1. Draw the circuit in Figure 10.4. If you are using the demo CD, use the ua741 opamps, which can be found in the **eval** library; otherwise, the LF411 opamp can be found in the **opamp** library. Performing a **Part Search** will result in

the LF411 opamp being sourced from different manufacturers. It does not matter which one you use. V1 is a V_{AC} source from the **source** library, used for an AC analysis, and the power symbols are VCC_CIRCLE symbols from **Place > Power**, renamed +12 V and −12 V, respectively.

NOTE

When running a Part Search for the LF411, make sure the search part is pointing to the <install dir> \Tools\ Capture > Library > PSpice library (see Chapter 5, Exercise 3).

---

2. You need to add a 5% tolerance to all the resistors. Hold down the control key and select the resistors R1, R2 and R3, then **rmb > Edit Properties**. In the **Property Editor** highlight the entire **Tolerance** Row (or column) as shown in Figure 10.5 and **rmb > Edit**. In the **Edit Property Values** box (Figure 10.6), type in **5%** and click on OK.

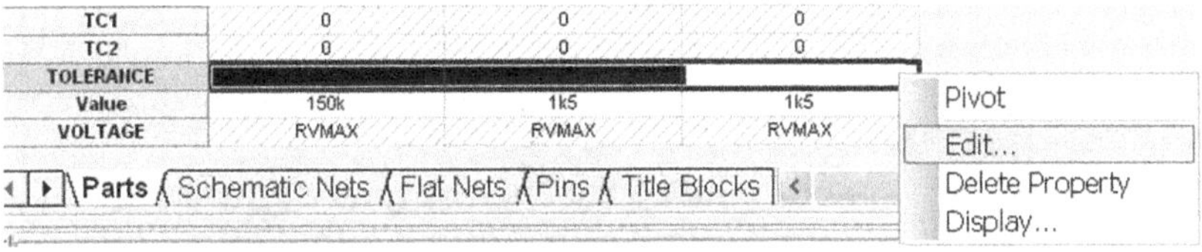

FIGURE 10.5
Select entire TOLERANCE row and select **Edit**.

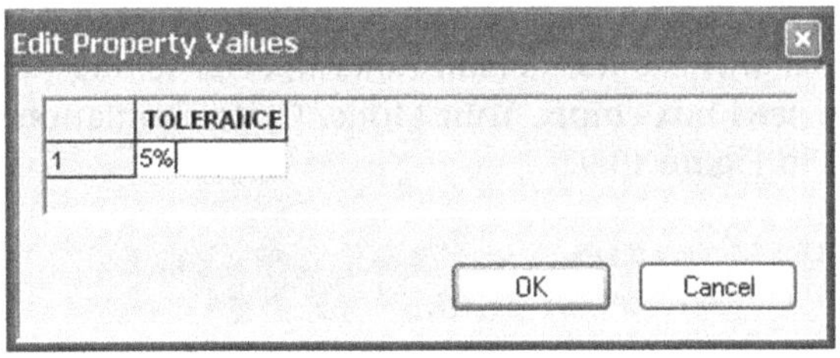

FIGURE 10.6
Adding a 5% tolerance to the resistors.

3. To display the component tolerances on the schematic, select the entire TOLERANCE row as in Figure 10.5, **rmb > Display** and in **Display Properties** (Figure 10.7), select **Value Only** and click on OK.

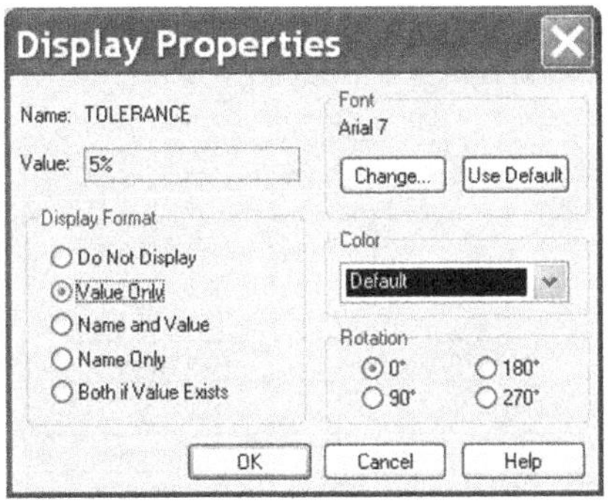

FIGURE 10.7
Displaying tolerance values.

4. Repeat Step 2 by assigning a 10% tolerance to capacitors C1 and C2.
5. Create a PSpice simulation profile called AC Sweep for an AC logarithmic sweep from 100 to 10 kHz with 50 points per decade (Figure 10.8). Select **Monte Carlo/Worst Case** in the **Options** box.

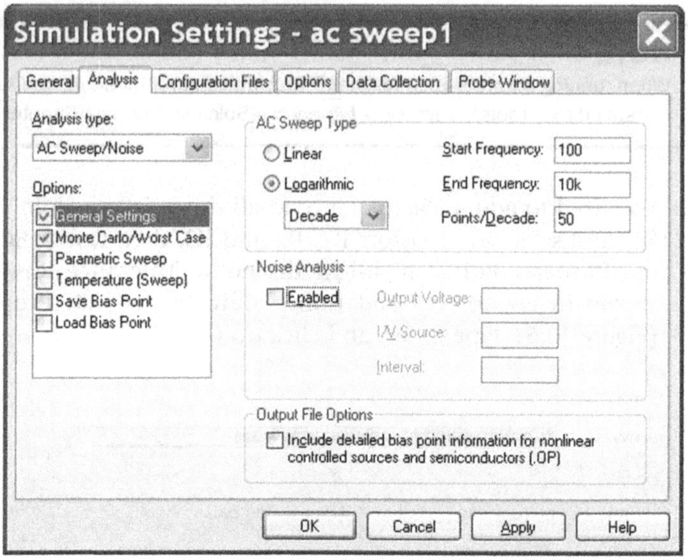

FIGURE 10.8 Simulation profile for an AC sweep.

The output variable is **V(out)** and the number of runs is 50. The distribution used will be uniform and you will use the default random seed setting by leaving the Random number seed box empty. Your Monte Carlo simulation settings should be as shown in Figure 10.9.

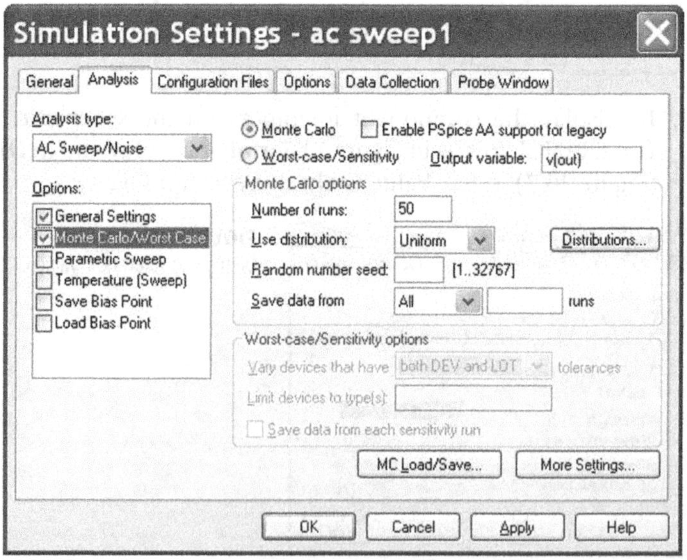

FIGURE 10.9 Monte Carlo simulation settings.

6. Place a dB voltage marker on node **out** (**PSpice > Markers > Advanced > dB Magnitude of Voltage**) and run the simulation.

NOTE

Monte Carlo runs can produce large data files. Since we are only interested in the output node, the simulation can be set up to collect data for display waveforms only for markers placed on the schematic. In the simulation profile, select the **Probe Window** tab and make sure that **All markers on open schematic** is selected.

---

7. When PSpice launches, a list of available runs is shown (Figure 10.10). Select **All** and click on OK.

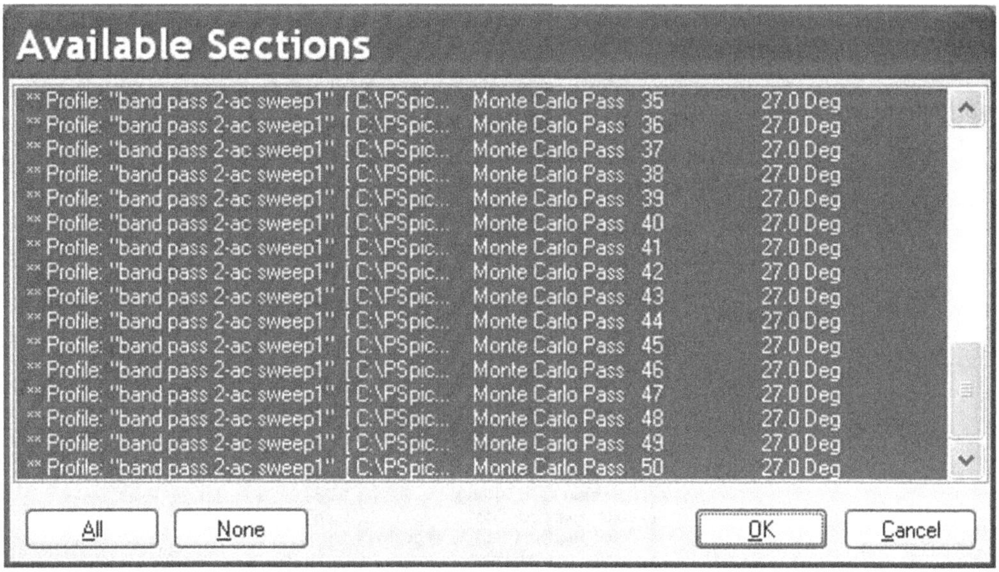

FIGURE 10.10
Available sections.

8. Figure 10.11 shows the band pass filter response showing a spread in the center frequency. Open the output file, **View > Output File**{ XE "**Output File**" } and scroll down to see the results for each Monte Carlo run. At the bottom of the file, you will see a summary of the statistical results and the deviation from the nominal for each run.
   However, to get a better picture of the spread of the center frequency, you can use a **Performance Analysis** to display histograms representing the statistically generated data. Performance analysis will be covered in more detail in Chapter 12, but for now, the steps required for a performance analysis for the bandpass filter are introduced in Exercise 2.

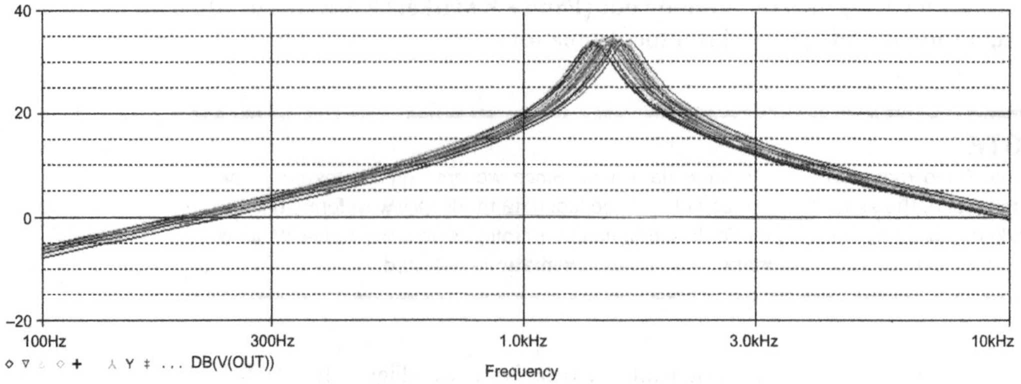

FIGURE 10.11
Monte Carlo band pass filter response.

Exercise 2

1. In Probe, from the top toolbar, select **Trace > Performance Analysis.** At the bottom of the **Performance Analysis** window, click on **Wizard,** then in the next window click on **Next.** You will be presented with a list of measurements as shown in Figure 10.12.

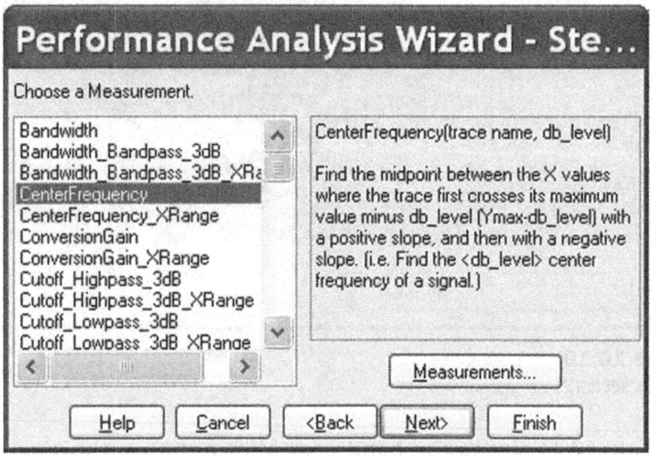

FIGURE 10.12
Available measurements.

2. Select the **CenterFrequency** measurement and click on **Next.**
3. In the **Measurement Expression** window (Figure 10.13), enter the name of the trace to search as **V(out)**. Alternatively, you can click on the **Name of trace to search** icon and search for the trace name, V(out).
   Enter a value of 3 for the **db level down for measurement** as shown. The center frequency of the filter will be measured 3 dB down either side of the maximum value for the Monte Carlo nominal run (no component tolerances applied). Click on **Next** in the **Performance Analysis Wizard.**

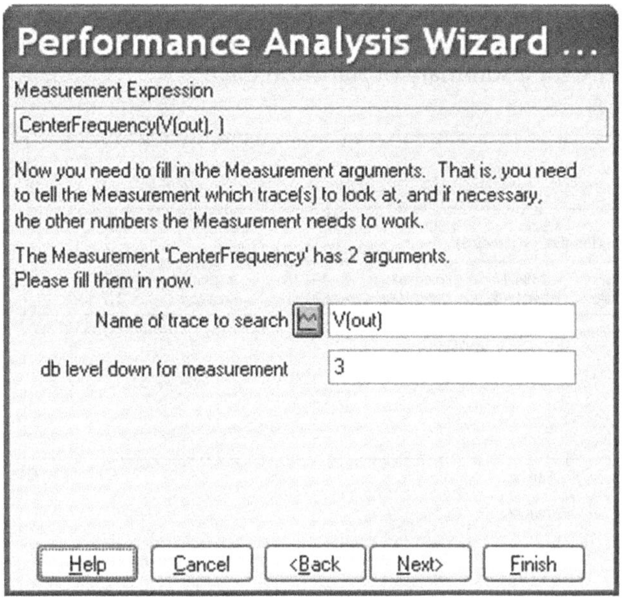

FIGURE 10.13
V(out) is selected as the trace variable with a −3 dB measurement down either side.

4. The nominal trace waveform of the band pass circuit will be displayed (Figure 10.14). This is the simulation response using the component nominal values. The trace shows two points where the measurements from the waveform will be taken. In this case, the center frequency is defined at 3 dB down from the maximum value. This is shown so that you can confirm whether you are seeing the correct circuit response and that the measurements are taken at the correct points on the waveform.

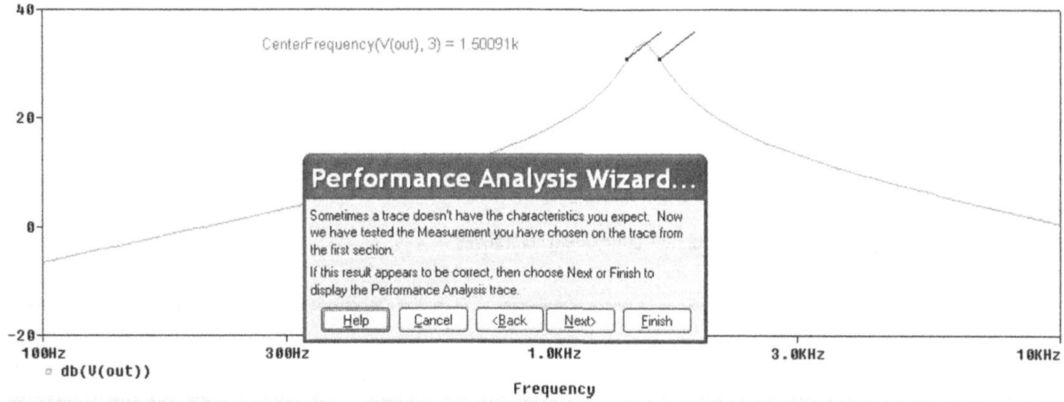

FIGURE 10.14
Nominal trace showing where the −3 dB measurement will be taken on the nominal circuit response waveform.

5. Click on **Next** in the **Performance Analysis Wizard**.
6. Histograms will now be displayed (Figure 10.15) representing the generated statistical data together with a summary of statistical data.

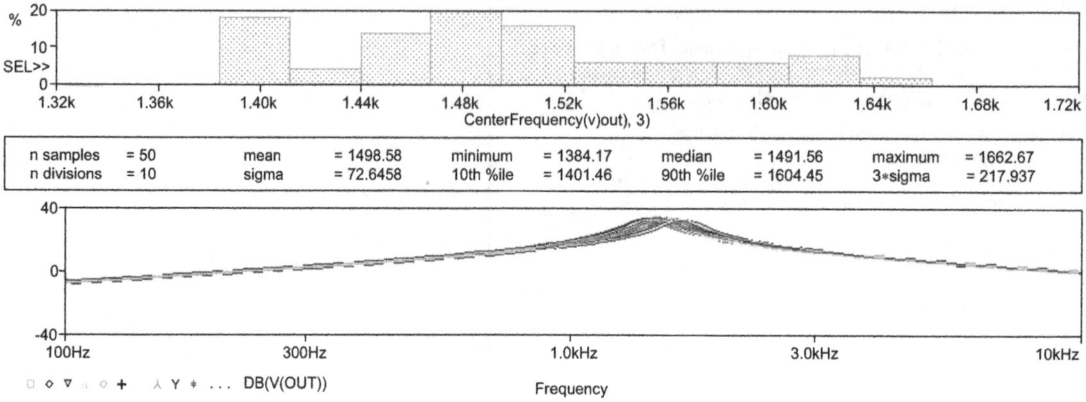

FIGURE 10.15
Histograms shows the possible spread in center frequency.

7. Change the resistor tolerances to 1% and the capacitor tolerances to 5% and re-run the simulation. Run a performance analysis as before using the center frequency measurement as above. The results are shown in Figure 10.16.

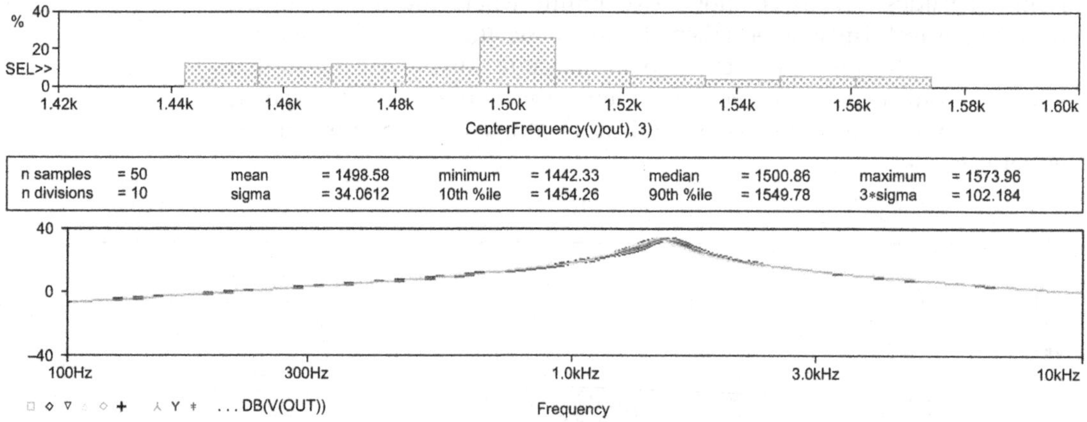

FIGURE 10.16
Statistical data for the center frequency with tighter component tolerances.

NOTE

Because the tolerances are displayed on the schematic, there is no need to use the Property Editor to change the tolerance values. Just double click on the tolerance values in the schematic and enter the new tolerance value.

---

8. Run another Performance analysis but this time, use the bandwidth measurement.

NOTE

The number of displayed histograms can be changed in Probe. Select **Tools > Options > Histogram Divisions**.

FILTER SPECIFICATIONS

The gain of the filter is given by:

$$\begin{aligned} |G| &= \frac{R3}{2R1} \\ |G| &= \frac{150 \times 10^3}{2 \times 1.5 \times 10^3} = 50 \end{aligned} \tag{10.1}$$

or in dB:

$$|G|_{dB} = 20log_{10}50 = 34 \text{ dB}$$

The band pass center frequency is given by:

$$f = \frac{1}{2\pi C\sqrt{\left(\frac{R1R2}{R1+R2}\right)R3}} \tag{10.2}$$

where C1 = C2 = C

$$f = \frac{1}{2\pi \times 10 \times 10^{-9}\sqrt{\left(\frac{1.5 \times 10^3 \times 1.5 \times 10^3}{1.5 \times 10^3 + 1.5 \times 10^3}\right) \times 150 \times 10^3}}$$

$$f = 1501 \text{ Hz}$$

The −3 dB bandwidth is given by:

$$\begin{aligned} f &= \frac{1}{\pi C R3} \\ f &= \frac{1}{\pi \times 10 \times 10^{-9} \times 150 \times 10^3} = 212.2 \text{ Hz} \end{aligned} \tag{10.3}$$

CHAPTER 11

Worst Case Analysis

Chapter Outline

Worst case analysis is used to identify the most critical components that will affect circuit performance. Initially, a sensitivity analysis is run on each individual component that has a tolerance assigned. The component value is effectively pushed towards both of its tolerance limits by a small percentage of its value to see which limit would have a greater effect on the worst case output. A worst case analysis is then performed by setting all the component values to their end tolerance limits which gave an indication of the worst case results. In order to reduce the number of simulation runs, collating functions can be used to detect differences from the nominal worst case output such as minimum, maximum or threshold differences.

In Figure 11.1, an equivalent inductor circuit known as a gyrator is implemented using U1B, R4, R5 and C2, which enables a high inductance value to be realized, 100 H in this example. The series connection of the equivalent inductor and C1 forms a series tuned circuit which determines the frequency of the notch filter. Using ideal component values, the initial simulation results will show the required notch filter response. However, components have tolerances that may affect the notch frequency. Adding component tolerances, a Worst Case analysis will determine which components are critical to the circuit's performance. The notch frequency can be determined by detecting the minimum output voltage of the filter. Therefore, a worst case analysis can be run using a collating function on the output voltage, which will only record the minimum output voltages.

Analog Design and Simulation using OrCAD Capture and PSpice. DOI: 10.1016/B978-0-08-097095-0.00011-8

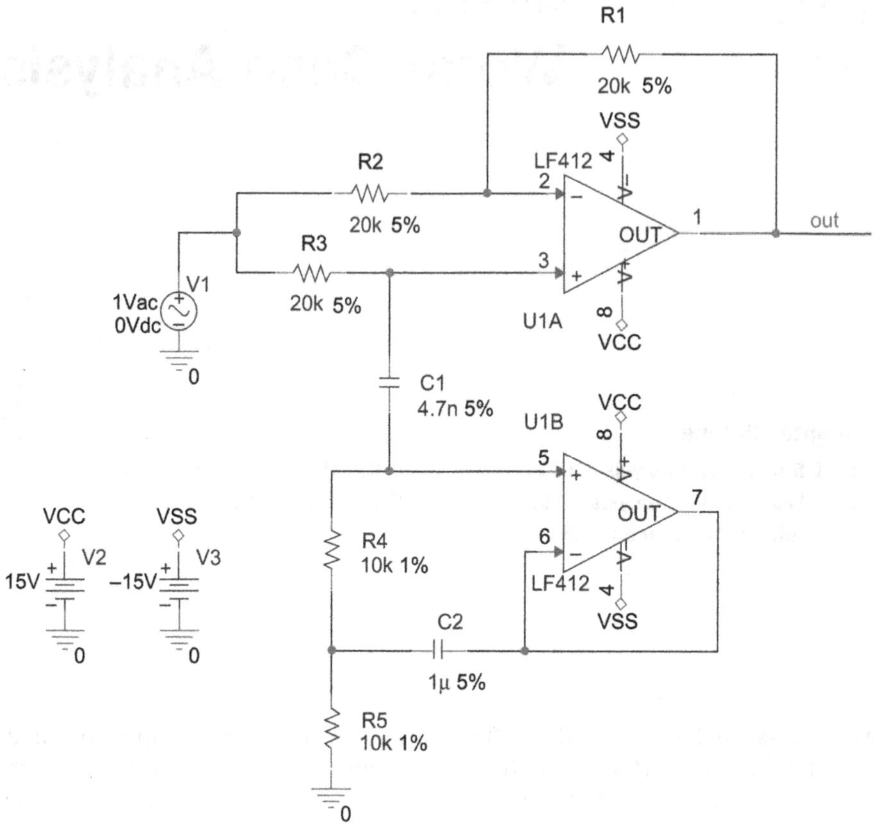

FIGURE 11.1 Notch filter using a gyrator circuit.

11.1 SENSITIVITY ANALYSIS

We are looking to investigate the effect that each component has on the notch frequency by recording the minimum output voltage of the filter, and hence to determine which components are critical to the notch frequency. A sensitivity analysis will be run on each component in turn, with its sensitivity value given by:

```
value = nominal value * (1 + RELTOL)
```

where RELTOL is the relative tolerance and can be found in the **PSpice Simulation Settings > Options** as shown in Figure 11.2. By default, RELTOL is 0.001 (0.1%).

For example, R1 has a value of 20 kΩ and a tolerance of 5% and so has an expected value between 19 and 21 kΩ. If RELTOL is set to 0.01 (±1%), the resistor value will be increased to 20 200 Ω and then decreased to 19 800 Ω. The direction in which the change in resistor value gave the worst case result

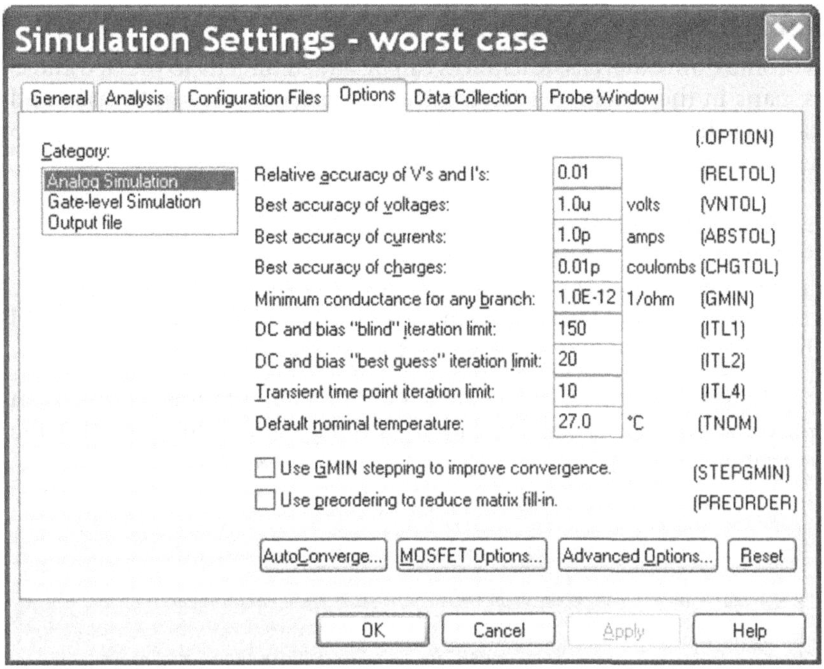

FIGURE 11.2
Simulation settings options.

(minimum voltage) determines which tolerance limit (upper or lower) to use in the worst case analysis. If the minimum output voltage was recorded with a 1% decrease in the resistor value, then the lower tolerance limit of 19 kΩ would used for R1 in the worst case analysis.

The value of R1 is then reset to its nominal value of 20 kΩ and a sensitivity analysis is performed on R2, varying its value towards its tolerance limits and recording the 'tolerance direction' that gave the minimum output voltage. The resistor values will then be set to their tolerance limit value that gave the worst case minimum output voltage.

It must be remembered that the above assumes that the output voltage changes continuously with a continuous change in component values and that there are no interdependencies with other components, which may not be the case.

11.2 WORST CASE ANALYSIS

Based upon the results of the sensitivity analysis, worst case analysis sets the component values to one of their tolerance limits. If R1, 20 kΩ −1% gave the minimum voltage, then R2 would be set to 20 kΩ −5%.

A worst case analysis is run in conjunction with a DC, AC or transient analysis and does not take into account the interdependence of the parameters. The results of the sensitivity and worst case analysis are written to the output file.

11.3 ADDING TOLERANCES

As with Monte Carlo analysis, tolerances can be added directly to the R, L and C Capture parts in the schematic. Alternatively, breakout parts can be used and the following **dev** or **lot** statements are added as used with a Monte Carlo analysis:

```
.model Rwcl RES R=1 dev=5% lot=2%
```

The dev statement causes the tolerance values of devices that share the same model name to vary independently from each other, while the lot statement will cause the tolerance values of devices to track together.

As with a Monte Carlo analysis, you have to define an output variable, which can be a node voltage or an independent current or voltage source. In Figure 11.3, the output variable is **V(out)**.

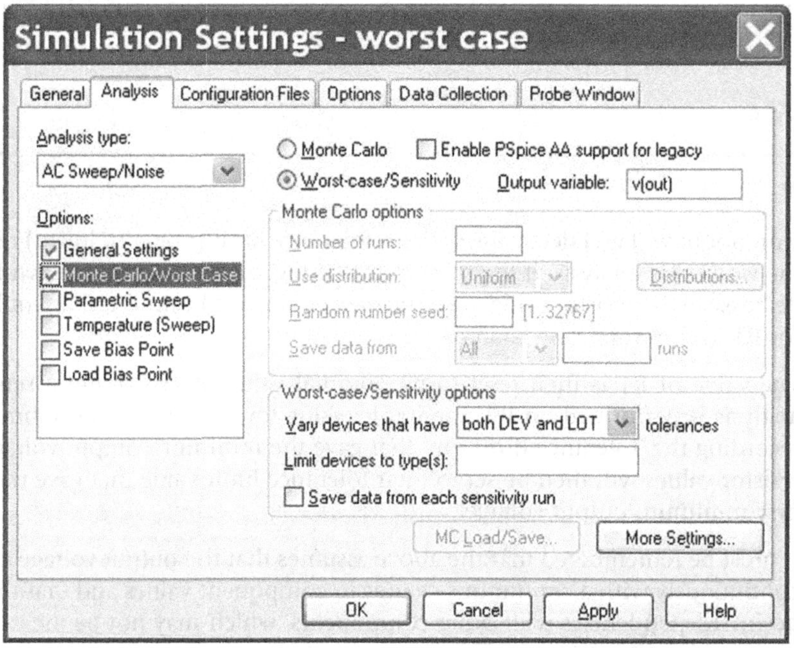

FIGURE 11.3
Worst case analysis simulation settings.

11.4 COLLATING FUNCTIONS

As with Monte Carlo, collating functions detect and compare the output response of the circuit with defined parameters. There are five functions that define the worst case results:

- YMAX finds the greatest distance in the *Y* direction in each waveform from the nominal run.
- MAX finds the maximum value of each waveform.

- MIN finds the minimum value of each waveform.
- RISE_EDGE(value) finds the first occurrence of the waveform crossing above the threshold (value). The function assumes that there will be at least one point that lies below the specified value followed by at least one above.
- FALL_EDGE(value) finds the first occurrence of the waveform crossing below the threshold (value). The function assumes that there will be at least one point that lies above the specified value followed by at least one below.

11.5 EXERCISE

1. Draw the circuit in Figure 11.4. The LF412 dual opamp is from the opamp library, or you can use any opamps.

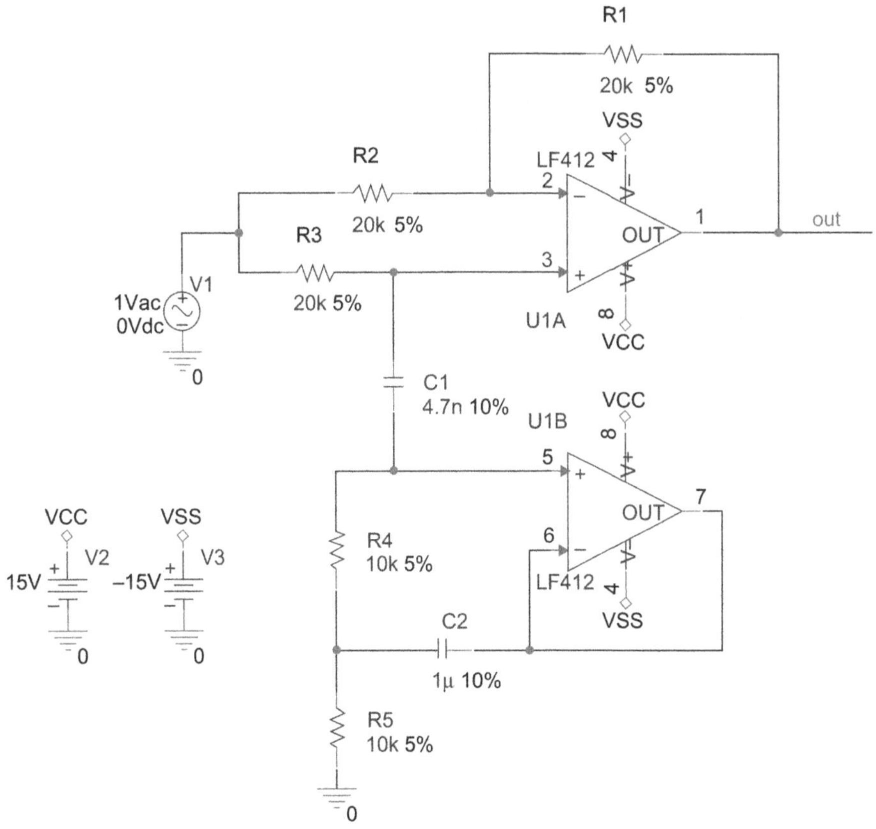

FIGURE 11.4 Notch filter.

2. Select all the resistors by holding down the control key and **rmb > Edit Properties**.
3. In the **Property Editor**, select and highlight the whole row (or column) for the **Tolerance** property and **rmb > Edit** (Figure 11.5).

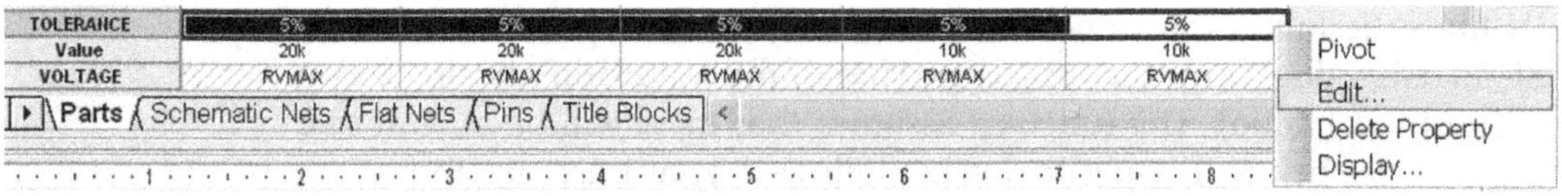

FIGURE 11.5
Selecting the tolerance property for all the resistors.

Add a 5% tolerance in the **Edit Property Values** window as shown in Figure 11.6, click on OK and then close the Property Editor.

FIGURE 11.6
Adding a 5% tolerance to the resistors.

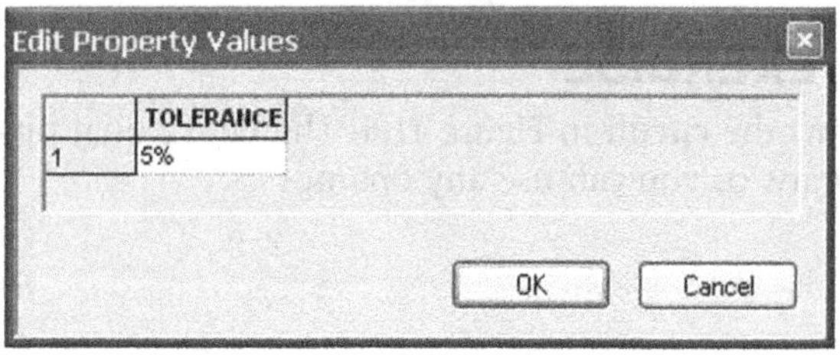

4. Repeat Steps 3 and 4 adding a 10% tolerance to the capacitors, C1 and C2.
5. Set up a simulation profile for an AC sweep from 10 Hz to 10 kHz performing a logarithmic sweep with 100 points per decade (Figure 11.7). Click on **Apply** but do **not** exit.

FIGURE 11.7
AC sweep simulation settings.

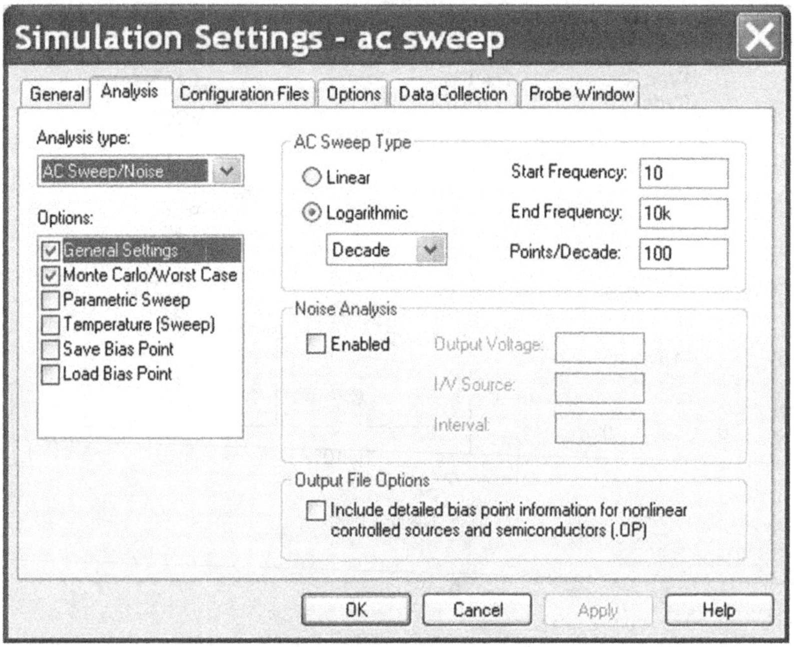

6. In the simulation profile, in **Options** select **Monte Carlo/Worst Case** and enter **V(out)** for the **Output variable** (Figure 11.8). Click on **Apply** but do **not** exit.

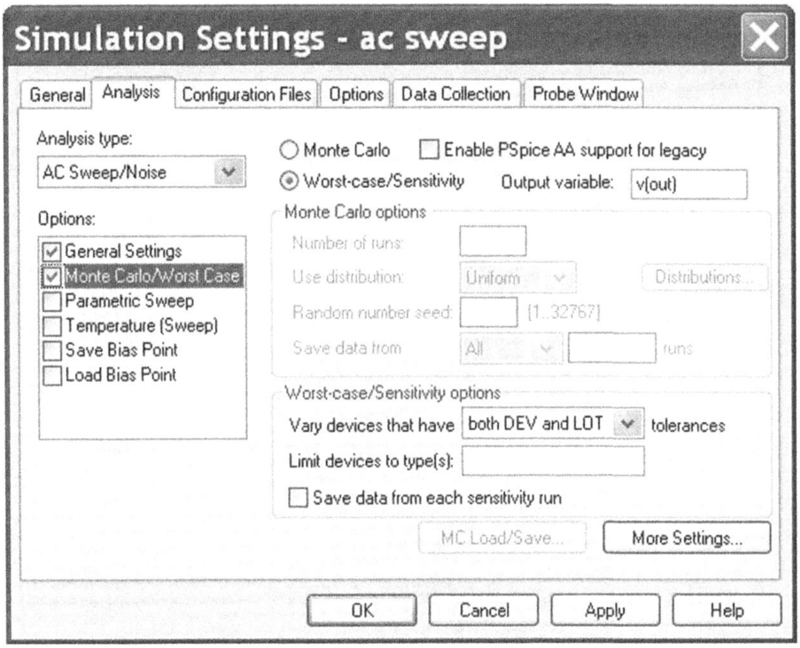

FIGURE 11.8
Setting up a worst case analysis.

7. Click on **More Settings** and in **Find**, select the **minimum (MIN)** for the Collating Function and select **Low** for the **Worst Case Direction** as shown in Figure 11.9. Click on OK but do **not** exit.

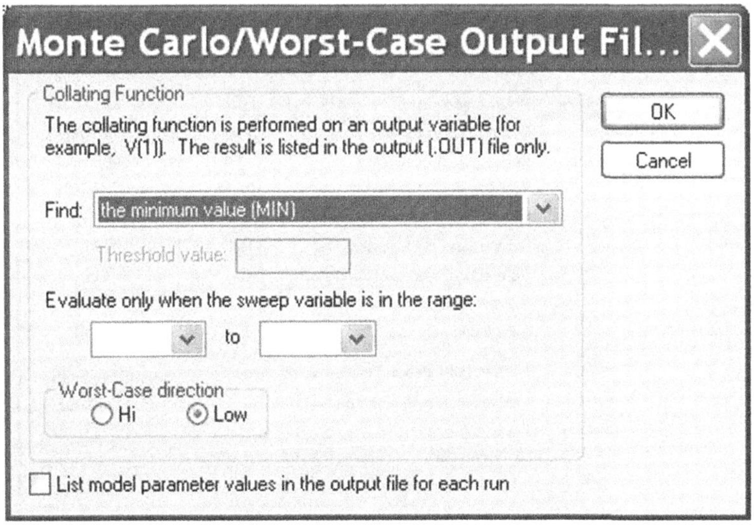

FIGURE 11.9
Selecting the MIN collating function.

8. Select the **Options** tab and select **Output File** and uncheck **Bias Point Node Voltages** and **Model parameter values** as shown in Figure 11.10. Click on OK but do **not** close the simulation profile window.

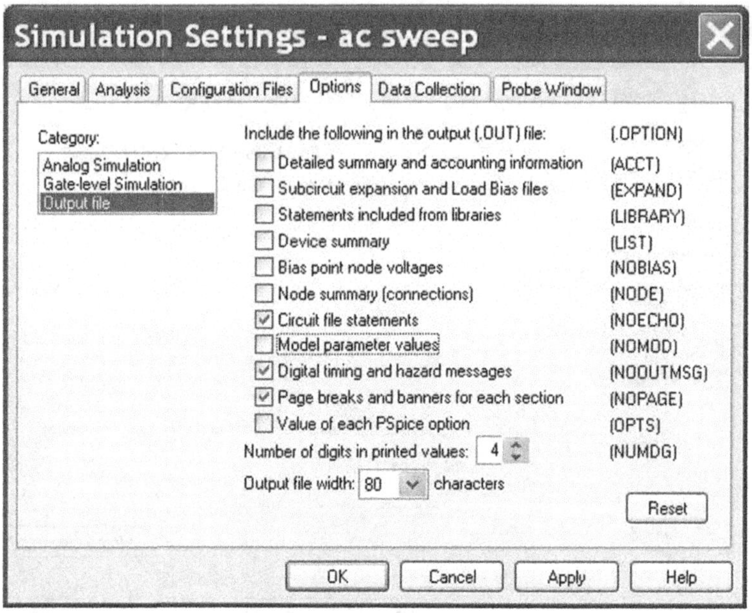

FIGURE 11.10
Output file options.

9. In the simulation profile select **Options** and change RELTOL to 0.01 (Figure 11.11). Click on OK to close the simulation settings profile.

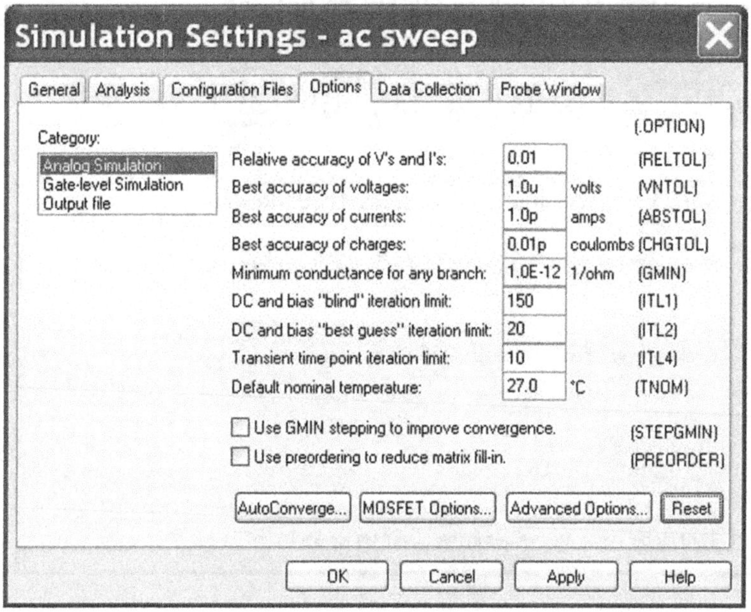

FIGURE 11.11
Changing RELTOL.

10. Place a Vdb marker on node **out** (**PSpice > Markers > Advanced > dB Magnitude of Voltage**).

11. Run the simulation .
12. In Probe you will see two output notch responses. The nominal response is at 234 Hz and the worst case response is at 199 Hz (Figure 11.12).

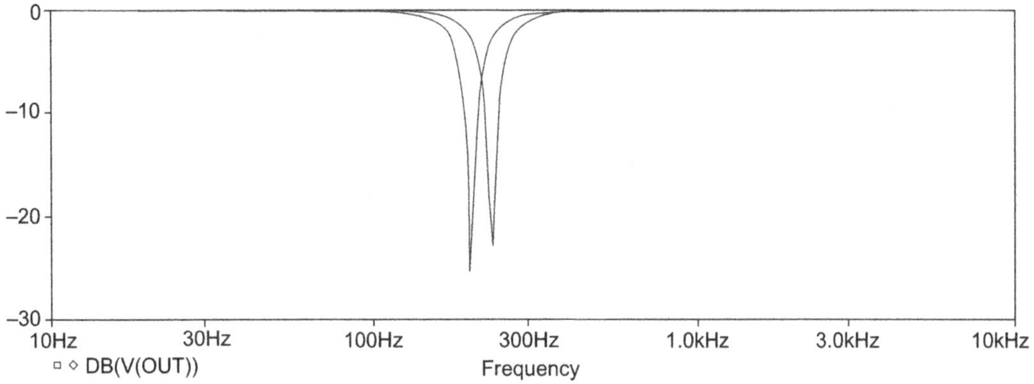

FIGURE 11.12
Ideal notch filter response at 234 Hz and worst case response at 199 Hz.

13. Open the **Output File** and scroll down to the list of Minimum values (Figure 11.13). Here, as expected, R4, R5, C1 and C2 have a greater effect on the notch output voltage compared to R1, R2 and R3.

```
 RUN                     MINIMUM VALUE

R_R2 R_R2 R               .0725 at F =   234.42
                        (      .2904% change per 1% change in Model Parameter)

R_R1 R_R1 R               .0725 at F =   234.42
                        (      .1786% change per 1% change in Model Parameter)

R_R3 R_R3 R               .0721 at F =   234.42
                        (    -.3169% change per 1% change in Model Parameter)

C_C2 C_C2 C               .0593 at F =   229.09
                        ( -37.543% change per 1% change in Model Parameter)

R_R4 R_R4 R               .0593 at F =   229.09
                        ( -37.601% change per 1% change in Model Parameter)

R_R5 R_R5 R               .0593 at F =   229.09
                        ( -37.616% change per 1% change in Model Parameter)

C_C1 C_C1 C               .0588 at F =   229.09
                        ( -38.143% change per 1% change in Model Parameter)
```

FIGURE 11.13
Minimum value.

Scroll down to **Worst Case All Devices**, as shown in Figure 11.14, where you can see, as the result of the sensitivity analysis, the tolerance direction in which the component values were set for the worst case analysis.

FIGURE 11.14
Table showing which tolerance limits were used.

Device	MODEL	PARAMETER	NEW VALUE	
C_C2	C_C2	C	1.1	(Increased)
C_C1	C_C1	C	1.1	(Increased)
R_R5	R_R5	R	1.05	(Increased)
R_R4	R_R4	R	1.05	(Increased)
R_R1	R_R1	R	.95	(Decreased)
R_R2	R_R2	R	.95	(Decreased)
R_R3	R_R3	R	1.05	(Increased)

14. At the bottom of the output file is the Worst Case Summary (Figure 11.15), showing that the worst case scenario is for the notch filter frequency at 199 Hz, which is a deviation of 7% from the nominal.

FIGURE 11.15
Worst case summary.

```
RUN                          MINIMUM VALUE

WORST CASE ALL DEVICES
                              .0522 at F =   199.53
                             (   7.0339% of Nominal)
```

15. Change the tolerance of the capacitors to 5% and resistors R4 and R5 to 1% and run the simulation to see whether this improves the predicted worst case response.
16. Figure 11.16 shows an improvement in the worst case notch frequency, which is given in the summary as 218 Hz (Figure 11.17), being closer to the nominal frequency of 234 Hz.

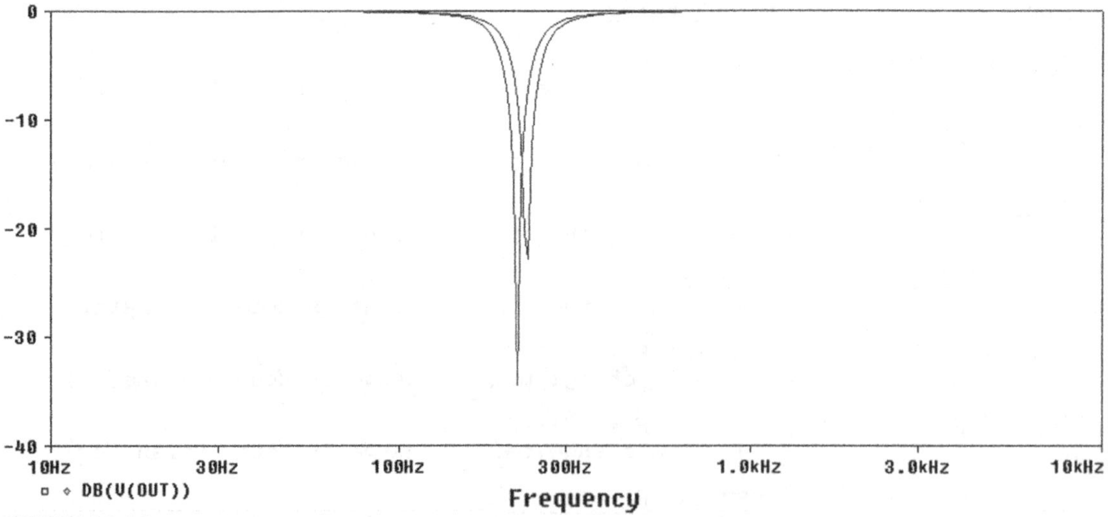

FIGURE 11.16
Improvement in worst case frequency deviation.

FIGURE 11.17
Improvement in worst case frequency.

```
RUN                          MINIMUM VALUE

WORST CASE ALL DEVICES
                              .019  at F =   218.78
                             (   4.8075% of Nominal)
```

CHAPTER 12
Performance Analysis

Chapter Outline

Performance analysis uses measurement definitions to scan a family of curves in Probe and to return a series of values based on the measurement definition. For example, after sweeping a voltage source connected to a CR network, a series of capacitor voltage charging curves will be obtained. Running a performance analysis using the rise time measurement definition on the resulting waveforms will generate a series of rise time values plotted against the swept source voltage.

12.1 MEASUREMENT FUNCTIONS

PSpice includes over 50 measurement definitions, some of which are listed in Table 12.1. A full list is given in the Appendix. The standard CenterFrequency and Bandwidth definitions were used in Chapter 10, Exercise 2. However, the measurement definitions also give you an option to define the range over which you want the measurement to be made. For example, the CenterFrequency_XRange definition gives you the option to measure the waveform over a specified *x*-range, which would be the frequency range. You can also custom design your own measurement definition.

12.2 MEASUREMENT DEFINITIONS

The measurement definitions can be viewed in PSpice from **Trace > Measurements**, which displays all the available measurements as well as the various options to create, view, edit and evaluate measurements (Figure 12.1).

Analog Design and Simulation using OrCAD Capture and PSpice. DOI: 10.1016/B978-0-08-097095-0.00012-X

Table 12.1 Some of the Measurement Definitions Available in PSpice

Definition	Description
Bandwidth	Bandwidth of a waveform (you choose dB level)
Bandwidth_Bandpass_3dB	Bandwidth (3 dB level) of a waveform
CenterFrequency	Center frequency (dB level) of a waveform
CenterFrequency_XRange	Center frequency (dB level) of a waveform over a specified *X*-range
ConversionGain	Ratio of the maximum value of the first waveform to the maximum value of the second waveform
Cutoff_Highpass_3dB	High pass bandwidth (for the given dB level)
Cutoff_Lowpass_3dB	Low pass bandwidth (for the given dB level)
DutyCycle	Duty cycle of the first pulse/period
Falltime_NoOvershoot	Fall time with no overshoot
Max	Maximum value of the waveform
Min	Minimum value of the waveform
NthPeak	Value of a waveform at its nth peak
Overshoot	Overshoot of a step response curve
Peak Value	Peak value of a waveform at its nth peak
PhaseMargin	PhaseMargin
Pulsewidth	Width of the first pulse
Q_Bandpass	Calculates *Q* (center frequency/bandwidth) of a bandpass response at the specified dB point
Risetime_NoOvershoot	Rise time of a step response curve with no overshoot
Risetime_StepResponse	Rise time of a step response curve
SettlingTime	Time from <begin_x> to the time it takes a step response to settle within a specified band
SlewRate_Fall	Slew rate of a negative-going step response curve

FIGURE 12.1
Available measurements.

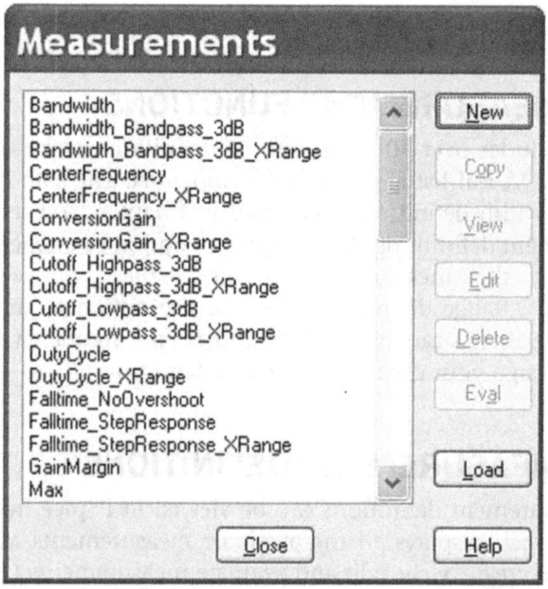

For example, Figure 12.2 shows the measurement definition for Risetime_NoOvershoot.

Rise time, by definition, is the time difference between the value of the voltage or current at 10% and 90% of the maximum value. So two measurements are needed, one when the voltage (or current) curve is at 10% of the maximum value (x1) and the other measurement when the curve is at 90% of the maximum value (x2).

In order to find the points at 10% and 90% of the curve, search commands are used:

Search forward level(10%,p) !1
Search forward level(90%,p) !2

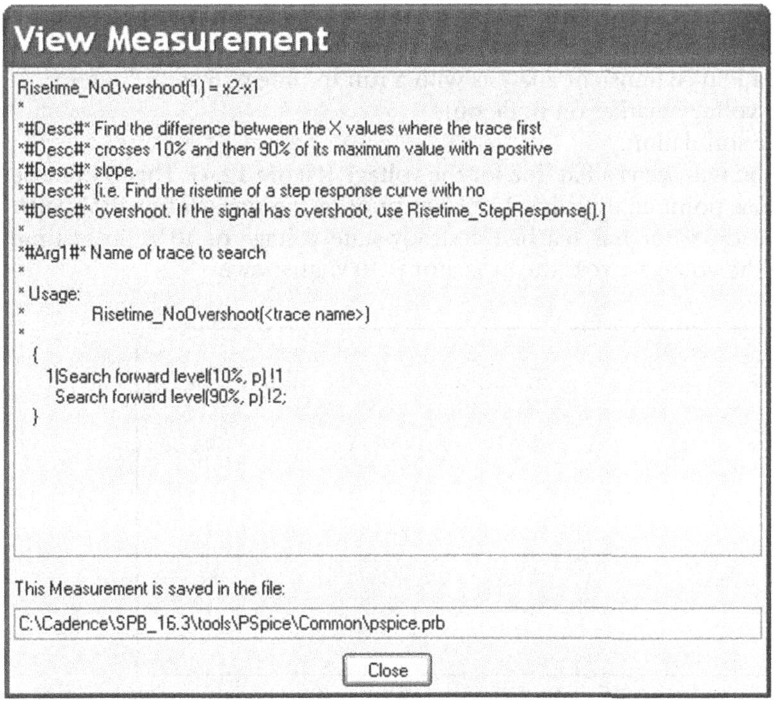

FIGURE 12.2
Risetime_NoOvershoot measurement definition.

In this example, search in the positive going direction (p) for the 10% level and return the first data point (x1), and also search in the positive going direction (p) for the 90% level and return the second data point (x2).

The first line in Figure 12.2 is called a marked point expression (has one value) and calculates the difference between x2 and x1 (x2 – x1). Any line starting with a # is a comment line and provides information to the user.

12.3 EXERCISES

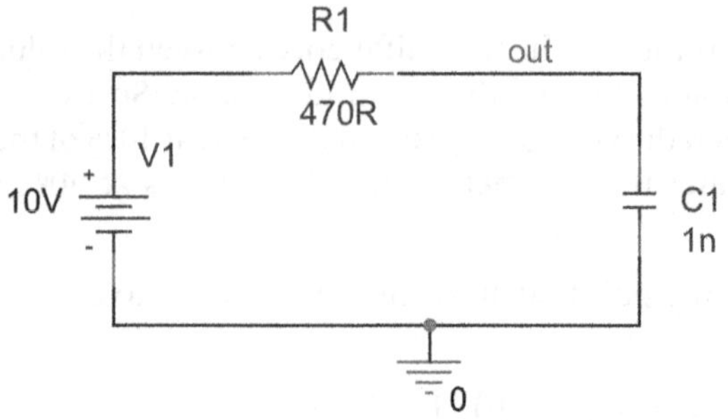

FIGURE 12.3
Measuring rise time.

Exercise 1

1. Create a project called Risetime and draw the CR circuit in Figure 12.3. Make sure you name the node **out** as shown.
2. Create a PSpice transient analysis with a run to time of 5 μs.
3. Place a voltage marker on node **out**.
4. Run the simulation.
5. What you will see is a flat line for the voltage (Figure 12.4). This is because a DC bias point analysis has been run prior to the transient analysis such that the capacitor has reached a steady-state voltage of 10 V. So at time $t = 0$ s, the voltage across the capacitor is 10 V, as shown.

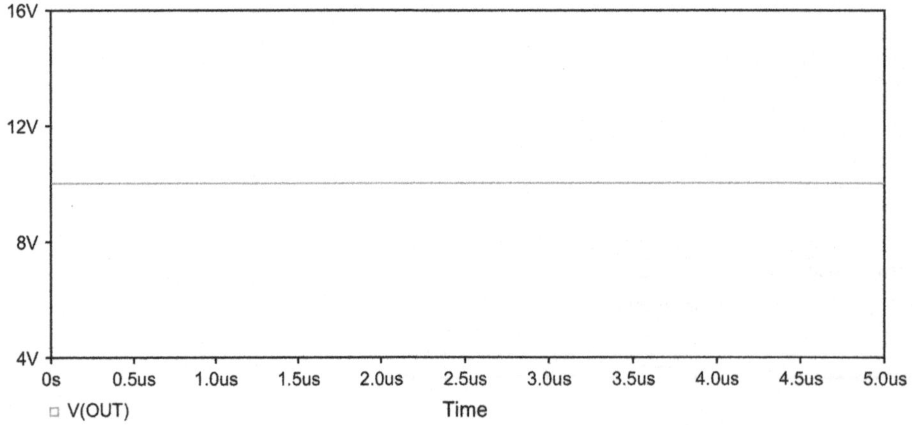

FIGURE 12.4
Capacitor voltage has reached steady state.

6. You can place an initial condition, IC1 part from the special library to ensure that at time $t = 0$ s, the voltage across the capacitor is 0 V, as in Chapter 7, Exercise 2. Alternatively, in this example, check the **Skip the initial transient bias point calculation** in the Simulation profile (Figure 12.5) and run the simulation.

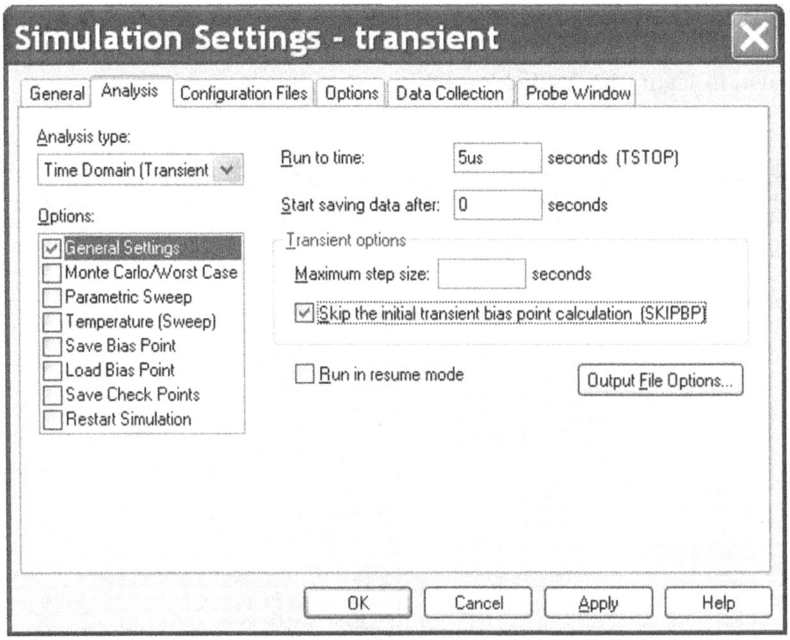

FIGURE 12.5
Skipping the initial bias point calculation.

7. You should see the familiar exponential voltage rise across the capacitor (see Figure 12.6).

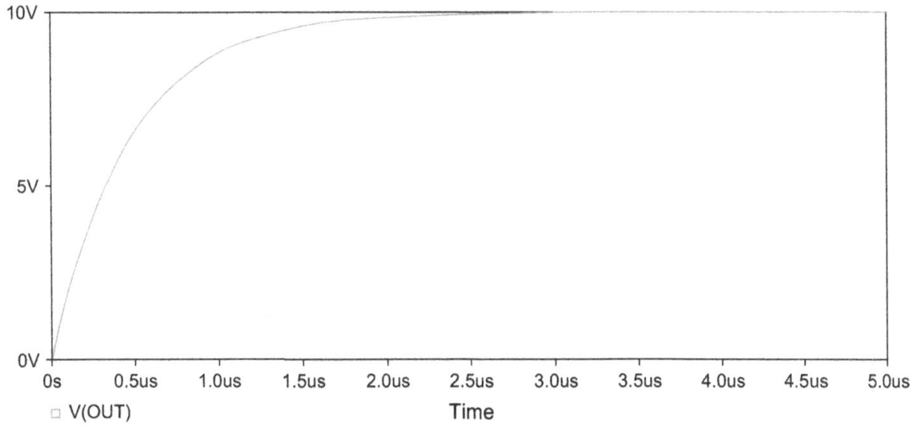

FIGURE 12.6
Voltage across capacitor initially rising exponentially.

8. In PSpice, select **Trace > Evaluate Measurement**, scroll down on the right-hand side (Measurements) and select Risetime_NoOvershoot(1). The (1) means that the function is expecting one parameter.

In the **Trace Expression** box, the cursor is automatically placed where the trace name will be entered. Select **V(out)** and the trace expression will be as shown in Figure 12.7. Click on OK.

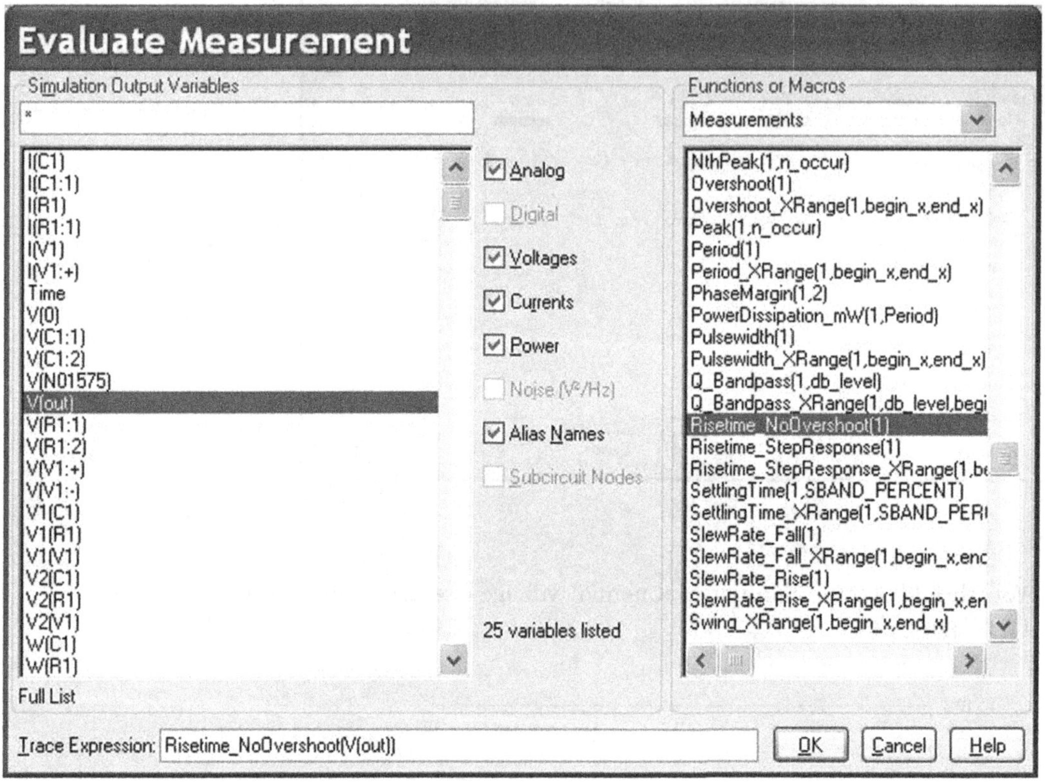

FIGURE 12.7
Selecting the Risetime_NoOvershoot measurement.

9. The Trace measurement result will be displayed as seen in Figure 12.8.

	Measurement Results			
	Evaluate	Measurement	Value	
▶	☑	Risetime_NoOvershoot(V(out))	1.03128u	
	Click here to evaluate a new measurement...			

FIGURE 12.8
Trace measurement evaluation.

10. Turn on the cursors, measure the voltage at 1 V (10%) and 9 V (90%) and confirm that the rise time is correct or .

Exercise 2

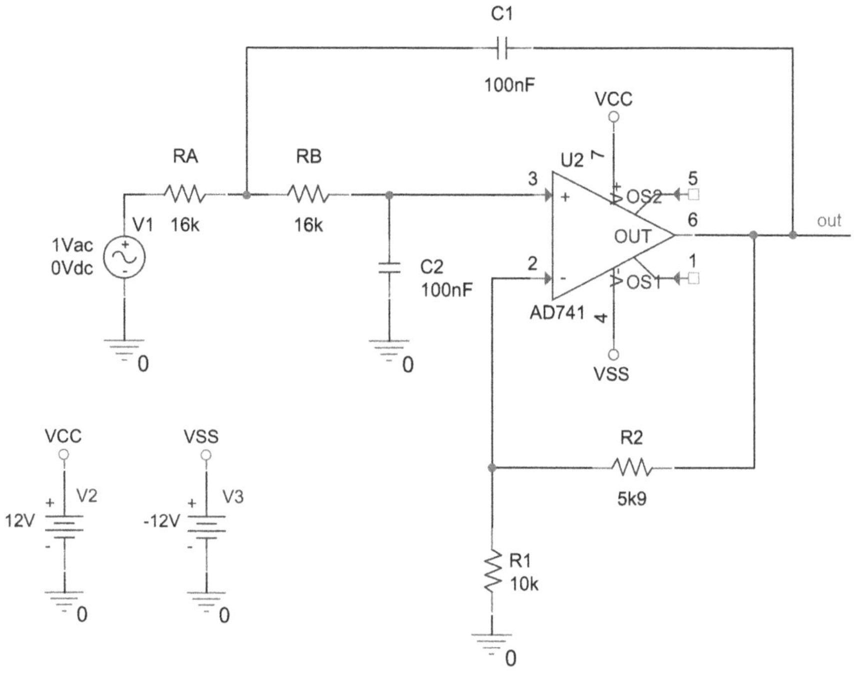

FIGURE 12.9
Sallen and Key low pass filter.

You will measure the low-pass cut-off frequency for a Sallen and Key filter.

1. Draw the Sallen and Key filter in Figure 12.9.
2. Set up an AC sweep for 1 Hz to 10 kHz, performing a logarithmic sweep with 20 points per decade.
3. Place a V_{dB} voltage marker on the output: **PSpice > Markers > Advanced > dB magnitude of Voltage**.
4. Run the simulation.
5. In PSpice, select **Trace > Evaluate Measurements > Cutoff_Lowpass_3dB()** and select V(out). The trace measurement should report a cut-off frequency of 99.6 Hz, as shown in Figure 12.10.

Measurement Results

	Evaluate	Measurement	Value
▶	☑	Cutoff_Lowpass_3dB(V(out))	99.62219
			Click here to evaluate a new measurement...

FIGURE 12.10
Cut off frequency measured as 99.6 Hz.

6. The frequency response is shown in Figure 12.11.

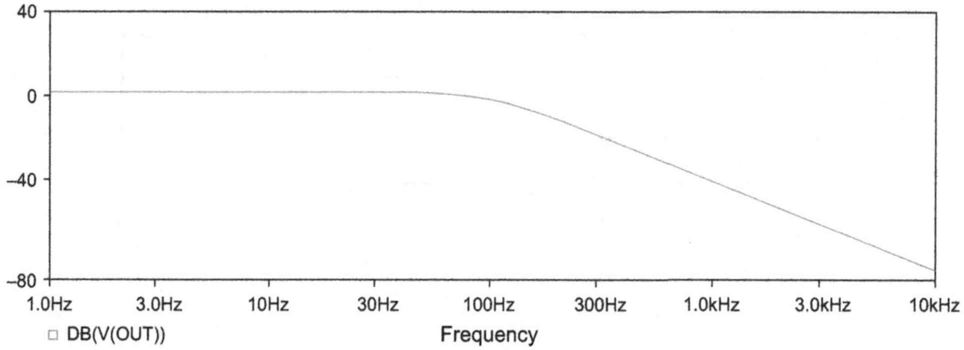

FIGURE 12.11
Frequency response of Sallen and Key filter.

CHAPTER 13

Analog Behavioral Models

Chapter Outline

Analog Behavioral Model (ABM) devices are extended versions of the traditional Spice voltage-controlled sources, the E device, which is a voltage-controlled voltage source (VCVS), and the G device, a voltage-controlled current source (VCCS). They provide transfer functions, mathematical expressions or look-up tables to describe the behavior of an electronic device or circuit. ABMs can provide a systems approach to designing electronic circuits. The electronic system is represented by a block diagram with each block represented by an ABM device which can reduce the total simulation time. If the system meets the required specifications, then each block can be successively replaced by its final electronic circuit. Alternatively, working electronic circuits can be replaced by an equivalent ABM block.

There are two types of ABM device: PSpice equivalent parts, which have a differential input and double-ended output; and control system parts, which have a single input and output pin. The standard E, F, G and H devices can be found in the analog library, whereas the ABM devices can be found in the ABM library.

13.1 ABM DEVICES

The extended sources provide five additional functions, which are defined as:

Value	Mathematical expression
Table	Look-up table

Analog Design and Simulation using OrCAD Capture and PSpice. DOI: 10.1016/B978-0-08-097095-0.00013-1

Freq	Frequency response
Chebyshev	Filter characteristics
Laplace	Laplace transform

Figure 13.1 shows a typical use of an ABM EValue device to implement a voltage doubler. The ABM has a differential input (IN+, IN−) and a double-ended output (OUT+, OUT−). When you first place the ABM, the default expression is given by:

```
V(%IN+, %IN-)
```

which calculates the difference between the voltages on the input pins IN+ and IN−. In order to multiply by a factor of 2, the expression is first enclosed in curly brackets (braces).

```
2*{V(%IN+, %IN-)}
```

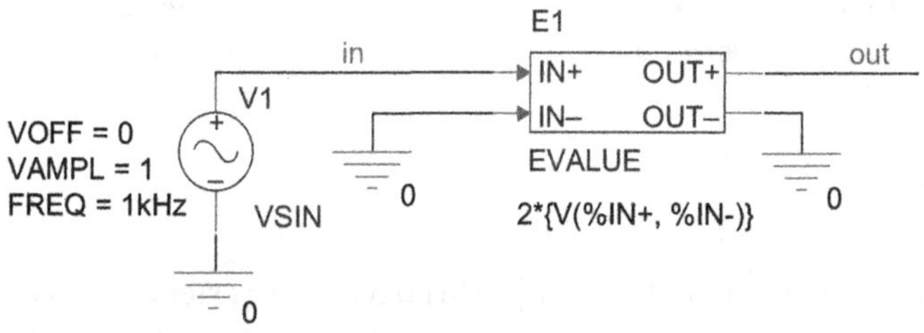

FIGURE 13.1
Amplitude doubler.

Figure 13.2 shows the output voltage when a 1 V sinewave is applied to the input pins.

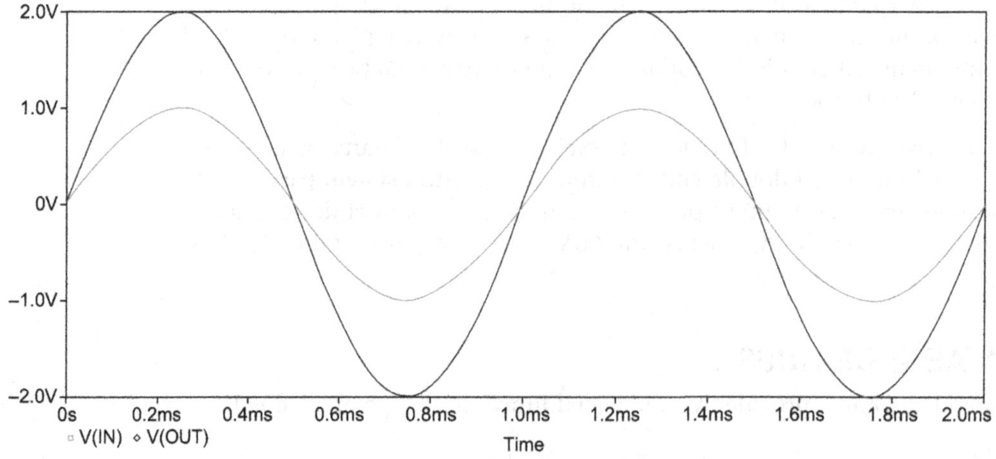

FIGURE 13.2
Input and output waveforms for the voltage doubler.

Other mathematical functions that can be applied to ABM expressions are shown in Table 13.1.

Table 13.1 Mathematical Functions

Function	Expression
ABS	Absolute value
SQRT	Square root $\sqrt{x}$
PWR	$\lvert x\rvert^{exp}$
PWRS	x^{exp}
LOG	Log base e $\ln(x)$
LOG10	Log base 10 $\log_{10}(x)$
EXP	e^x
SIN	sin
COS	cos
TAN	tan
ATAN	$\tan^{-1}$
ARCTAN	$\tan^{-1}$

Conditional statements can also be applied to ABM parts. For example, in Figure 13.3, **if** the input voltage is greater than 4 V, **then** output 0 V **else** output 5 V. This effectively is a comparator. The resulting waveform is shown in Figure 13.4.

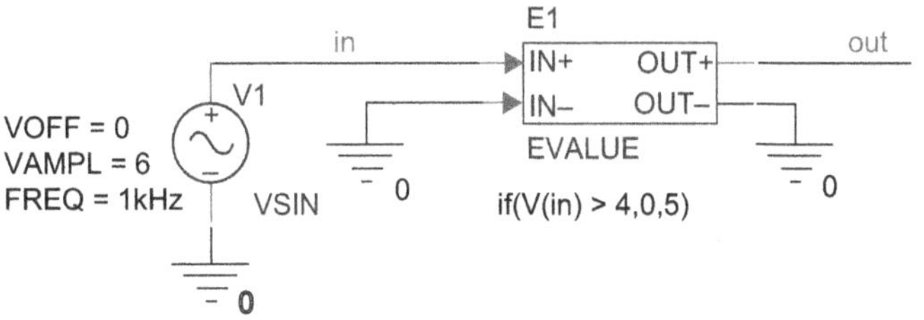

FIGURE 13.3
ABM comparator.

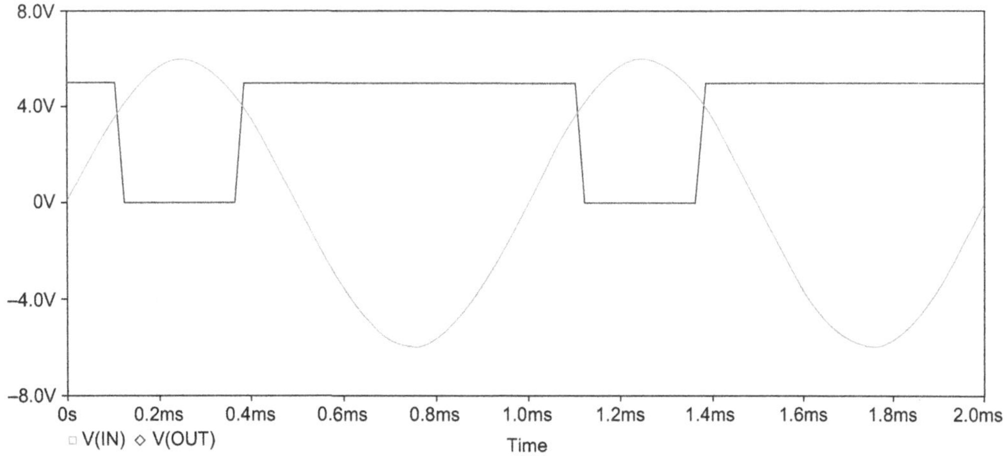

FIGURE 13.4
ABM comparator.

The first order 159 Hz low-pass filter in Figure 13.5 can be defined by the transfer function:

$$\frac{V_{\text{out}}}{V_{\text{in}}} = \frac{1}{1 + j\dfrac{\omega}{\omega_c}} \tag{13.1}$$

where the cut-off frequency is given by:

$$\omega_c = \frac{1}{CR} = 2\pi f \tag{13.2}$$

$$f = \frac{1}{2\pi CR} = \frac{1}{2\pi \times 10^{-6} \times 10^{3}} = 159 \text{ Hz}$$

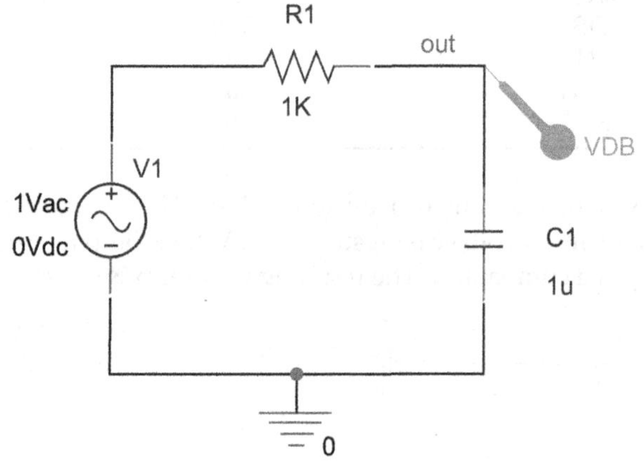

FIGURE 13.5
Low-pass filter.

Figure 13.6 shows the frequency response of the filter.

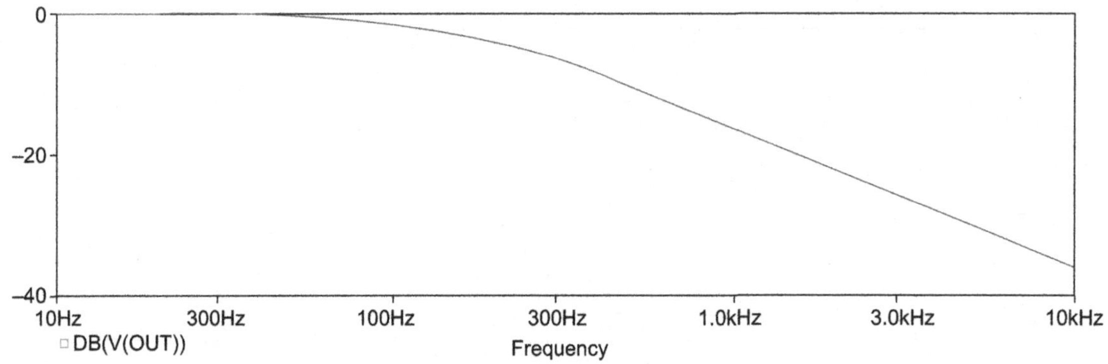

Measurement Results

Evaluate	Measurement	Value
☑	Cutoff_Lowpass_3dB(V(out))	159.40481

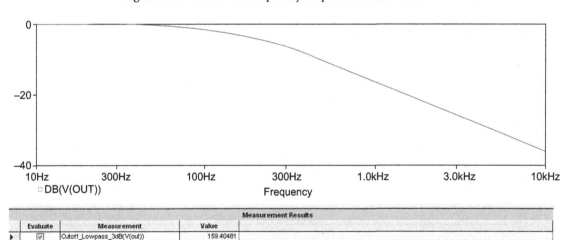

FIGURE 13.6
Low-pass filter frequency response.

Laplace ABMs can be used to implement filter circuits where the transfer function is transformed to the s domain, where $s = j\omega$ and $\omega = 2\pi f$, so the transfer function for the low-pass filter is defined in the s domain by:

$$\frac{V_{\text{out}}}{V_{\text{in}}} = \frac{1}{1 + s\tau} \tag{13.3}$$

where $\tau = \text{CR} = 10^{-6} \times 10^{3}$; $\tau = 10^{-3}$ or 0.001 s.

The filter circuit is redrawn in Figure 13.7 with the transfer function given as:

$$\frac{V_{out}}{V_{in}} = \frac{1}{1 + 0.001 * s} \tag{13.4}$$

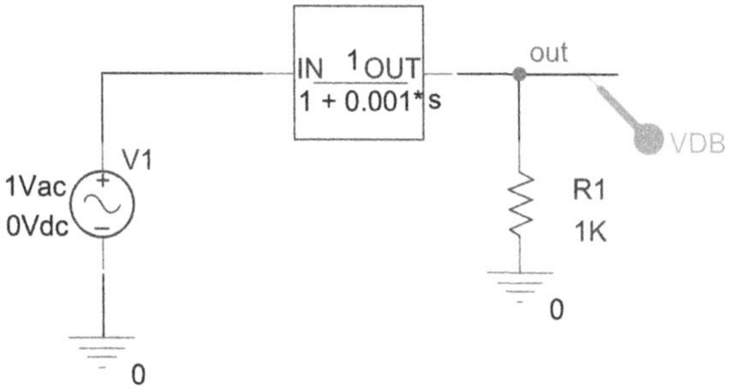

FIGURE 13.7
Laplace low-pass filter.

NOTE

When you enter the coefficient for s, make sure you that enter the multiplier, i.e. 1+0.001*s and not 1+0.001 s.

Figure 13.8 shows the low-pass frequency response for the Laplace filter.

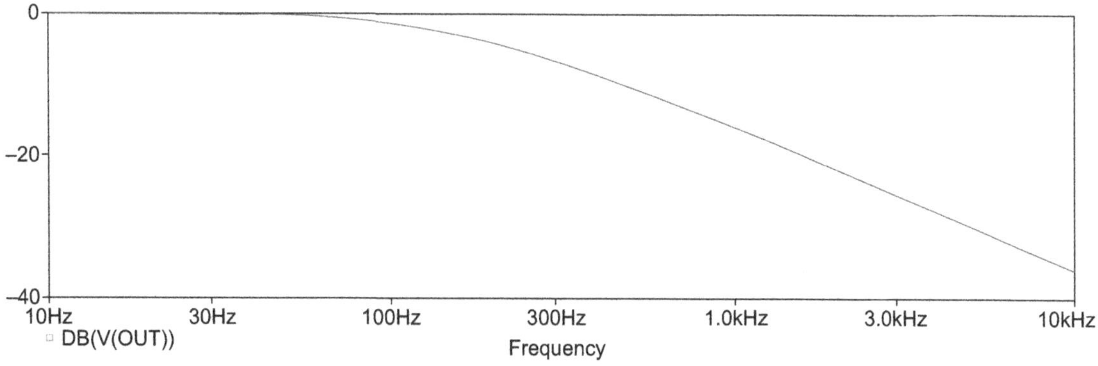

Measurement Results

Evaluate	Measurement	Value
☑	Cutoff_Lowpass_3dB(V(out))	159.40481

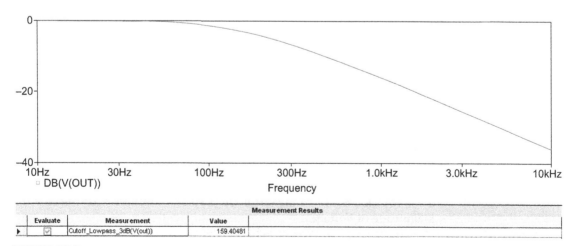

FIGURE 13.8
Low-frequency response for the Laplace filter.

In the circuit of Figure 13.9, both ABM inputs are grounded but the ABM expression is referencing a net name called 'source'. This is useful to reduce the wiring complexity of a circuit, especially if there are multiple ABMs being driven by a single source. Note that since a GValue ABM part is being used, the output is a current and therefore cannot be left unconnected, hence the output resistor, R1, provides a DC path to 0 V.

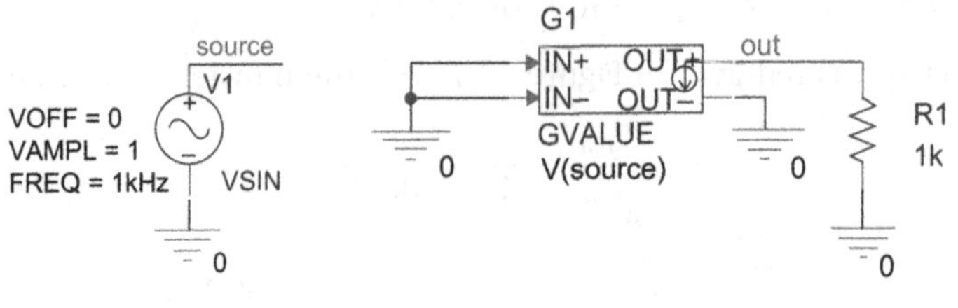

FIGURE 13.9
Referencing the **source** node in the circuit.

13.2 EXERCISES

Exercise 1

In Figure 13.10, an EFreq ABM part is being used to model a bandpass filter using a table of parameters defining frequency, magnitude and phase.

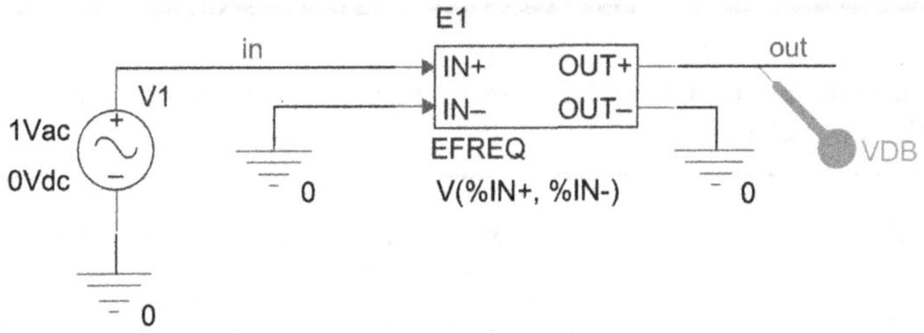

FIGURE 13.10
Using a frequency table.

1. Draw the circuit in Figure 3.10 using an EFreq part from the ABM library. The V_{AC} source (V1) is from the source library.
2. Double click on EFreq to open the Property Editor, scroll to the **Table** property and enter:

   (0.1,-40,170)(1k,-40,160)(2k,-20,140)(3k,-0,100)(6k,-0,-100)(10k,-20,-140)(20k,-40,-160)(30k,-40,-170)

3. Set up a simulation profile for an AC sweep from 1 Hz to 100 kHz and place a V_{dB} voltage marker on the **out** node (**PSpice > Markers > Advanced > dB Magnitude of Voltage**).
4. Run the simulation. You should see the bandpass response as shown in Figure 13.11.

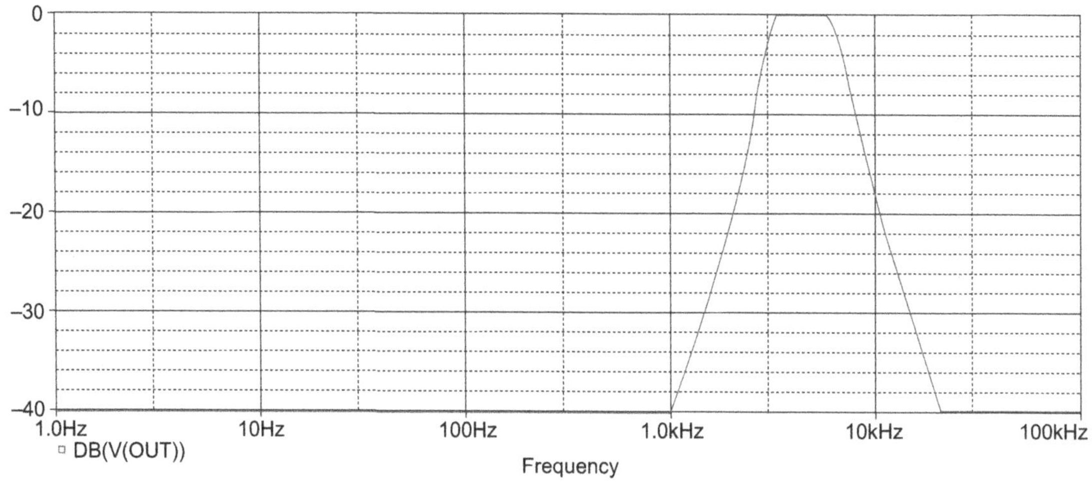

FIGURE 13.11
Bandpass filter response using a Freq ABM.

Exercise 2

1. Draw the ABM rectifier as shown in Figure 13.12. The net name **in** is referenced inside the ABM expression brackets.

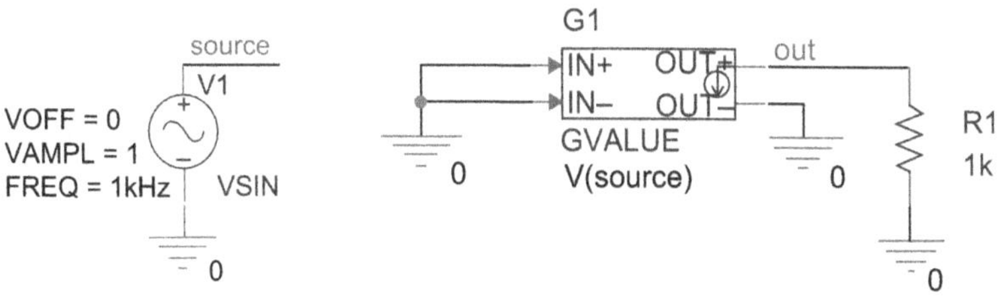

FIGURE 13.12
Behavioral implementation of a rectifier.

2. Set up a simulation profile for a transient sweep with a **run to time** of 4 ms.
3. Run the simulation.
4. Investigate the use of the EFREQ ABM device in modeling the 1500 Hz bandpass filter discussed in Chapter 10.

CHAPTER 14

Noise Analysis

Chapter Outline

Noise analysis is run in conjunction with an AC analysis and calculates the output noise and equivalent input noise in a circuit. The output noise, at a specified output node, is the root mean square (RMS) sum of the noise generated by all the resistors and semiconductors in the circuit. If the circuit is considered as noiseless, then the equivalent input noise is defined as the noise required at the input in order to generate the same output noise. This is the same as dividing the output noise by the gain of the circuit in order to obtain the equivalent input noise.

14.1 NOISE TYPES

14.1.1 Resistor Noise

Johnson noise or thermal noise is due to the random thermal agitation of electrons in a conductor, which increases with frequency and temperature. In PSpice, the thermal noise contribution from a resistor is represented by a current source in parallel with a noiseless resistor. Because of its random nature, the noise current source is represented as a mean square value given by:

$$\overline{i^2} = \frac{4\mathrm{k}T\Delta f}{R} \quad (\mathrm{A}^2/\mathrm{Hz}) \tag{14.1}$$

where:

k – Boltzmann's constant: $1.38\mathrm{e}^{-23}$ (J K^{-1}),

Analog Design and Simulation using OrCAD Capture and PSpice. DOI: 10.1016/B978-0-08-097095-0.00014-3

T – absolute temperature (K),
R – resistance (Ω),
Δf – frequency bandwidth (Hz).

14.1.2 Semiconductor Noise

Semiconductor noise is generally made up of thermal, shot and flicker noise. Thermal noise is generated by the intrinsic parasitic resistances for the device. Shot noise, however, is a randomly fluctuating noise current generated when a current flows across a PN junction and is given by:

$$\overline{i^2} = 2qI \ (\mathrm{A^2/Hz}) \tag{14.2}$$

where:

q – electron charge, 1.602×10^{-19} C,
I – current through the device (A).

Flicker noise is a phenomenon not widely understood but has been attributed to imperfections in semiconductor channels and the generation and recombination of charge carriers. However, what is known is that flicker noise occurs at low frequencies and that the noise current decreases with frequency exhibiting a $1/f$ characteristic. Flicker noise current is given by:

$$\overline{i^2} = \frac{KF \times Id^{AF}}{\Delta f} \ (\mathrm{A^2/Hz}) \tag{14.3}$$

where:

KF – flicker noise coefficient,
Id – current through the device,
AF – flicker noise exponent,
Δf – frequency bandwidth.

14.2 TOTAL NOISE CONTRIBUTIONS

After a noise analysis is run, the thermal, shot and flicker noise contributions for resistors and semiconductors are made available as trace variables in Probe. Table 14.1 shows the available noise variables for some of the devices.

The noise spectral density for NTOT(device) is measured in units of $\mathrm{V^2/Hz}$.

The total circuit noise is represented as either NTOT(ONOISE) in units of $\mathrm{V^2/Hz}$ or the RMS summed output, V(ONOISE) in units of $\mathrm{V/\sqrt{Hz}}$.

The equivalent input noise is V(INOISE) and is calculated from $\left(\frac{V(ONOISE)}{gain\ of\ circuit}\right)$ in units of $\mathrm{V/\sqrt{Hz}}$ or $\mathrm{A/\sqrt{Hz}}$. If the input source is a current source, the units are $\mathrm{A/\sqrt{Hz}}$. If the input source is a voltage source, the units are $\mathrm{V/\sqrt{Hz}}$.

Table 14.1 Noise Output Variables Available in Probe

Device	Output Variable	Noise
Resistor	NTOT	Thermal noise
Diode	NRS	Parasitic thermal noise for RS
	NSID	Shot noise
	NFID	Flicker noise
	NTOT	Total of noise contributions
Bipolar transistor	NRB	Parasitic thermal noise for RB
	NRC	Parasitic thermal noise for RC
	NRE	Parasitic thermal noise for RE
	NSIB	Shot noise for base current
	NSIC	Shot noise for collector current
	NFIB	Flicker noise
	NTOT	Total of all noise contributions
MOSFET	NRD	Parasitic thermal noise for RD
	NRG	Parasitic thermal noise for RG
	NRS	Parasitic thermal noise for RS
	NRB	Parasitic thermal noise for RB
	NSID	Shot noise
	NFID	Flicker noise
	NTOT	Total of all noise contributions

14.3 RUNNING A NOISE ANALYSIS

You have to run an AC analysis in order to run a noise analysis. In the simulation settings for an AC sweep, there is an option to enable the Noise Analysis as shown in Figure 14.1. In this example, the output voltage node V(out) has

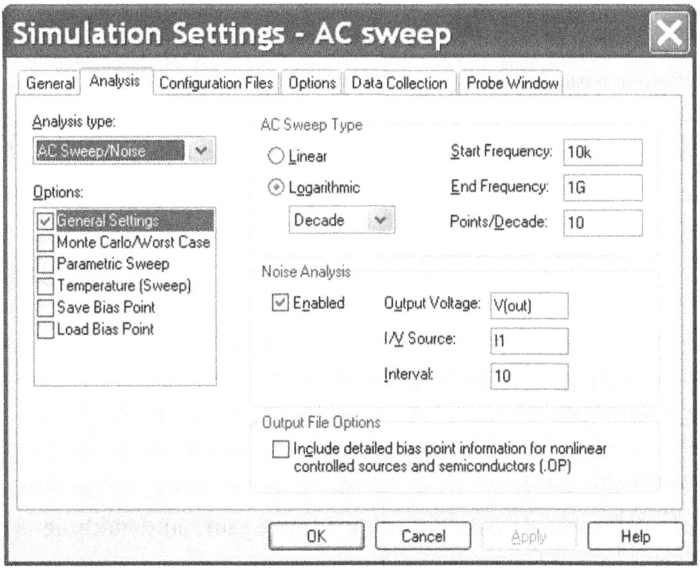

FIGURE 14.1
AC sweep and noise analysis settings.

been specified as the node at which the total output noise will be calculated. The **I/V Source** references the current or voltage source used as an input to the circuit, which is usually a V_{AC} or I_{AC} source. In this example, the reference designator is V1.

The **Interval** is an integer that specifies how often results are written to a table which is generated in the output file. Each table entry will be determined by the frequency **Interval**, which specifies the *n*th interval of the range set in the AC sweep. For example, in Figure 14.1, the AC sweep is from 10 kHz to 1 GHz with 10 points per decade; the decade frequencies will be 10 kHz, 100 kHz, 1 MHz, 10 MHz, 100 MHz and 1 GHz. So the 10th frequency interval will be 100 kHz, the 20th interval 1 MHz, and so on. If the number of points/decade was set at 5, with the same interval integer of 10, then the frequencies written to the output file would be 10 kHz, 1 MHz and 100 MHz. If no interval number is entered, then no table will be generated in the output file. This is not to be confused with the frequency interval of the waveform data points in Probe, which is determined by the AC frequency sweep settings.

14.4 NOISE DEFINITIONS

The instantaneous value of noise voltage at any time t, is given by $v_n(t)$. As the noise voltage is statistical in nature, the root mean square (rms) value is given by:

$$E_n = \sqrt{\overline{v_n(t)^2}} \quad (V) \tag{14.4}$$

Similarly for noise current, the rms value is given by:

$$I_n = \sqrt{\overline{i_n(t)^2}} \quad (A) \tag{14.5}$$

The rms noise voltage across a resistor is given by:

$$E_n = \sqrt{4kTR\Delta f} \quad (V) \tag{14.6}$$

and the rms noise current for a resistor is given by:

$$I_n = \sqrt{\frac{4kTR\Delta f}{R}} \quad (A) \tag{14.7}$$

where:

k – Boltzman's constant 1.38×10^{-23} ($J\ K^{-1}$),
T – absolute temperature (K),
R – resistance (Ω),
Δf – frequency bandwidth (Hz).

In PSpice, noise calculations are made assuming a unity gain bandwidth, ie $\Delta f = 1$Hz.

The rms noise power for a resistor is given by:

$$P_n = \frac{E_n^2}{R} = I_n^2 R \quad (W) \tag{14.8}$$

The noise power spectral density, S, is given by:

$$S = \frac{P_n}{\Delta f} \quad (W/Hz) \tag{14.9}$$

Substituting for Pn from (14.8) into (14.9):

$$S = \frac{\left(\frac{E_n^2}{R}\right)}{\Delta f} = \frac{\left(\frac{4kTR}{R}\right)}{\Delta f}$$

$$S = 4kT \quad (W/Hz) \tag{14.10}$$

Noise voltage spectral density, e_n, is given by:

$$e_n = \frac{E_n}{\Delta f} = \frac{\sqrt{4kTR\Delta f}}{\Delta f}$$

$$e_n = \frac{\sqrt{4kTR}}{\Delta f}\frac{\sqrt{\Delta f}}{}$$

$$e_n = \frac{\sqrt{4kTR}}{\Delta f}\frac{\sqrt{\Delta f}}{}\frac{\sqrt{\Delta f}}{\sqrt{\Delta f}}$$

$$e_n = \frac{\sqrt{4kTR}}{\sqrt{\Delta f}} \quad (V/\sqrt{Hz}) \tag{14.11}$$

Similarly the noise current spectral density, i_n, is given by:

$$i_n = \frac{\sqrt{\frac{4kT}{R}}}{\sqrt{\Delta f}} \quad (A/\sqrt{Hz}) \tag{14.12}$$

If the rms quantities, En and In, are uncorrelated (independent of each other), then noise sources can be added together such that:

$$E_n^2 = E_{n1}^2 + E_{n2}^2$$

or

$$E_n = \sqrt{E_{n1}^2 + E_{n2}^2} \tag{14.13}$$

In PSpice, the noise contributions for a bipolar transistor are represented by the thermal noise sources for the intrinsic base, emitter and collector resistances and the shot and flicker noise contributions for the base and collector currents. Each noise source is represented by the following spectral power densities assuming a unit frequency bandwidth ($\Delta f = 1Hz$).

Collector parasitic resistance thermal noise:

$$Ic^2 = \frac{4kT}{\left(\frac{RC}{AREA}\right)} \quad (A^2/Hz) \tag{14.14}$$

Base parasitic resistance thermal noise:

$$Ib^2 = \frac{4kT}{RB} \quad (A^2/Hz) \tag{14.15}$$

Emitter parasitic resistance thermal noise:

$$Ie^2 = \frac{4kT}{\left(\frac{RE}{AREA}\right)} \quad (A^2/Hz) \tag{14.16}$$

Base shot and flicker noise currents:

$$Ib = 2qIb + \frac{KF \times Ib^{AF}}{\Delta f} \quad (A/Hz) \tag{14.17}$$

Collector shot noise current:

$$Ic = 2qIc \quad (A/Hz) \tag{14.18}$$

where:

AREA – area scaling factor. By default this is set to 1,
AF – flicker noise exponent,
KF – flicker noise coefficient.

14.5 EXERCISE

A simple transistor circuit is used to illustrate the contribution of component noise in a circuit. A large value resistor is used for RB in order to compare its noise contribution to the noise contribution from the transistor.

1. Draw the circuit in Figure 14.2. The Q2N304 transistor is from the bipolar library and the V_{AC} source V1 is from the source library.

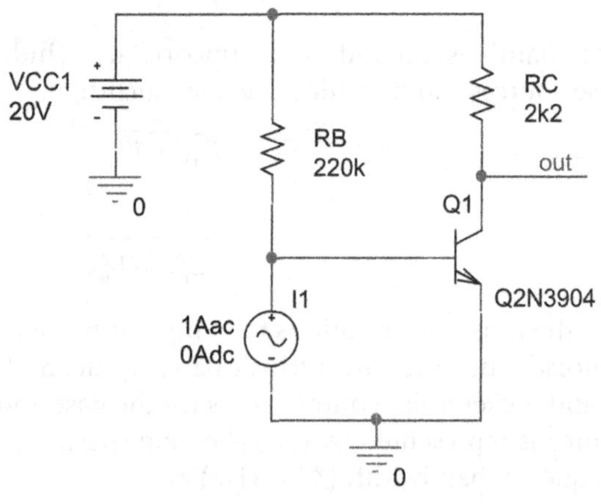

FIGURE 14.2
A simple transistor amplifier.

2. Create a PSpice simulation profile and set up an AC sweep from 10 kHz to 1 GHz using a logarithmic sweep using 10 points/decade. Enable **Noise Analysis**, defining **V(out)** as the output node and I1 as the source (Figure 14.3).

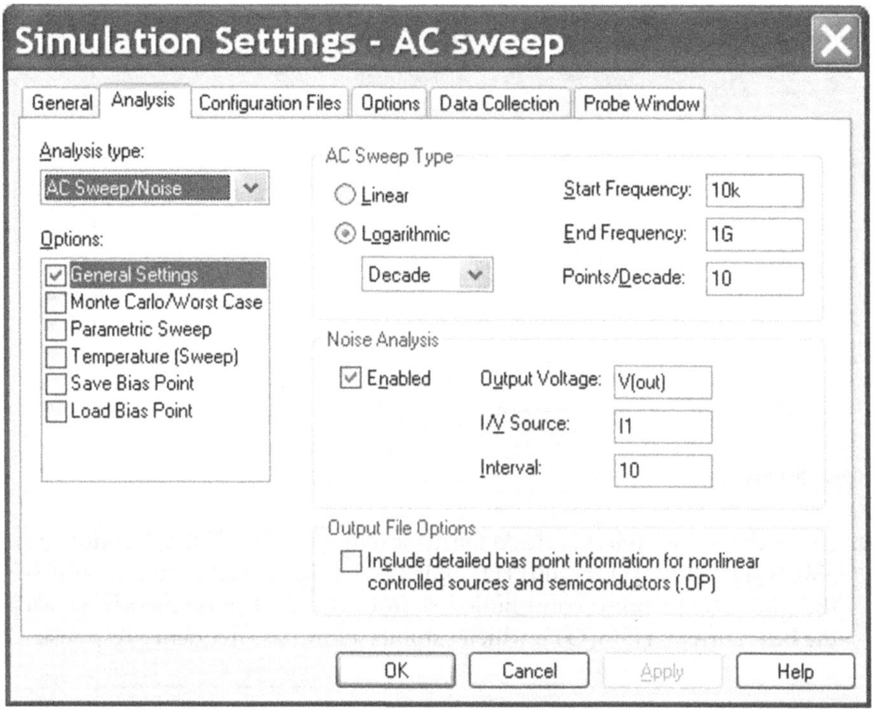

FIGURE 14.3
AC with noise-enabled analysis.

3. Run the simulation.
4. In PSpice, add the trace (**Trace > Add**) for the noise contribution from the collector resistor, NTOT(RC) (Figure 14.4). Units are V^2/Hz.

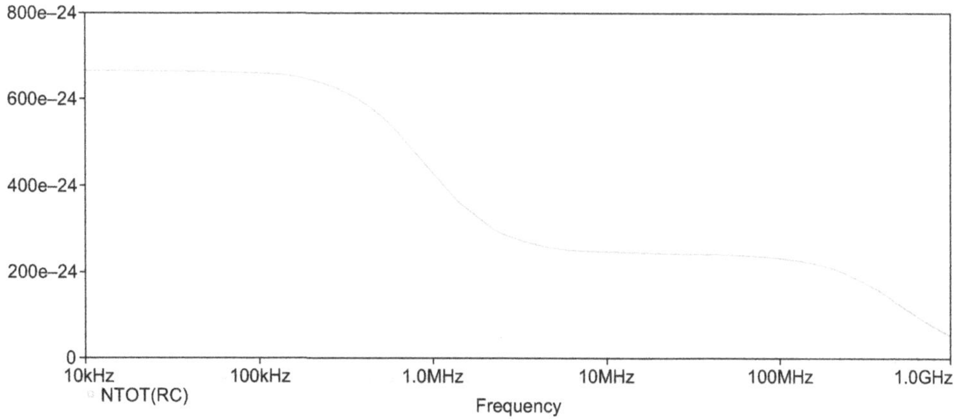

FIGURE 14.4
Noise spectral density for the collector resistor.

5. Add the trace for the noise contribution from the base resistor, NTOT(RB). Figure 14.5 shows that the larger base resistor contributes a larger noise contribution compared to the collector resistor.

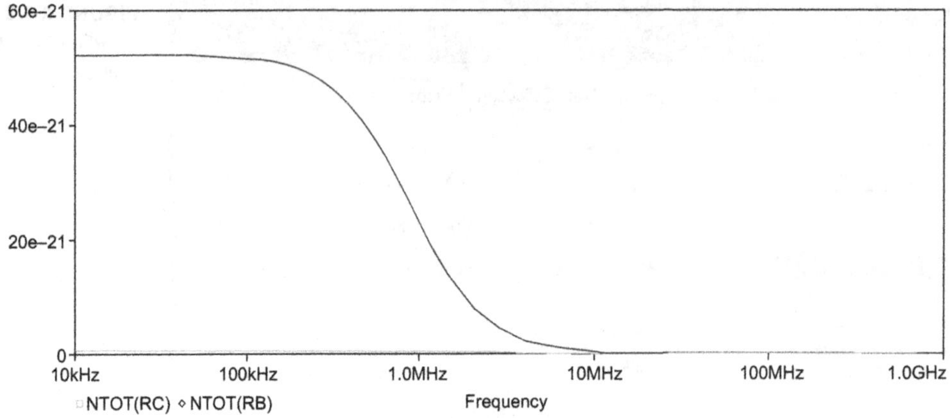

FIGURE 14.5
A greater noise contribution is from the base resistor.

6. Delete the resistor traces and add the noise traces for Q1: NFIB(Q1), NRB(Q1), NRC(Q1), NRE(Q1), NSIB(Q1), NSIC(Q1) (Figure 14.6). You should see that the greatest noise contribution is from the shot noise associated with the base current, NSIB(Q1), which exhibits a low pass frequency response.

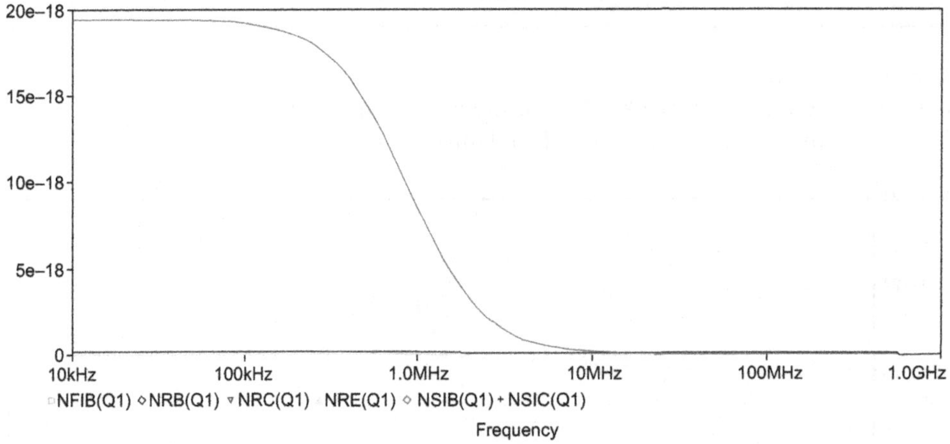

FIGURE 14.6
Transistor noise traces.

NOTE

To delete a trace, select the trace name, which turns red, and press the delete key.

---

7. Delete the collector current shot noise trace NSIC(Q1) and you will see the noise contributions from the intrinsic transistor resistances.
8. Delete all the traces, **Trace > Delete All Traces**, and add the noise trace for the base resistor NTOT(RB) and the total noise trace for Q1, NTOT(Q1). As you can see in Figure 14.7, the transistor contributes the greatest noise in the circuit.

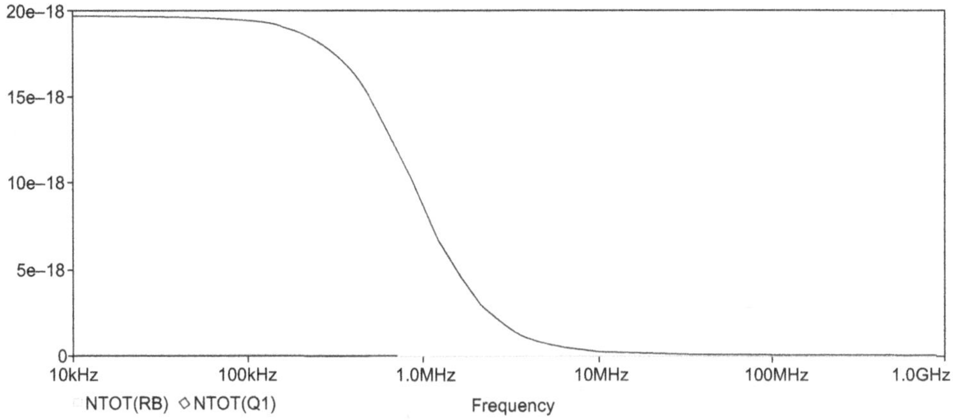

FIGURE 14.7
The transistor contributes the greatest noise in the circuit.

9. Delete all the traces.
10. Add the NTOT(ONOISE) trace, which is the total output noise for the circuit in units of V^2/Hz as shown in Figure 14.8.

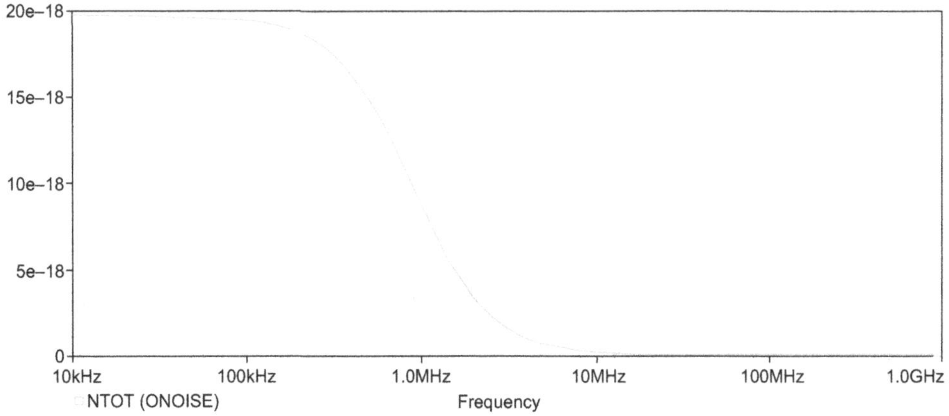

FIGURE 14.8
Total output noise in the circuit.

11. Delete NTOT(ONOISE) and add the trace for the equivalent input noise I(NOISE). You should see the noise increase with an increase in frequency due to the current gain of the transistor being proportional to frequency (Figure 14.9). As the input source is a current source, the units are $A/\sqrt{Hz}$.

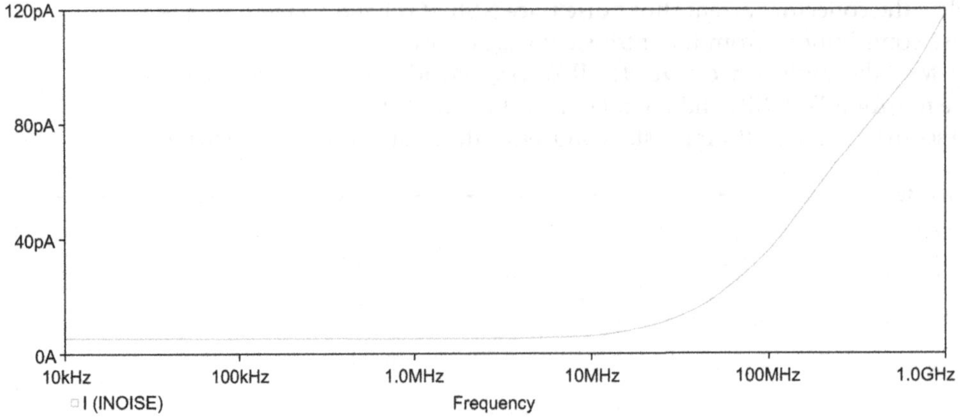

FIGURE 14.9
Increase in equivalent input noise with frequency.

12. Delete I(NOISE) and add the trace for V(ONOISE) (Figure 14.10). The units are $V/\sqrt{Hz}$.

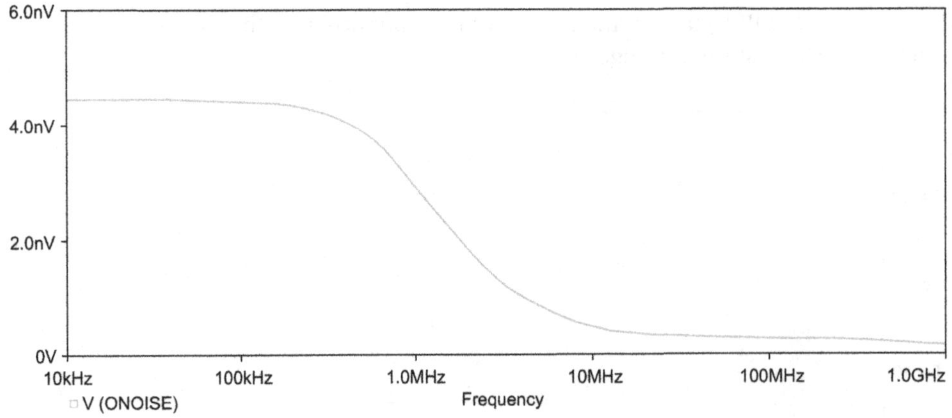

FIGURE 14.10
Output noise of the circuit.

13. A table of results will be generated for the noise analysis and can be seen in the **Output File** for decade frequencies of 10 kHz, 100 kHz, 1 MHz, 10 MHz, 100 MHz and 1 GHz. Select **View > Output File** and scroll down to the start of the noise analysis, which is 10 kHz (Figure 14.11).

The first section of the Output File states the frequency used for the noise calculation. The calculated noise contributions from the transistor and the two resistors are shown in units of V^2/Hz. Next, the total circuit noise was calculated and shown in units of V^2/Hz for NTOT(ONOISE) and $V/\sqrt{Hz}$ for V(ONOISE).

The transfer function was then calculated giving, in the above case, a circuit gain of 832.9 at 10 kHz, from which the equivalent input noise I(NOISE) was calculated and shown in units of $A/\sqrt{Hz}$, as the input source was a current source.

```
****     NOISE ANALYSIS                    TEMPERATURE =   27.000 DEG C
******************************************************************************
    FREQUENCY =   1.000E+07 HZ

**** TRANSISTOR SQUARED NOISE VOLTAGES (SQ V/HZ)

         Q_Q1

RB       1.845E-26

RC       1.649E-20

RE       0.000E+00

IBSN     1.513E-19

IC       6.411E-20

IBFN     0.000E+00

TOTAL    2.319E-19

**** RESISTOR SQUARED NOISE VOLTAGES (SQ V/HZ)

         R_RC       R_RB

TOTAL    2.452E-22  4.060E-22

**** TOTAL OUTPUT NOISE VOLTAGE         =  2.325E-19 SQ V/HZ

                                        =  4.822E-10 V/RT HZ

     TRANSFER FUNCTION VALUE:

       V(OUT)/I_I1                     =  7.340E+01

     EQUIVALENT INPUT NOISE AT I_I1 =  6.569E-12 A/RT HZ
```

FIGURE 14.11 Output file showing noise calculations.

CHAPTER 15

Temperature Analysis

Chapter Outline

A change in temperature can affect the performance and characteristics of a circuit. The components most affected by a change in temperature include semiconductors, resistors, capacitors and inductors. All of these components have an inbuilt temperature dependence model parameter such that performing a temperature sweep will change component and subsequent circuit behavior.

15.1 TEMPERATURE COEFFICIENTS

For a resistor, the change in its nominal value due to a change in temperature is defined as:

$$R = R(\text{nom}) * (1 + \text{TC1} * (T - T_{\text{nom}}) + \text{TC2} * (T - T_{\text{nom}})^2) \qquad (15.1)$$

where:

TC1 – linear temperature coefficient (ppm/°C),

TC2 – quadratic temperature coefficient ($\text{ppm}/{}^\circ\text{C}^{-2}$),

T – simulation temperature (°C),

T_{nom} – nominal temperature (°C), set by default to 27°C.

There is a TCE – exponential coefficient which if specified gives the resistor value as:

$$R = R(\text{nom}) * 1.01^{\text{TCE}*(T - T_{\text{nom}})} \qquad (15.2)$$

Manufacturers normally specify the linear coefficient TC1 for resistors.

Analog Design and Simulation using OrCAD Capture and PSpice. DOI: 10.1016/B978-0-08-097095-0.00015-5

The temperature coefficients specified for resistors are given in parts per million per degree Celsius (ppm/°C). So for a 10 kΩ resistor with a linear temperature coefficient of +200 ppm/°C, TC1 = 0.0002 and with no TC2 specified, a 20°C rise in temperature will give:

$$R = 10\,000 \times (1 + (0.0002 * 20))$$

Therefore, for a 20°C rise in temperature, $R = 10\,040\ \Omega$.

Similarly, for inductors and capacitors the component values are given by:

$$\mathrm{L} = \mathrm{L(nom)} * (1 + \mathrm{TC1} * (T - T_{\mathrm{nom}}) + \mathrm{TC2} * (T - T_{\mathrm{nom}})^2)$$

$$\mathrm{C} = \mathrm{C(nom)} * (1 + \mathrm{TC1} * (T - T_{\mathrm{nom}}) + \mathrm{TC2} * (T - T_{\mathrm{nom}})^2)$$

There is no TCE exponential coefficient for the inductors and capacitors.

In previous versions of OrCAD, the temperature coefficients were not readily available on the Capture parts therefore to add TC1 and TC2. Breakout parts as used for Monte Carlo analysis are used where the temperature coefficients are added to the PSpice model definition.

For example, to add a linear temperature coefficient, TC1 = 0.02 (ppm/°C), place an Rbreak resistor from the Breakout library, **rmb > Edit PSpice Model** and edit the PSpice model from:

```
.model Rbreak RES R=1
```

to

```
.model Rtemp RES R=1 TC1=0.02
```

15.2 RUNNING A TEMPERATURE ANALYSIS

An AC, DC or transient analysis is normally run using the default nominal temperature (TNOM) of 27°C, which is set in the simulation profile under the **Options** tab (Figure 15.1). TNOM is the default nominal temperature and is also the temperature at which model parameters were measured.

If you want to run a transient analysis at a different temperature then you need to specify the simulation temperature by selecting **Temperature (Sweep)** in the simulation profile and then enter either a single simulation temperature or a list of temperatures values as shown in Figure 15.2.

In the example above, three transient analyses will be run at the specified temperatures of 27, 55 and 125°C and plotted, one for each temperature, on one graph in Probe, the PSpice waveform viewer.

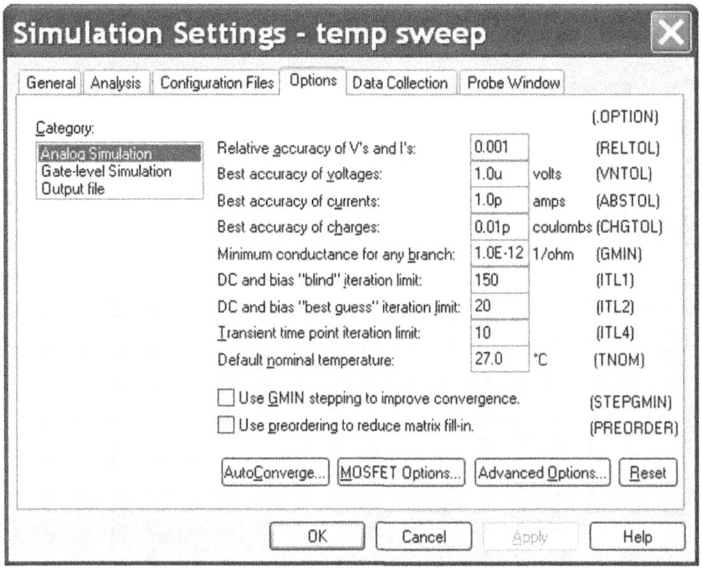

FIGURE 15.1
Default simulation options.

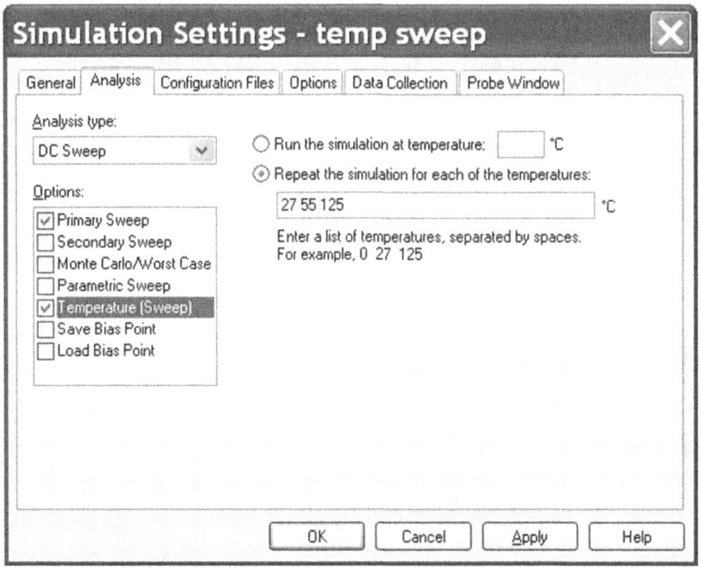

FIGURE 15.2
Setting the simulation temperature.

If you want to use temperature as a swept variable, i.e. for temperature to appear on the *x*-axis, you run a DC sweep and then select temperature as the sweep variable, as shown in Figure 15.3. The temperature will be swept from 0 to 50°C in 1°C increments. The change in temperature is represented by $T - T_{\text{nom}}$.

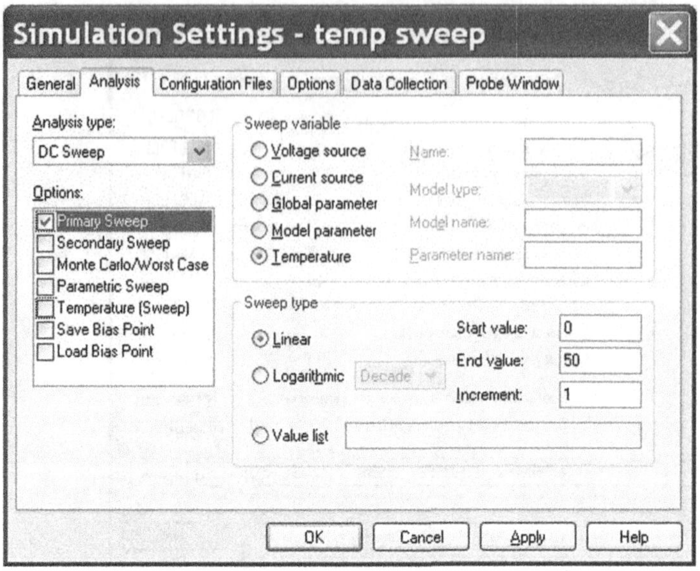

FIGURE 15.3
Performing a temperature sweep.

15.3 EXERCISES

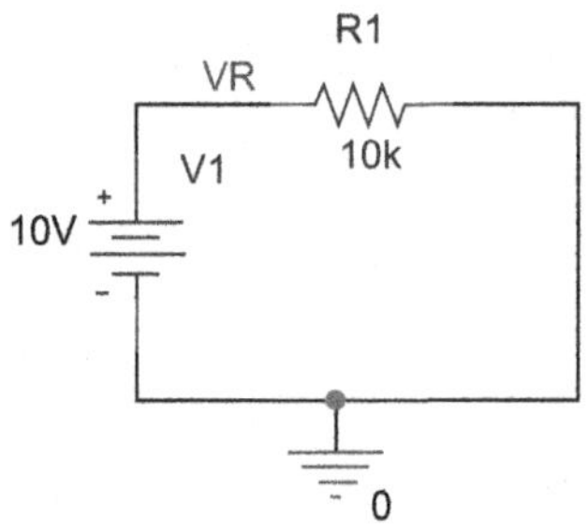

FIGURE 15.4
Adding resistor temperature coefficients.

Exercise 1

The specification for a 10 kΩ resistor is given as 200 ppm/°C. Therefore, TC1 = 0.0002.

1. Draw the circuit shown in Figure 15.4 and double click on the resistor to open the Property Editor. Enter a linear coefficient value of 0.0002 in TC1 and **Display** both the property **Name and Value** of TC1 as shown in Figure 15.5. Make sure you name the node as **VR**, **(Place > Net Alias)**.

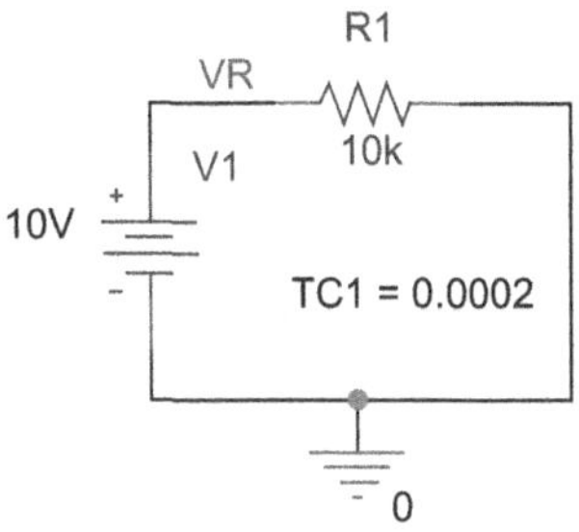

FIGURE 15.5
Displaying TC1.

2. Set up a DC linear temperature sweep from 0°C to +50°C in steps of 1°C as shown in Figure 15.6.

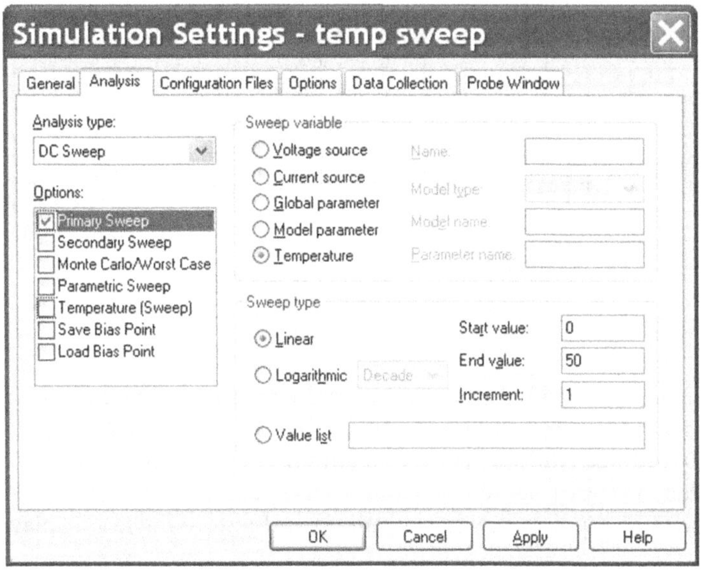

FIGURE 15.6
DC temperature sweep.

3. Run the simulation.
4. In order to plot resistance, you need to plot the ratio of the voltage across R1 divided by the current flowing through R1. In Probe, select **Trace > Add**  or .
5. At the bottom of the **Add Traces** window, there is an expression field in which you will define the resistance ratio. Select the V(VR) trace from the list of **Simulation Output Variables**; then, in the right-hand side **Functions or Macros**, select the divider symbol '/' and then select I(R1) from the **Simulation Output Variables**. You should see the expression as in Figure 15.7.

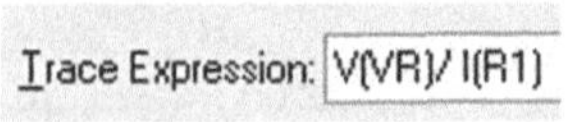

FIGURE 15.7
Trace expression for resistance of R1.

Note that you can also type in the expression rather than select variables or functions. Probe will then display the expected increase in resistance with temperature (Figure 15.8).

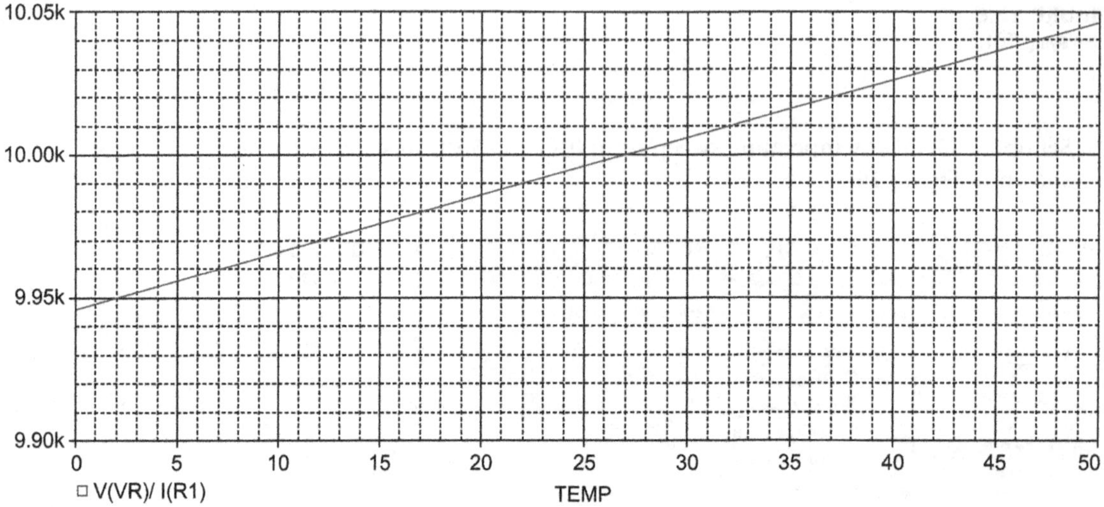

FIGURE 15.8
Resistance change with a change in temperature.

Turn the cursor on and verify that the value of the resistor is 10 kΩ at 27°C and 10 040 Ω at 47°C, which represents a rise in temperature of 20°C.

6. Reset TC1 to 0 and add and display the quadratic temperature coefficient, TC2 with a value of 0.001 (Figure 15.9) and resimulate; but this time, rather than add in the trace expression, you can call up and restore the last display automatically.

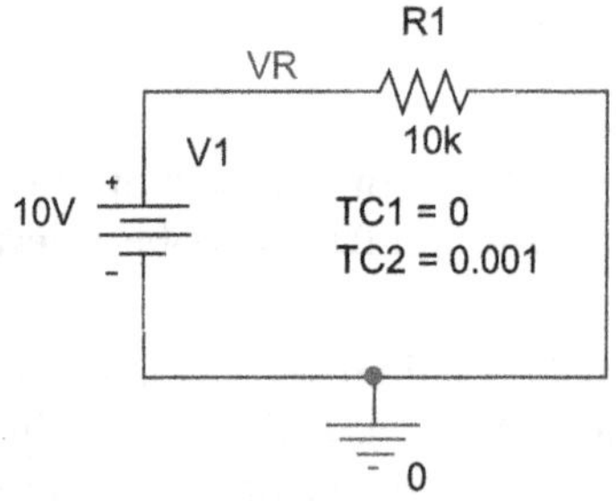

FIGURE 15.9
Adding a quadratic temperature coefficient.

7. In Probe, select **Window > Display Control > Last Session > Restore**. You should see the quadratic response as shown in Figure 15.10.

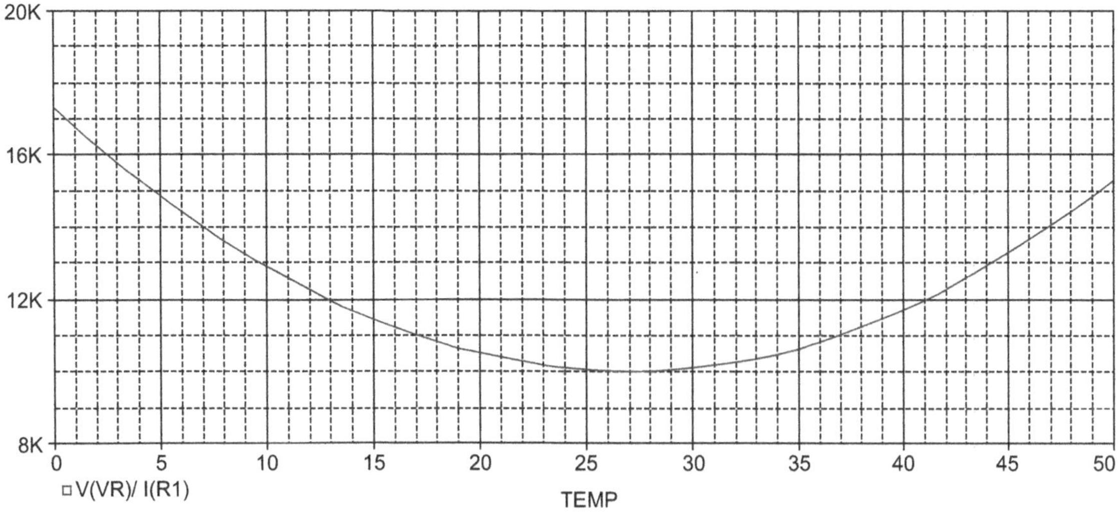

FIGURE 15.10
Variation of resistance using quadratic temperature coefficient TC2.

Exercise 2

1. Draw the circuit shown in Figure 15.11. The D1N914 diode can be found in either the diode or eval library.

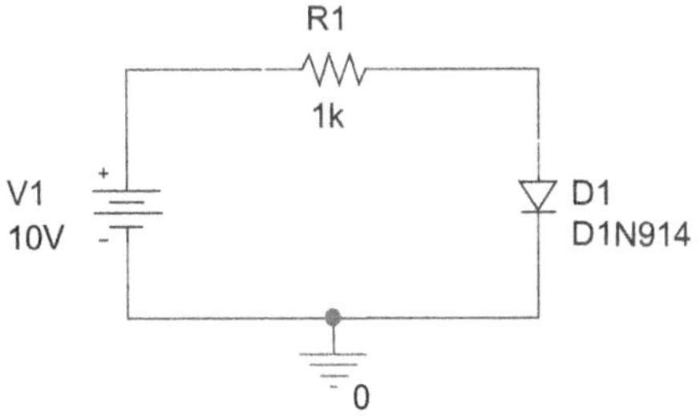

FIGURE 15.11
Using a nested sweep to determine temperature effects on the diode.

2. Set up a nested DC sweep. For the primary sweep, sweep V1 from 0 V to +10 V in steps of 0.01 V (Figure 15.12). For the secondary sweep, sweep the temperature from −55°C to +75°C in steps of 10°C (Figure 15.13).

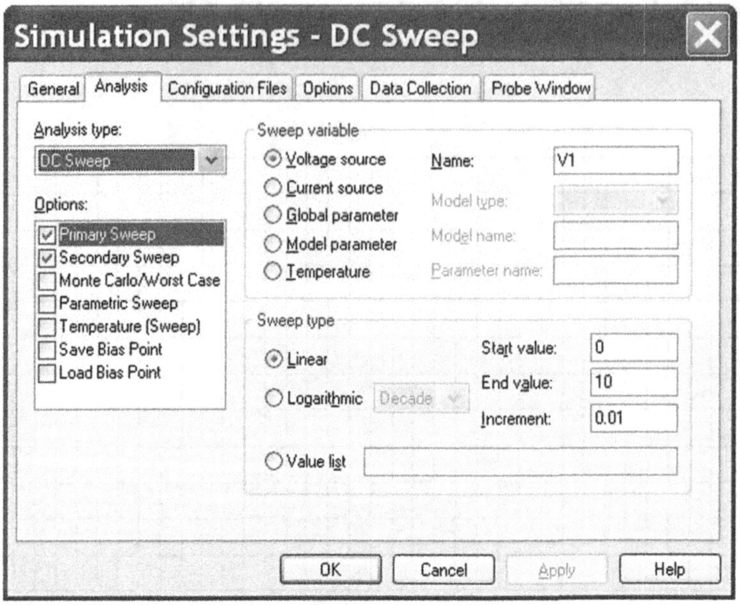

FIGURE 15.12
Primary sweep.

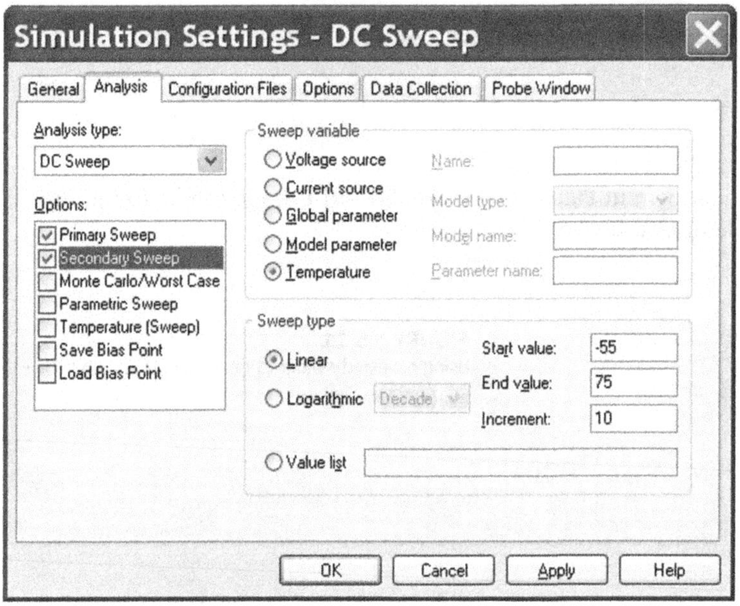

FIGURE 15.13
Secondary sweep.

3. We need to display the current through the diode versus the voltage across the diode. Select **Plot > Axis Settings > X Axis**. Select **Axis Variable**... and select V1(D1) (Figure 15.14). Click on OK.

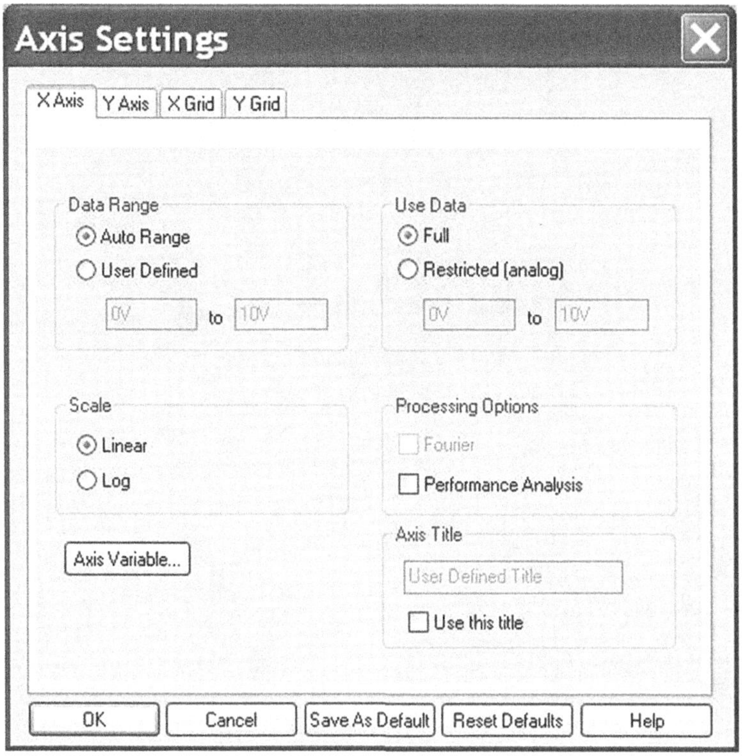

FIGURE 15.14
Changing axis variable.

4. Add the trace for the current through the diode. **Trace > Add > I(D1)**. Figure 15.15 shows the family of temperature current–voltage curves for D1.

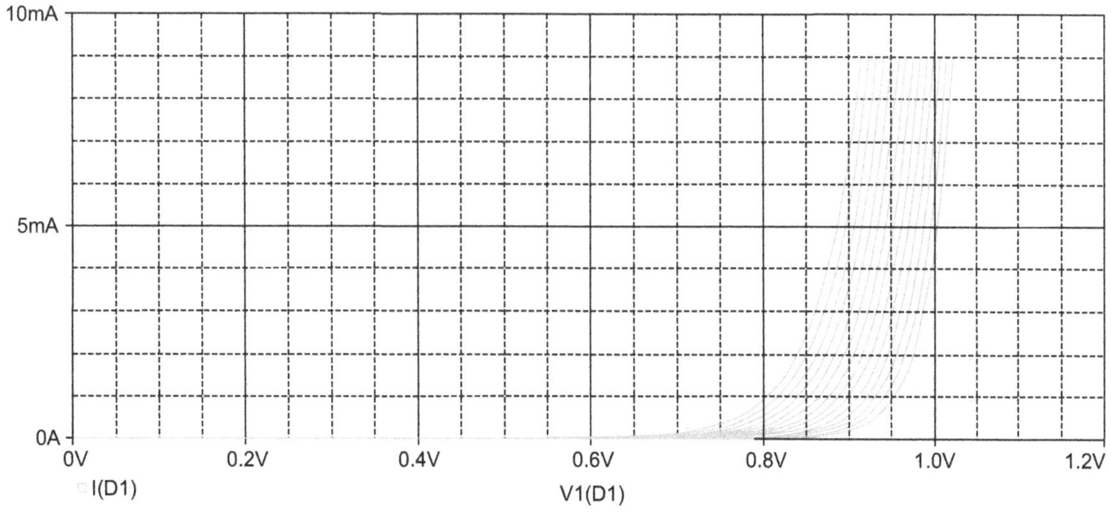

FIGURE 15.15
Family of temperature current–voltage curves for the D1N914 diode.

CHAPTER 16
Adding and Creating PSpice Models

Chapter Outline

PSpice models can be created and edited in the PSpice Model Editor, which can be started in standalone mode from the **Start** menu, **PSpice > Simulation Accessories > Model Editor**, or by highlighting a PSpice part in the schematic in Capture, **rmb > Edit PSpice Model**. When you edit a PSpice part from Capture, a copy of the PSpice model is created in a library file, which will have the same name as the project, i.e. <project name>.lib. This is so that the original PSpice model does not get modified. The copied library is written to the project file and can be seen as one of the Configured PSpice libraries in the Project Manager. Whenever you create a new Capture part for a PSpice model, you need to reference the library in which the model can be found by providing the path to the library file in the simulation profile under the **Configuration Files > Category > Library**.

16.1 CAPTURE PROPERTIES FOR A PSPICE PART

For PSpice simulation, a Capture part needs to have four specific properties attached. These are the:

- Implementation: name of the model
- Implementation Path: left blank as model is searched for in the configured libraries in the simulation profile

Analog Design and Simulation using OrCAD Capture and PSpice. DOI: 10.1016/B978-0-08-097095-0.00016-7

- Implementation Type: PSpice Model
- PSpiceTemplate: provides the Capture part interface to the model or subcircuit.

The above properties are automatically added when a Capture part is created either in Capture or in the Model Editor. Figure 16.1 shows the attached properties for a Q2N3904 transistor.

The PSpiceTemplate is defined as:

```
Q^@REFDES %c %b %e @MODEL
```

where the first character Q defines a bipolar transistor. Other common device types are shown in Table 16.1. A complete table of device types is shown in Appendix 2.

The ^ is used by the netlister to define the hierarchical path to the device. The ^ is effectively replaced by the hierarchical path to the device. For example, the netlist in Figure 16.2 is that of a hierarchical design consisting of a **Top** level design containing a **Bottom** hierarchical block. At the bottom level, there are two resistors R1 and R2, their hierarchical paths being defined by R_Bottom_R1 and R_Bottom_R2. There is also a resistor at the top level, R1.

FIGURE 16.1
Q2N3904 attached properties.

	A
	SCHEMATIC1 : PAGE1 : Q1
Color	Default
COMPONENT	2N3904
Designator	
Graphic	Q2N3904.Normal
ID	
Implementation	Q2N3904
Implementation Path	
Implementation Type	PSpice Model
Location X-Coordinate	360
Location Y-Coordinate	110
Name	INS960
Part Reference	Q1
PCB Footprint	TO92
Power Pins Visible	
Primitive	DEFAULT
PSpiceTemplate	Q^@REFDES %c %b %e @MODEL
Reference	Q1
Source Library	C:\CADENCE\SPB_16.3\TOOLS\C
Source Package	Q2N3904
Source Part	Q2N3904.Normal
Value	Q2N3904

Table 16.1 PSpice Implementation Definitions

Character	Device Type	Pin Order
R	Resistor	1,2
C	Capacitor	1,2
L	Inductor	1,2
D	Diode	Anode, cathode
Q	Transistor	Collector, base, emitter
M	MOSFET	Drain, gate, source, bulk
Z	IGBT	Collector, gate, emitter
I	Current source	+ve node, −ve node
V	Voltage source	+ve node, −ve node
X	Subcircuit	Node 1, node 2, ...node n

```
**** INCLUDING Top.net ****
* source HIERARCHY
R_Bottom_R1          N00522 N00469  1k TC=0,0
R_Bottom_R2          0 N00522  1k TC=0,0
V_V1          N00469 0 10V
R_R1          0 N00522  1k TC=0,0
```

FIGURE 16.2
Hierarchical netlist.

@REFDES is the reference designator such as Q1, R2, etc. The % defines the pin names, the order of which is defined in Table 16.1 and the @MODEL references the PSpice model name.

16.2 PSPICE MODEL DEFINITION

The basic definition for a PSpice model is given by:

```
.MODEL <model name> <model type>
+ ([<parameter name> = <value>
```

The model name must start with one of the PSpice device characters as shown in Table 16.1 and can be up to eight characters long. The model type is specific to the model; for example an NPN type is specific only to a bipolar transistor. Table 16.2 shows the associated model types for the PSpice models.

For example, the model definition for a Q2N3904 transistor is given by:

```
.model Q2N3904    NPN(Is=6.734f Xti=3 Eg=1.11 Vaf=74.03 Bf=416.4 Ne=1.259
+          Ise=6.734f Ikf=66.78m Xtb=1.5 Br=.7371 Nc=2 Isc=0 Ikr=0 Rc=1
+          Cjc=3.638p Mjc=.3085 Vjc=.75 Fc=.5 Cje=4.493p Mje=.2593 Vje=.75
+          Tr=239.5n Tf=301.2p Itf=.4 Vtf=4 Xtf=2 Rb=10)
*          National    pid=23          case=TO92
*          88-09-08 bam      creation
*$
```

Table 16.2 PSpice Model Types

Model	Device Type	Device
Qname	NPN	NPN bipolar
	PNP	PNP bipolar
	LPNP	Lateral PNP
Dname	D	Diode
Cname	CAP	Capacitor
Kname	CORE	Non-linear magnetic core
Lname	IND	Inductor
Mname	NMOS	N-channel MOSFET
	PMOS	
Jname	NJF	N-channel JFET
	PJF	P-channel JFET
Rname	RES	Resistor
Tname	TRN	Transmission line
Bname	GASFET	N-channel GAsFET
Zname	IGBT	N-channel IGBT
Nname	DINPUT	Digital input device
Oname	DOUTPUT	Digital output device
Wname	ISWITCH	Current-controlled switch
Uname	UADC	Multibit ADC
	UDAC	Multibit DAC
	UDLY	Digital delay line
	UEFF	Edge-triggered flip-flop
	UGATE	Standard gate
	UIO	Digital I/O model
	UTGATE	Tristate gate
Sname	VSWITCH	Voltage-controlled switch

The model name, Q2N3904, starts with a letter Q to signify that this is a bipolar transistor model. The model type is that of NPN and the parameter list is enclosed in { } brackets. The comment lines start with an asterisk, *, and gives information such as the semiconductor vendor, date of creation and the printed circuit board (PCB) footprint.

Most devices are defined using the basic model definition. However, the complete PSpice model definition is given by:

```
.MODEL <model name> [AKO: <reference model name>]
+ <model type>
+ ([<parameter name> = <value> [tolerance specification]]*
+ [T_MEASURED=<value>] [[T_ABS=<value>] or
+ [T_REL_GLOBAL=<value>] or [T_REL_LOCAL=<value>]])
```

A Kind Of (AKO) is used when you want to create a model based upon another model (referenced) but change some of the model parameters. For example, if you want to create a 2N3904 transistor model which has a minimum BF of 75 but want to keep the other model parameters the same, then the model definition would be:

```
.model Q2N3904_minBF AKO:Q2N3904 NPN (BF=75)
```

In this way you can build up a family of transistors based on the original transistor but with different parameters.

In the model definition there are three parameters that relate to the temperature at which the model parameter values were calculated or measured:

T_MEASURED
T_ABS
T_REL_GLOBAL or T_REL_LOCAL

T_MEASURED is the temperature at which model parameters were measured or derived. Default is 27°C and, if set, this overrides TNOM.

T_ABS sets the absolute device temperature. No matter what the circuit temperature, the device temperature will be equal to T_ABS.

T_REL_GLOBAL is used to provide a temperature offset from the circuit temperature. The device temperature will be equal to the difference between the circuit temperature and T_REL_GLOBAL.

T_REL_GLOBAL is useful if you have a transistor that is going to be operating at a higher temperature than other transistors. For example, if you have a 2N3904 transistor that is going to be located near a heat source which is running at 5°C higher than ambient temperature, then you would add T_REL_GLOBAL=5°C to the end of the parameter list as shown below:

.model Q2N3904 NPN(Is=6.734f Xti=3 Eg=1.11 Vaf=74.03 Bf=416.4 Ne=1.259

+ Ise=6.734f Ikf=66.78m Xtb=1.5 Br=.7371 Nc=2 Isc=0 Ikr=0 Rc=1

+ Cjc=3.638p Mjc=.3085 Vjc=.75 Fc=.5 Cje=4.493p Mje=.2593 Vje=.75

+ Tr=239.5n Tf=301.2p Itf=.4 Vtf=4 Xtf=2 Rb=10 **T_REL_GLOBAL=5**)

* National pid=23 case=TO92

* 88-09-08 bam creation

*$

So when you run a temperature analysis at −55°C, 27°C and 125°C, the Q2N3904 will be simulated at −50°C, 32°C and 130°C.

16.3 SUBCIRCUITS

PSpice devices can also be represented as a network of components, which is the usual case for operational amplifiers (opamps) and voltage regulators. These devices are classed as subcircuits and the first letter in the PSpiceTemplate is an X. The PSpiceTemplate for the LF411 opamp is shown below:

```
X^@REFDES %+ %- %V+ %V- %OUT @MODEL
```

The order of the pins must match the PSpice subcircuit definition for the LF411 as shown below:

```
* connections:     non-inverting input
*                  | inverting input
*                  | | positive power supply
*                  | | | negative power supply
*                  | | | | output
*                  | | | | |
.subckt LF411      1 2 3 4 5
  c1    11 12 4.196E-12
  c2     6  7 10.00E-12
  css   10 99 1.333E-12
  dc     5 53 dy
  de    54  5 dy
  dlp   90 91 dx
  dln   92 90 dx
  dp     4  3 dx
  egnd 99   0 poly(2),(3,0),(4,0) 0 .5 .5
  fb     7 99 poly(5) vb vc ve vlp vln 0 31.83E6 -1E3 1E3 30E6 -30E6
  ga     6  0 11 12 251.4E-6
  gcm    0  6 10 99 2.514E-9
  iss   10  4 dc 170.0E-6
  hlim 90   0 vlim 1K
  j1    11  2 10 jx
  j2    12  1 10 jx
  r2     6  9 100.0E3
  rd1    3 11 3.978E3
  rd2    3 12 3.978E3
  ro1    8  5 50
  ro2    7 99 25
  rp     3  4 15.00E3
  rss   10 99 1.176E6
  vb     9  0 dc 0
  vc     3 53 dc 1.500
  ve    54  4 dc 1.500
  vlim   7  8 dc 0
  vlp   91  0 dc 25
  vln    0 92 dc 25
.model dx D(Is=800.0E-18 Rs=1m)
.model dy D(Is=800.00E-18 Rs=1m Cjo=10p)
.model jx NJF(Is=12.50E-12 Beta=743.3E-6 Vto=-1)
.ends
*$
```

NOTE

When you create a Capture part for a subcircuit, a rectangular box is drawn such that you can edit the box and draw your own graphics for that part using the Part Editor in Capture.

16.4 MODEL EDITOR

The Model Editor is used to view text model definitions and to display graphical model characteristics and model parameters. Figure 16.3 shows the forward current versus voltage curve for a diode. When the Model Editor first loads the PSpice library, the model text description is shown. To see the graphical model characteristics, select **View > Extract Model**. Figure 16.3 also shows a table in which data from a manufacturer's datasheet can be entered such that a new model can be created based upon the manufacturer's empirical data.

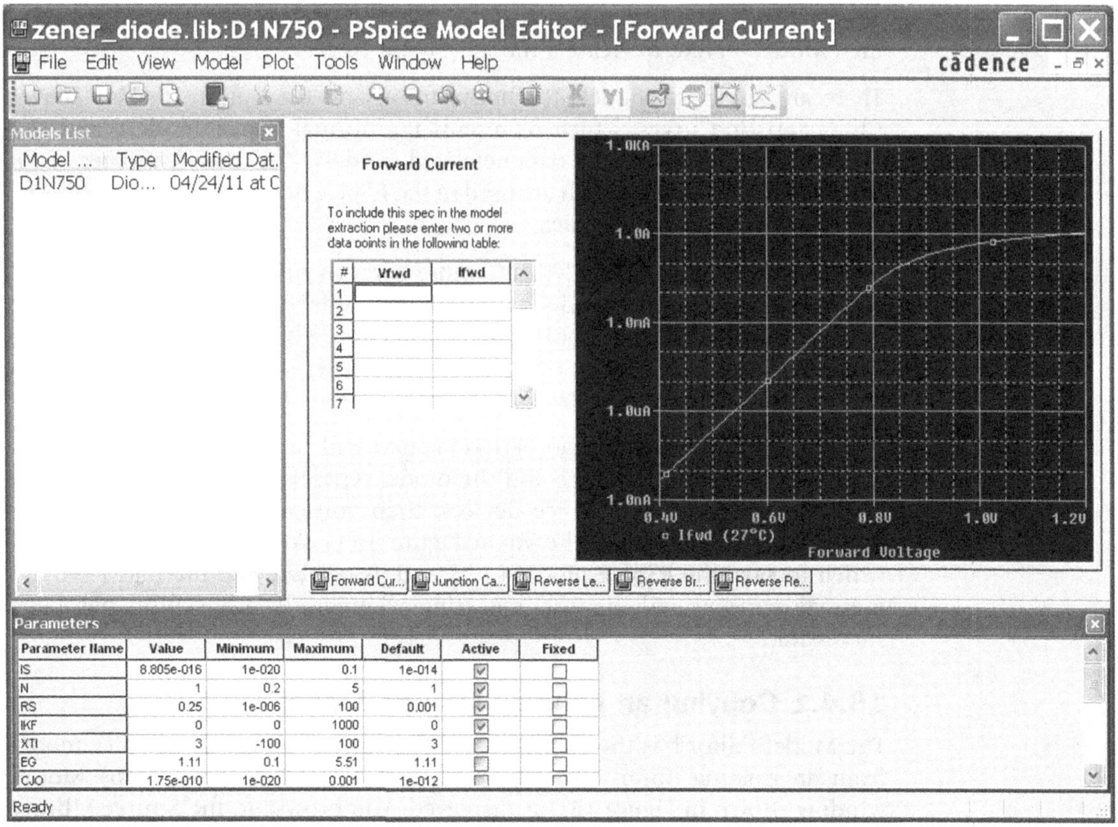

FIGURE 16.3
Model Editor.

FIGURE 16.4
Creating a new model.

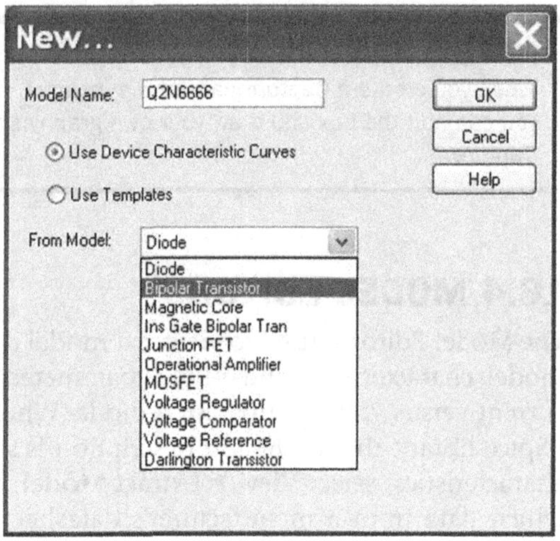

New models can be created by first selecting **File > New**, to create a library file, then **Model > New**, to create a model (Figure 16.4).

There are 11 device models to choose from, and the option for **Use Device Characteristic Curves**, which represents the normal PSpice models, and **Use Template**, which relates to parameterized models. The Use Templates is for parameterized models, which are used in the PSpice Advanced Analysis software and are not covered in this text.

Once a model has been created, a Capture part can automatically be generated. The first character of the model name and the model type defined in Table 16.2 determine which type of Capture part to generate. Before you generate a Capture part, you need to make sure that the correct schematic editor is selected. In **Tools > Options** > **Schematic Editor**, select Capture (Figure 16.5).

As mentioned previously, the part generated will be determined by the first character in the model name and the model type. If the part to be generated is one of the standard PSpice devices, then you can use **File > Export to Capture Part Library**. As shown in Figure 16.6, you select the libraries in which to save the PSpice model (.lib) and associated Capture part (.olb). A message window will appear, reporting whether the translation has been successful.

16.4.1 Copying an Existing PSpice Model

The Model Editor has the facility to make a copy of an existing PSpice model from an existing library. Selecting **Model > Copy From**, the **Copy Model** window shown in Figure 16.7 is displayed. You browse to the **Source Library** and select the model from the displayed model list and enter the **New Model** name.

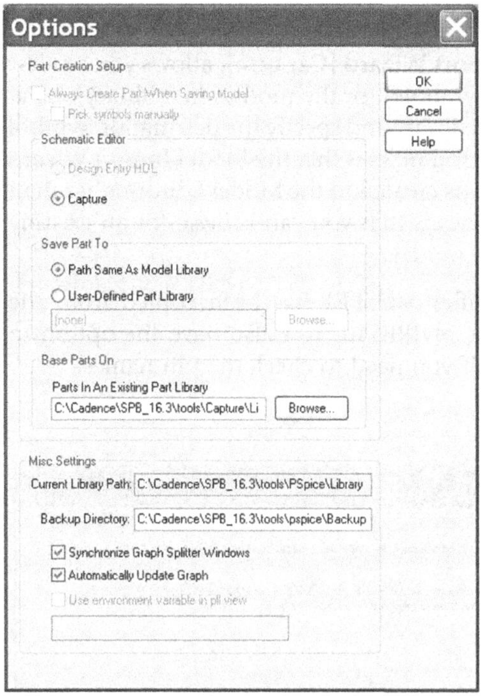

FIGURE 16.5
Selecting the Capture schematic Editor in Options.

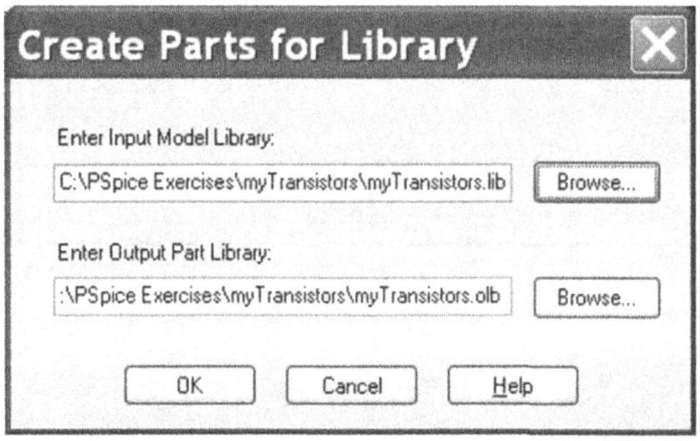

FIGURE 16.6
Creating and saving the PSpice model and Capture part.

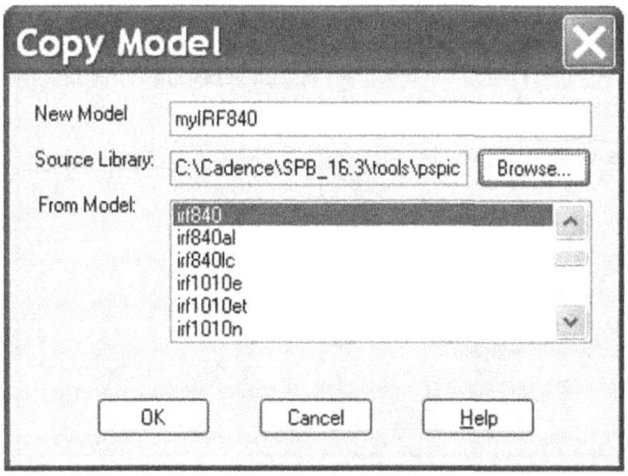

FIGURE 16.7
Copying a PSpice model from an existing library.

16.4.2 Model Import Wizard

The Model Import Wizard, **File > Import Wizard [Capture]**, allows you to view and select or replace Capture parts (symbols) for the models in a library one at a time. You browse to the input model library and specify the destination symbol library as shown in Figure 16.8. The good news is that the Model Import Wizard will display opamp symbols for opamps created in the Model Editor rather than rectangular boxes. The other good news is that you can browse for an existing Capture symbol to associate with the model.

In Figure 16.9 an international rectifier MOSFET has been copied from the IRF.lib library, modified and saved as myIRF540. You also have the option to view the model text, which is useful if you need to check the pin names.

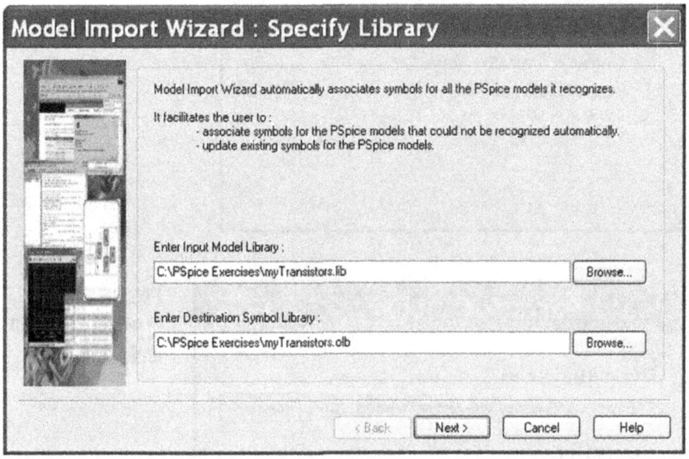

FIGURE 16.8
Model Import Wizard.

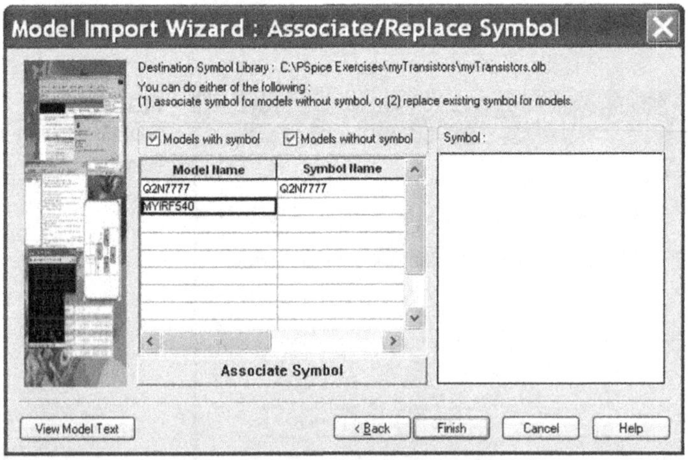

FIGURE 16.9
Associate or replace a symbol to the model.

As no symbol exists for the MOS device yet, clicking on **Associate Symbol** will allow you to browse to a suitable library which contains a symbol for an NMOS transistor (Figure 16.10).

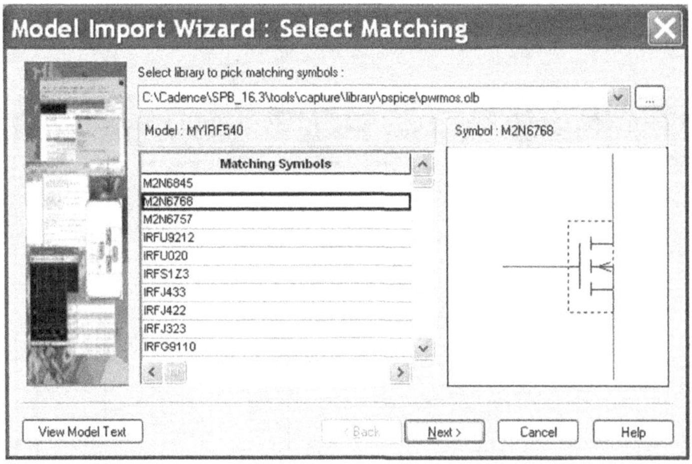

FIGURE 16.10
Finding a matching symbol for the model.

NOTE

The MOS device is modeled as a subcircuit; if generated in the Model Editor without using the Model Import Wizard, it would be displayed as a rectangular box in Capture, which would then have to be edited.

In this example, the Capture symbol for an NMOS symbol was found in the pwrmos.olb library (Figure 16.10). However, you now have to associate the model terminals (1, 2, 3) with the Capture symbol pins (d, g, s) (Figure 16.11). All the symbol pin names will be listed.

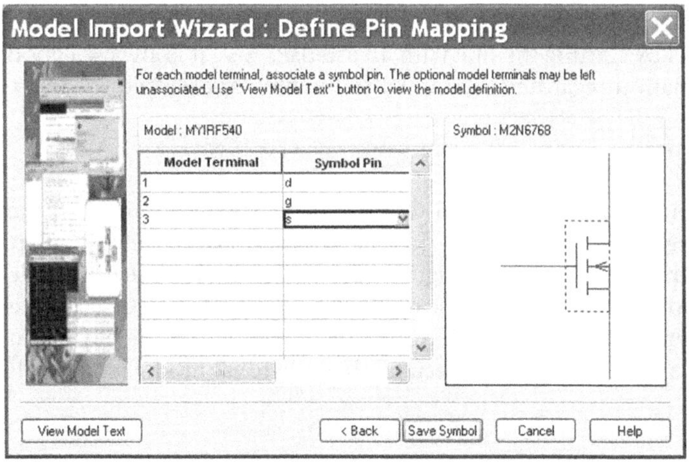

FIGURE 16.11
Defining model terminals and pins.

When you select **Save Symbol** in Figure 16.11, the models listed in the PSpice myTransistors.lib library file are shown (Figure 16.12). Note that as all models have an associated capture symbol, the option to **Replace Symbol** is now shown. If there is no associated symbol, the option **Associate Symbol** is shown.

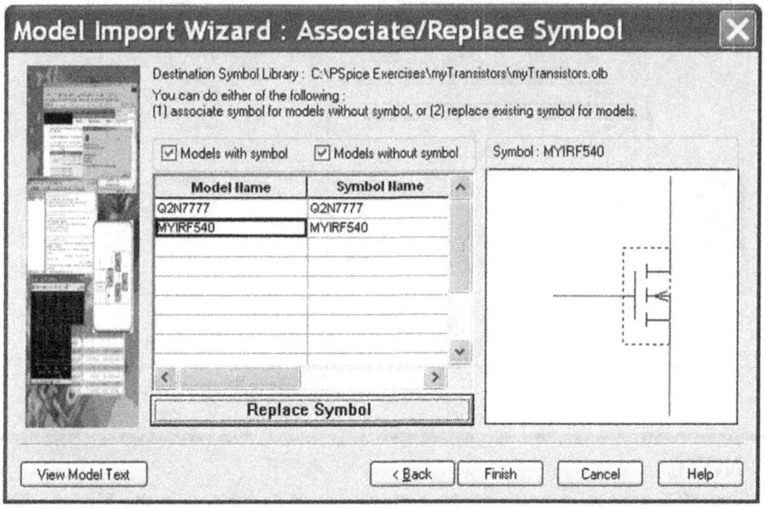

FIGURE 16.12
Summary of models in myTransistors PSpice library.

An example of using the Model Import Wizard can be found in Chapter 21, Exercise 2.

16.4.3 Downloading Models from a Vendor Website

The Model Editor is useful for displaying the characteristic curves for models, especially if the PSpice model has been downloaded from a vendor's website. By default, the Model Editor ignores the first line in a PSpice model file, so make sure the **.model** statement starts on the second line. Comment lines can be added to the model by starting the line with an asterisk, *, so it is always a good idea to add information regarding the model to the file, for example:

```
* Q2N7777 transistor downloaded from semiconductor website 23.4.2011
```

A Capture part can also be generated in Capture from a PSpice model. In the Project Manager highlight the design file (.dsn) and select **Tools > Generate Part** to open the **Generate Part** window (Figure 16.13). Select **Netlist/source file type** to **PSpice Model Library** and browse to the PSpice model in **Netlist/source file**. Then in **Implementation name** select the model name from the file. If you have downloaded a library of models, then all the individual model names will be available.

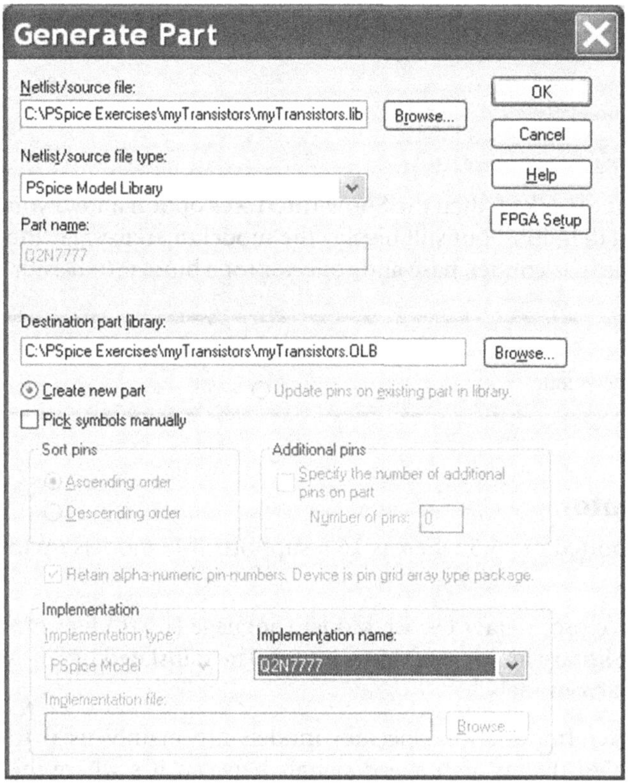

FIGURE 16.13
Generating a Capture part from a PSpice library model.

16.4.4 Encryption

There is an encryption facility in the Model Editor that allows you to encrypt PSpice models or libraries such that the models can be used for simulation but the model definitions cannot be viewed. To encrypt a model library, select **File > Encrypt Library** and in the **Library Encryption** window (Figure 16.14) browse to the library to be encrypted and the folder where the library will be saved to.

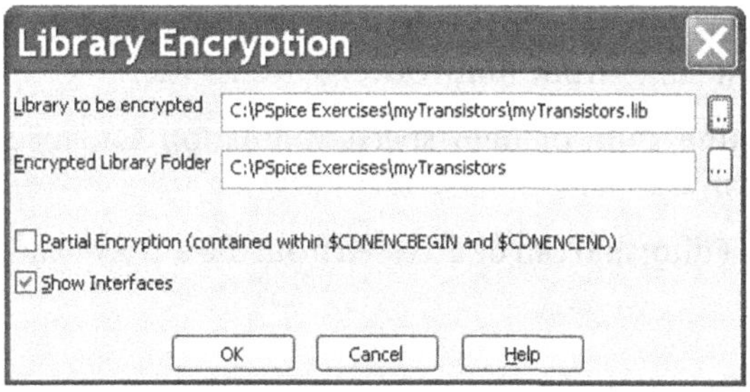

FIGURE 16.14
Library encryption.

If you only want to encrypt part of the library, i.e. two of the PSpice library models, then add the following text to the beginning and end of the model definitions:

```
$CDNENCSTART beginning of model text
$CDNENCFINISH end of model text
```

If **Partial Encryption** is not checked, then the **Show Interfaces** option allows you to encrypt the model text definition but still display the model interfaces, i.e. the model connecting pins such as emitter, base and collector for a bipolar transistor.

NOTE

Comment lines are not encrypted.

16.4.5 IBIS Translator

IBIS 1.1 models are supported in 16.5 whereas 16.6 supports IBIS models up to version 5.0.

In 16.6, IBIS and Cadence proprietary Device Model Language (DML) files can be translated into PSpice library files in the Model Editor. The translated IBIS I/O buffers are defined as macromodels.

IBIS (Input Output Buffer Information Specific) models are mainly used to analyse the transmission of digital high speed signals between ICs where the transmitted medium, usually copper traces, are modelled as transmission lines. The input and output buffers of ICs are characterised by tables of measured Voltage/Current and Voltage/Time data to provide a representation of I/O buffer behaviour without disclosing any proprietary information regarding the internal implementation of the I/O buffers. IBIS models based on these V/I and V/T tables can be used to analyse the Signal Integrity of the transmission of data by predicting the circuit response to mismatched impedance lines, crosstalk from adjacent transmission lines, overshoot, undershoot, ground bounce and simultaneous switching of digital high speed signals.

IBIS models provide a relatively accurate simulation model as they take into account structures for ESD protection and inherent parasitics associated with die bonding to the IC package pins. Compared to standard SPICE models, IBIS simulations run faster and do not suffer from non-convergence issues.

DML files are used by the existing suite of high speed simulation Cadence software tools for high speed analysis.

The IBIS translator is in the Model Editor and can be accessed from the top tool bar.

```
Model > IBIS Translator
```

The IBIS translator is not available in the demo Lite CD.

16.5 EXERCISES

Exercise 1

The breakdown voltage for a zener diode will be modified using the Model Editor.

1. Create a project called **zener_diode** and draw the circuit in Figure 16.15. The 4V7 zener diode can be found in either the eval library or the diode library.

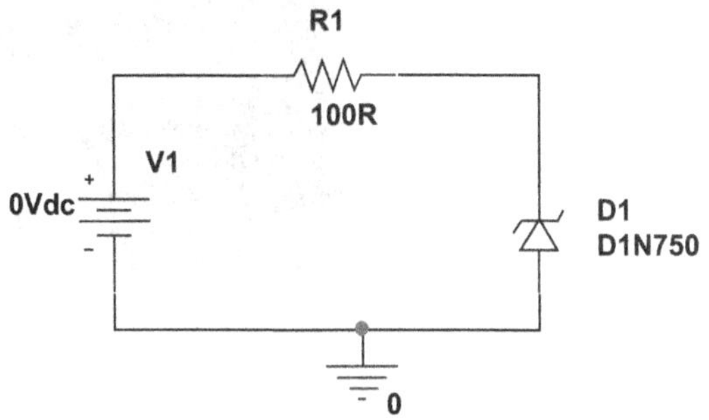

FIGURE 16.15
Verifying the zener breakdown voltage.

2. Create a **PSpice > New Simulation Profile** and select the **Analysis type** to DC sweep. Create a simulation profile for a DC sweep for V1 from 1 V to 10 V in steps of 0.1 V (Figure 16.16) to confirm the 4.7 V zener diode breakdown voltage. It does not matter if V1 has a voltage of 0 V as shown in the schematic, as a DC sweep is being performed.

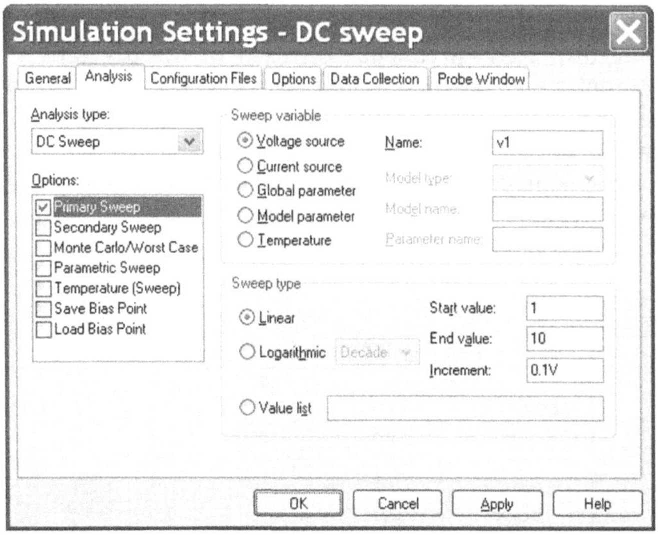

FIGURE 16.16
DC sweep simulation profile.

3. Select the diode and **rmb > Edit PSpice Model** to open the Model Editor. Note the library name at the top of the Model Editor, **zener_diode.lib** (Figure 16.17).

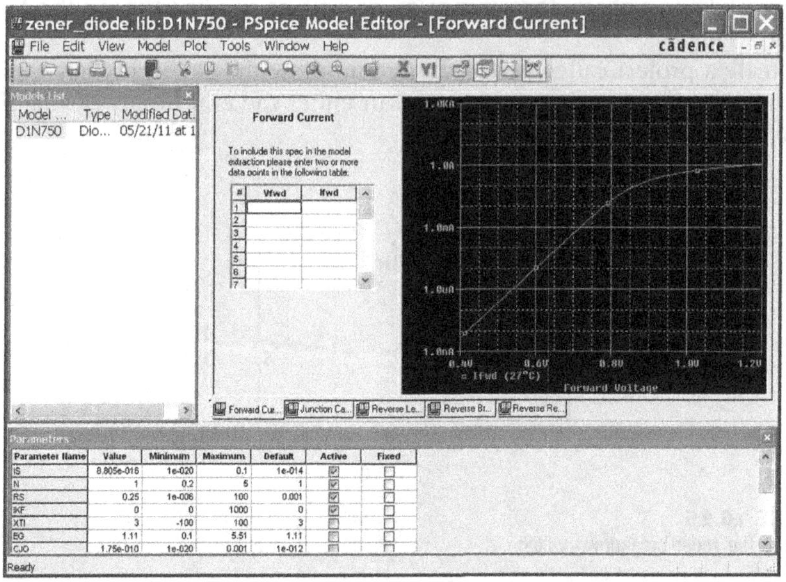

FIGURE 16.17
Creating a zener.lib file.

4. From the top toolbar, select **View > Extract Model.** In the first PSpice Model Editor window, click on **Yes**, then click on OK in the next window regarding parameters. The Model Editor will display the forward current characteristic for the diode as shown in Figure 16.17.
5. Under the displayed curve, click on each of the five tabs to view the diode model characteristics. When you select a curve, the **Parameter** window (Figure 16.18) displays an **Active** check against each parameter that is associated with the displayed characteristic curve. Note the diode characteristic for the **Reverse Breakdown** curve and the active parameters.

FIGURE 16.18
Model Editor showing extracted model parameter curves.

Parameters

Parameter Name	Value	Minimum	Maximum	Default	Active	Fixed
FC	0.5	0.001	10	0.5	☐	☐
ISR	1.859e-009	1e-020	0.1	1e-010	☐	☐
NR	2	0.5	5	2	☐	☐
BV	4.7	0.1	1000000	100	☑	☐
IBV	0.020245	1e-009	10	0.0001	☑	☐
TT	5e-009	1e-016	0.001	5e-009	☐	☐

6. In the Parameters section at the bottom of the Model Editor, scroll down and locate the diode breakdown parameter, BV. Change the value from 4.7 V to 8.2 V and click on the Fixed box (Figure 16.19). Select the **Reverse Breakdown** tab under the curve, which should have changed to 8.2 V.

Parameters

Parameter Name	Value	Minimum	Maximum	Default	Active	Fixed
FC	0.5	0.001	10	0.5	☐	☐
ISR	1.859e-009	1e-020	0.1	1e-010	☐	☐
NR	2	0.5	5	2	☐	☐
BV	8.2	0.1	1000000	100	☑	☑
IBV	0.020245	1e-009	10	0.0001	☑	☐
TT	5e-009	1e-016	0.001	5e-009	☐	☐

FIGURE 16.19
Editing the breakdown voltage.

7. Save the library file, **File > Save** and close the Model Editor.
8. Edit the Simulation Profile. Select **Configuration Files > Library** and note that the zener_diode.lib has been added to the profile. The attached design icon indicates that the file is local only to the design (Figure 16.20). The world icon, , indicates that the file is global and can be seen by all designs. The nom.lib contains all the PSpice library files and hence is global to all designs.

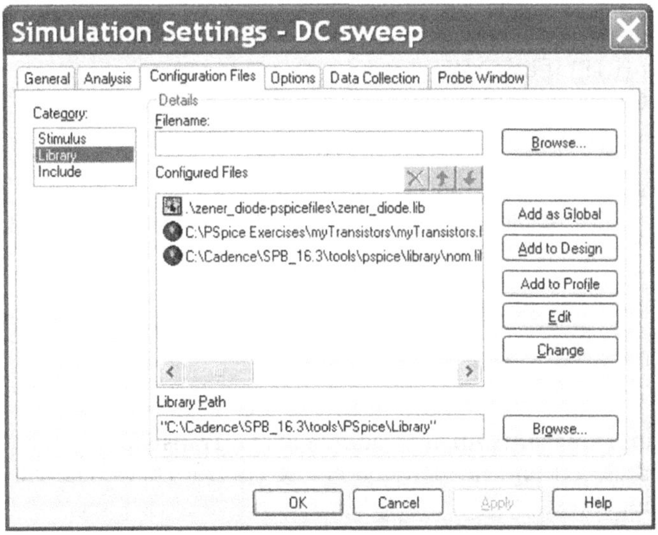

FIGURE 16.20
Zener_diode.lib library added as local to design.

9. Select the zener_diode.lib file and click on **Edit**. The Model Editor will open up. Select the D1N750 in the **Models List** and the diode model parameter will appear as before. This enables model library files to be quickly viewed. Close the Model Editor and the simulation profile.
10. Rerun the simulation and confirm that the diode breakdown voltage is now 8.2 V.

Exercise 2

Whenever you create a new PSpice model and Capture part it is recommended that you create a new directory for your model. Do not install the new libraries in the Capture or PSpice folders. If a new OrCAD release is installed, then the PSpice and Capture libraries will be reinstalled and so any new models created will be lost. For this exercise we will assume that a PSpice model for a transistor has been downloaded from a semiconductor's website. In order to recreate this scenario we will copy an existing transistor model from the bipolar.lib library to a new file, myTransistors.lib.

1. Using a text editor such as WordPad or Notepad, browse to the installed PSpice libraries, **<install path>Orcad 16.3\Tools\PSpice\Library**, and select the bipolar.lib or eval.lib PSpice library. Make sure you select **Files of type** to **All Files**.
2. In the library file, scroll down and select the Q2N3904 model definition as shown below (use Control F to find the Q2N3904):

```
.model Q2N3904         NPN(Is=6.734f Xti=3 Eg=1.11 Vaf=74.03 Bf=416.4 Ne=1.259
+                Ise=6.734f Ikf=66.78m Xtb=1.5 Br=.7371 Nc=2 Isc=0 Ikr=0 Rc=1
+                Cjc=3.638p Mjc=.3085 Vjc=.75 Fc=.5 Cje=4.493p Mje=.2593 Vje=.75
+                Tr=239.5n Tf=301.2p Itf=.4 Vtf=4 Xtf=2 Rb=10)
*                Nationalpid=23          case=TO92
*                88-09-08 bam    creation
*$
```

3. Select the model text and copy and paste it into a new text file. Do not use rich text format (RTF) if using WordPad.
4. Edit the transistor model name to Q2N7777 and add a comment line to the first line 1 as shown below:

```
* example of a downloaded transistor model
.model Q2N3904         NPN(Is=6.734f Xti=3 Eg=1.11 Vaf=74.03 Bf=416.4 Ne=1.259
+                Ise=6.734f Ikf=66.78m Xtb=1.5 Br=.7371 Nc=2 Isc=0 Ikr=0 Rc=1
+                Cjc=3.638p Mjc=.3085 Vjc=.75 Fc=.5 Cje=4.493p Mje=.2593 Vje=.75
+                Tr=239.5n Tf=301.2p Itf=.4 Vtf=4 Xtf=2 Rb=10)
*                Nationalpid=23          case=TO92
*                88-09-08 bam    creation
*$|
```

5. Save the file as **myTransistors.lib** in a folder called **myTransistors**. Make sure the file is saved as text and not RTF, otherwise control characters will be added to the model text.
6. Create a new PSpice project called myTransistors.
7. In the Project Manager make sure the myTransistors.dsn file is highlighted and select **Tools > Generate Part**. In the **Generate Part** window (Figure 16.21) select:

   In **Netlist/source file:** type: select **PSpice Model Library**
   In **Netlist/source file:** browse to the myTransistors.lib file.
   In **Destination part library:** browse to the same folder where myTransistors.lib is.
   In **Implementation name:** there will only be one entry Q2N7777.
   Click on **OK**.

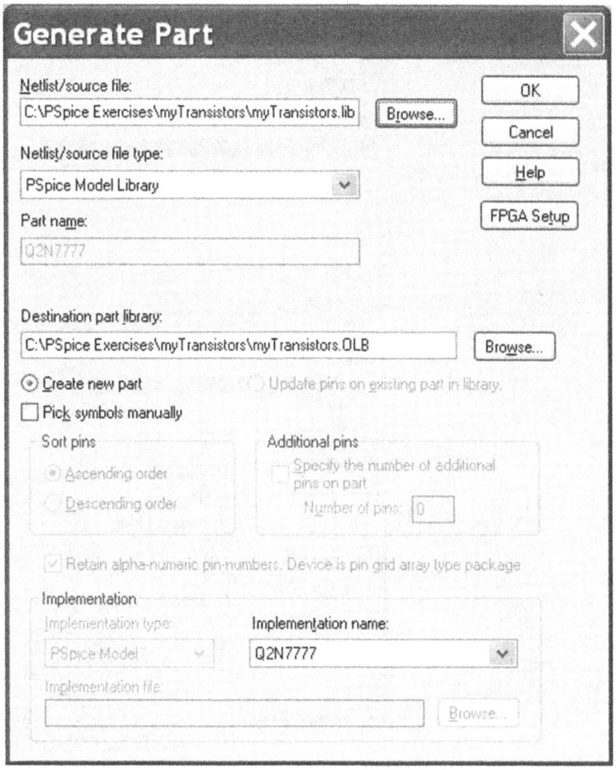

FIGURE 16.21
Generating a Capture part.

8. The myTransistors.olb Capture library will be created and added to the library folder in the Project Manager (Figure 16.22). Expand the library to see the Q2N7777 part.

FIGURE 16.22
myTransistors library with Q2N7777.

9. Open the schematic page and select **Place > Part** or press P on the keyboard. The library **myTransistors** has automatically been added to the libraries and the **Part List** contains the transistor graphical symbol for an NPN transistor (Figure 16.23).
   Also, the PSpice icon appears, indicating that the transistor has a PSpice model attached. You now need to make the myTransistors.lib file available for simulation in the Simulation Profile.

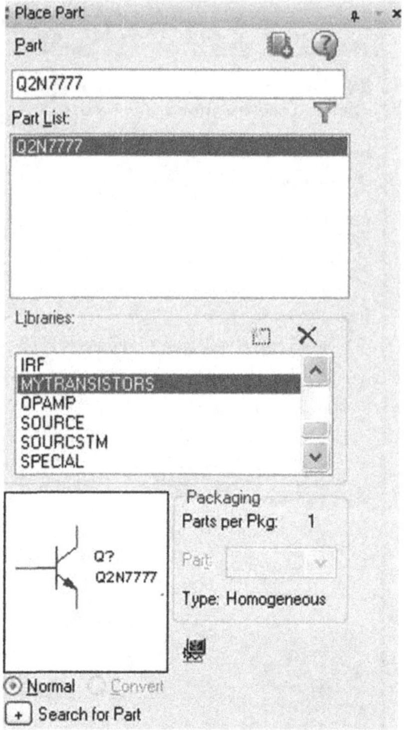

FIGURE 16.23
myTransistors.olb library available in library list.

10. Create a new simulation profile, **PSpice > New Simulation Profile**. Under the **Configurations Files** tab, select **Category > Library** and browse to the folder where myTransistors.lib is. As mentioned in Exercise 1 Step 8, you can add the library files Global to the design, local to the design or to the Profile. In this exercise, add the file as Global (Figure 16.24) and click on OK. The transistor is ready to be simulated and the myTransistors.lib library will be available for every new design.

Exercise 3

In Exercise 2, the model definition for the Q2N7777 has a Q for the first character in the model name, so this is recognized as a transistor and has an NPN as the model type. Hence, an NPN transistor symbol was generated. If the model name starts with an X for a subcircuit, a rectangular box is drawn which can then be edited in Capture using the Part Editor. You highlight the part in the library, **rmb > Edit Part**. Many of the power MOSFET models, for example, are defined as subcircuits.

However, in the Model Editor, the **Model Import Wizard** can be used to select an existing Capture graphic symbol for the model rather than having to edit the graphics in the Part Editor.

1. Open the Model Editor from the **Start** menu: **All Programs > Cadence (or OrCAD) > PSpice > Simulation Accessories > Model Editor**

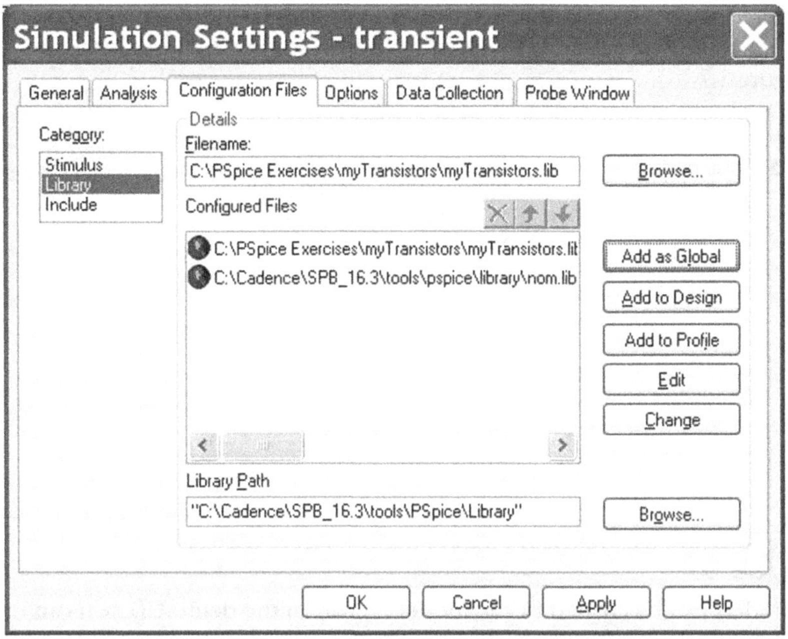

FIGURE 16.24
Adding myTransistors.lib PSpice library file as Global.

2. Select **File > Open** and browse to the myTransistors.lib library file. Click on Q2N7777 in the **Models List** and the model text will be displayed. Note that the first line which you added is not displayed.
3. Select **View > Extract model** and click on **Yes** in the Model Editor window. There will be eight characteristic curves for the transistor model.
4. Select **File > Export to Capture Part Library**. If the Capture option is not available, select **Tools > Options** and select Capture as the Schematic Editor (see Figure 16.5).

In the **Create parts for Library** window, the output library folder will show the same location as the PSpice library file, as shown in Figure 16.25. Click on OK and if a window appears asking to save the library, click on **Yes**. The translator window will appear and hopefully report no errors.

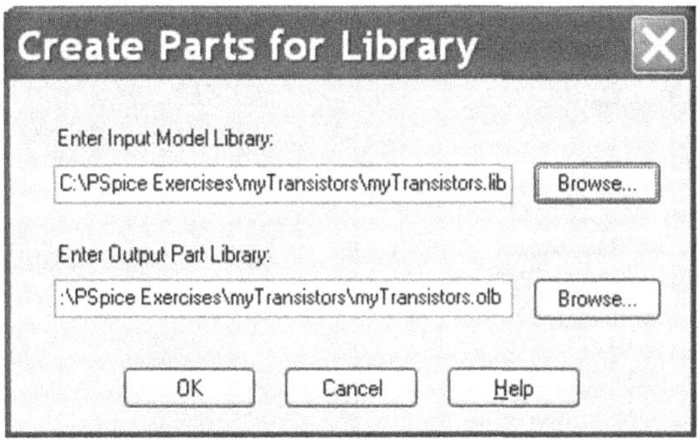

FIGURE 16.25
Folder locations for Capture part and PSpice model library.

Exercise 4

You can also copy PSpice models from existing models in the Model Editor.

1. In the Model Editor, open the **bipolar.lib** library from the PSpice library folder.

2. Select **Model > Copy From** and in the **Copy Model** window enter Q2N3906X in **New Model**, select Q2N3906 from the library list and click on OK (Figure 16.26).

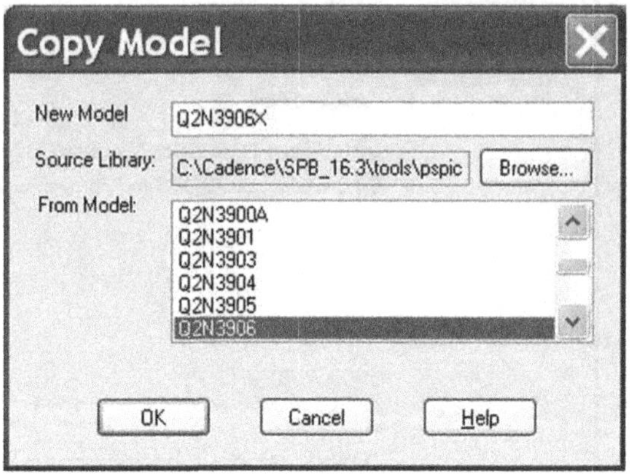

FIGURE 16.26
Copying a PSpice model.

Exercise 5

New models can be created in the Model Editor. In the demo CD you can only create new diode models in the Model Editor.
The Use Templates is for parameterized models, which are used in the PSpice Advanced Analysis software and are not covered in this text. For this exercise, select **Use Device Characteristic Curves.**

1. In the Model Editor, select **File > New.**
2. Select **Model > New**, enter the model name and select one of the Model types from the pull-down menu as shown in Figure 16.27.
3. Investigate the other different model types.

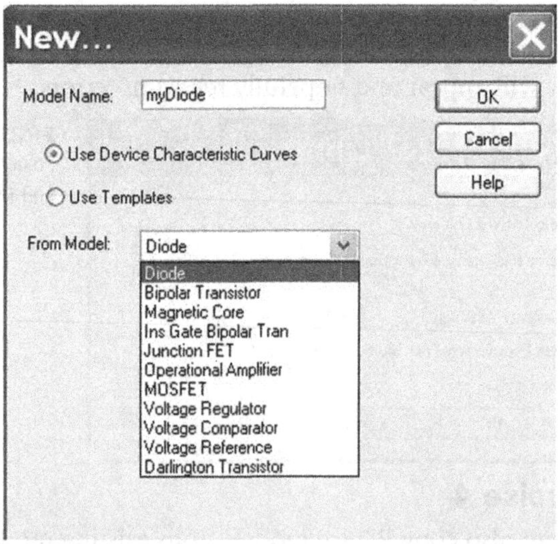

FIGURE 16.27
Creating new PSpice models.

CHAPTER 17

Transmission Lines

Chapter Outline

The signal integrity of high-speed signals via transmission lines is dependent on the frequency-related signal and dispersion losses of the transmission lines. Signal power loss is attributed to the increase in conductor resistance (skin effect) and the increase in dielectric conductance (dielectric loss) with an increase in frequency. Dispersion is the distortion of the signal wave shape resulting from delays introduced by the distributed frequency-dependent inductance and capacitance of the transmission line. Any reflected signals, due to impedance mismatch, will also exhibit loss and dispersion and subsequently degrade the performance of the transmission line.

Ideal and lossy transmission lines are modeled in PSpice using Tline distributed models and TLUMP lumped line segment models.

17.1 IDEAL TRANSMISSION LINES

The parameters required for an ideal transmission line are the characteristic impedance (Z0) and either the transmission line delay (TD) or the normalized line length (NL) which is the number of wavelengths along the line at a given frequency. You cannot enter TD and NL together. If you do not specify the frequency for NL, then the frequency defaults to 0.25 which represents the quarter wave frequency.

Analog Design and Simulation using OrCAD Capture and PSpice. DOI: 10.1016/B978-0-08-097095-0.00017-9

The time delay, TD, along a transmission line is given by:

$$TD = \frac{LEN}{v_p} \tag{17.1}$$

where TD is the transmission delay (s), LEN is the length of the transmission line (m), and v_p is the velocity of the propagating wave (propagating velocity) (m s^{-1}).

For transmission lines, the propagation velocity is expressed as a percentage of the speed of light, such that:

$$v_p = c \times VF \tag{17.2}$$

where VF is the velocity factor which has a value between 0 and 1, and c is the speed of light at 3×10^8 m s^{-1}.

The normalized transmission line length is given by:

$$NL = \frac{LEN}{\lambda} \tag{17.3}$$

From $v = f\lambda$, the wavelength is given as:

$$\lambda = \frac{v_p}{f} \tag{17.4}$$

Equation 17.3 can be therefore rewritten as:

$$NL = LEN\frac{f}{v_p} \tag{17.5}$$

where f is frequency (Hz) and λ is wavelength (m).

PSpice uses a T device from the **analog** library to model an ideal transmission line. Figure 17.1a shows the Capture part for the T device and Figure 17.1b the associated properties in the Property Editor.

So, for an ideal transmission line, if you do not know the delay time (TD) then you can enter values for NL and f and, as mentioned above, if you do not enter the frequency, then the default value of 0.25 is used, which represents the quarter wave frequency.

Initial conditions for voltage and current can be applied to the transmission line.

17.2 LOSSY TRANSMISSION LINES

Transmission lines can be considered to consist of a number of identical sections known as RLCG lumped line segments, as shown in Figure 17.2. The R represents the line resistance, L the line inductance, C the dielectric capacitance and G the dielectric conductance. For long transmission lines, one solution would be to use a number of lumped RLCG segments connected together. PSpice provides lumped line segments of up to 128 segments in the TLine library. However, lumping together large line segments can lead to long simulation times.

(a)

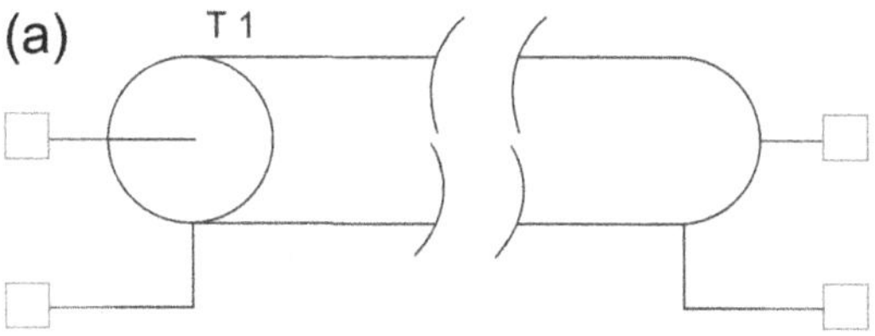

(b)

Designator	
F	
Graphic	T.Normal
ID	
Implementation	
Implementation Path	
Implementation Type	<none>
Location X-Coordinate	130
Location Y-Coordinate	420
Name	INS8246
NL	
Part Reference	T1
PCB Footprint	
Power Pins Visible	
Primitive	DEFAULT
PSpiceOnly	TRUE
PSpiceTemplate	T^@REFDES %A+ %A- %B
Reference	T1
Source Library	C:\CADENCE\SPB_16.3 ...
Source Package	T
Source Part	T.Normal
TD	
Value	T
Z0	

FIGURE 17.1
Ideal transmission line Tline: (a) Capture part: T device; (b) associated Tline properties.

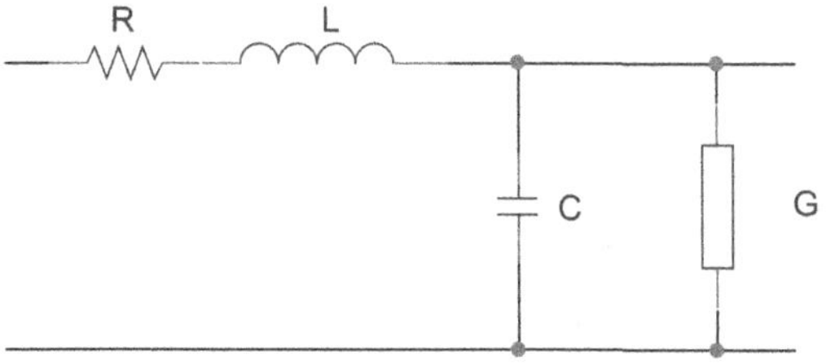

FIGURE 17.2
RLCG lumped line segment of a transmission line.

Simpler RC transmission line models are also available in the TLine library, as are over 40 coaxial cable models and twisted wire pair models.

An alternative approach for lossy transmission lines is to use a distributed model which relies on an impulse response convolution method to determine the transmission line response. Figure 17.3 shows the TLOSSY PSpice device and the associated properties in the Property Editor.

The length of the transmission line is represented by the LEN property and the R, L, C and G properties are specified as per unit length.

(a)

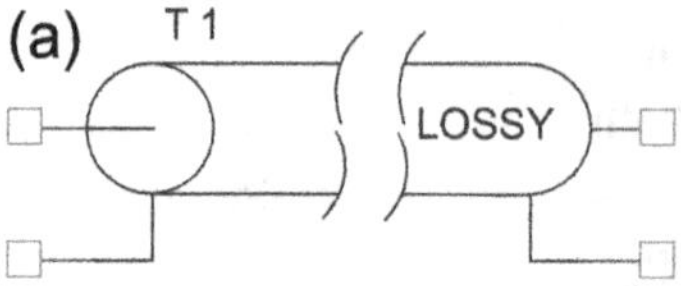

(b)

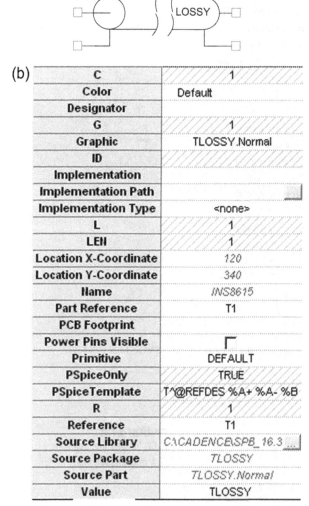

C	1
Color	Default
Designator	
G	1
Graphic	TLOSSY.Normal
ID	
Implementation	
Implementation Path	
Implementation Type	<none>
L	1
LEN	1
Location X-Coordinate	120
Location Y-Coordinate	340
Name	INS8615
Part Reference	T1
PCB Footprint	
Power Pins Visible	
Primitive	DEFAULT
PSpiceOnly	TRUE
PSpiceTemplate	T^@REFDES %A+ %A- %B
R	1
Reference	T1
Source Library	C:\CADENCE\SPB_16.3
Source Package	TLOSSY
Source Part	TLOSSY.Normal
Value	TLOSSY

FIGURE 17.3
Lossy transmission line TLOSSY: (a) Capture part: TLOSSY device; (b) associated TLOSSY properties.

NOTE

The maximum internal time step generated for distributed transmission line models is limited to one half of the transmission line delay, TD. Therefore, for short transmission lines, the simulation time may be considerably longer for distributed line models compared to using a lumped line model for a short transmission line.

17.3 EXERCISES

The following exercises demonstrate the basic transmission line characteristics for different load terminations.

Exercise 1

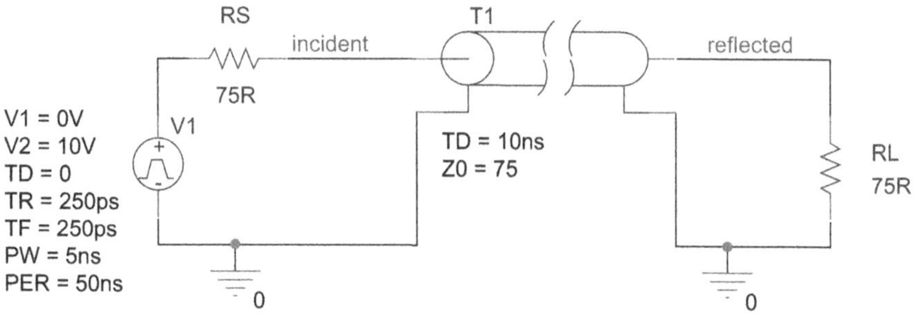

FIGURE 17.4
Matched source and load impedance transmission line.

MATCHED LOAD FOR RL

1. Draw the circuit in Figure 17.4. The transmission device T is from the **analog** library and the voltage pulse is from the **source** library. When you place the load resistor, RL, on the schematic, by default, pin 1 is on the left-hand side. Rotate resistor RL three times such that pin 1 is at the top, which connects to T1. By convention, current flowing into pin 1 is defined as positive, such that a measured negative current at pin 1 represents a current flowing out of pin 1.
2. Double click on T1 to open the Property Editor and add and display the property values as shown in Figure 17.5.

TD	10ns
Value	T
Z0	75

FIGURE 17.5
Adding property values for TD and Z0.

Highlight both TD and Z0 by holding down the control key, select **Display** and in the **Display** Properties window, select **Display > Name and Value** as shown in Figure 17.6.

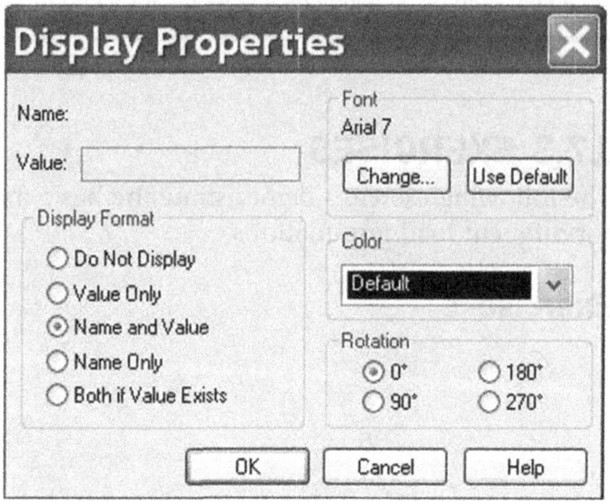

FIGURE 17.6
Display the TD and Z0 property values.

3. Create a transient analysis simulation profile with a Run to time of 50 ns and a Maximum step size of 100 ps (Figure 17.7).

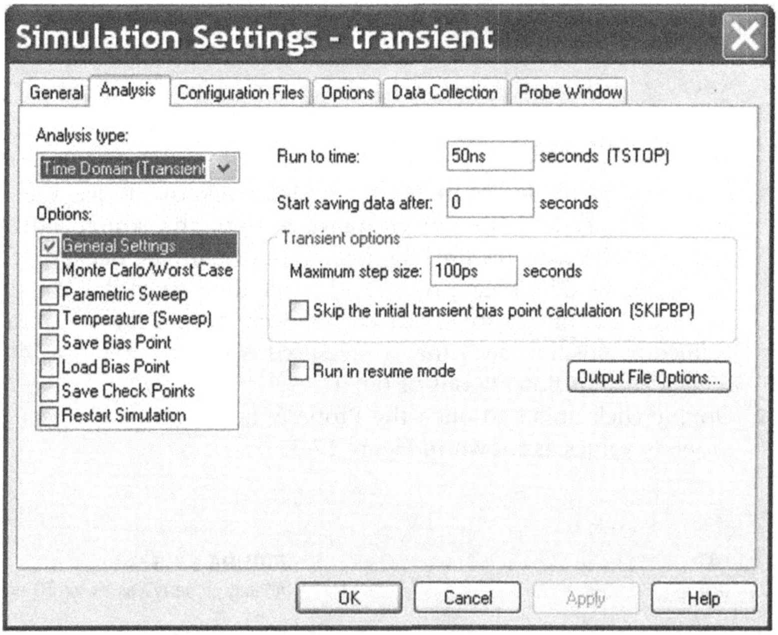

FIGURE 17.7
Transient analysis simulation settings.

4. Add voltage markers on the incident and reflected nodes (Figure 17.8) and run the simulation.

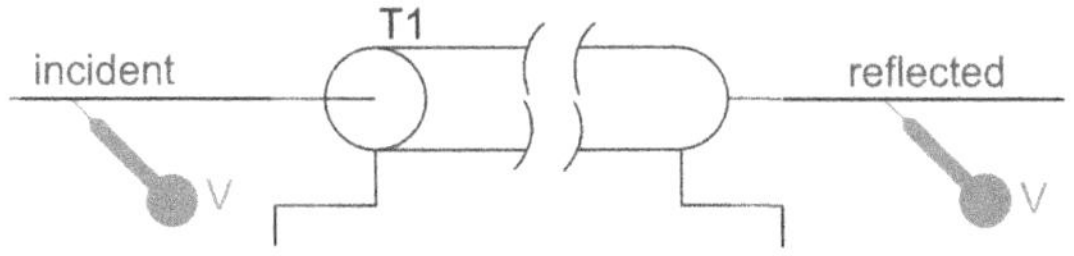

FIGURE 17.8
Placing voltage markers.

5. Figure 17.9 shows the source pulse and the delayed load pulse after 10 ns. Initially, the source resistor and transmission line act as a potential divider, so the voltage amplitude divides down to 5 V.

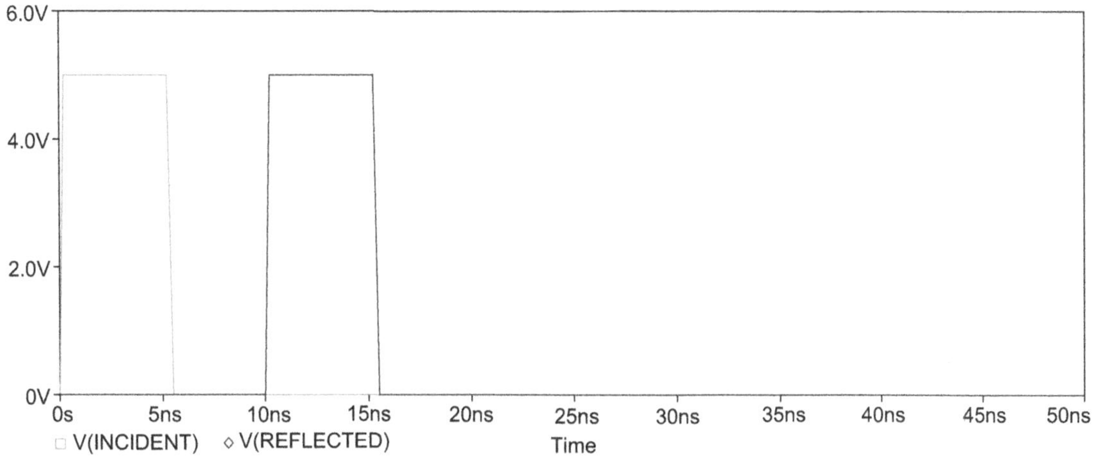

FIGURE 17.9
Matched transmission line voltage waveforms.

6. Delete the voltage markers in Capture and place current markers on the input pin (incident) to T1 and the top pin of the load resistor, RL (Figure 17.10).

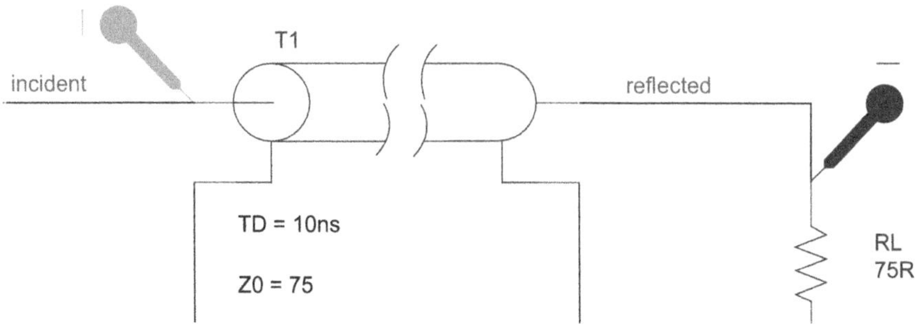

FIGURE 17.10
Placing current markers.

There is no need to rerun the simulation as the waveform display in Probe will automatically be updated. You should see that both currents have the same magnitude and are separated by the specified 10 ns delay (Figure 17.11).

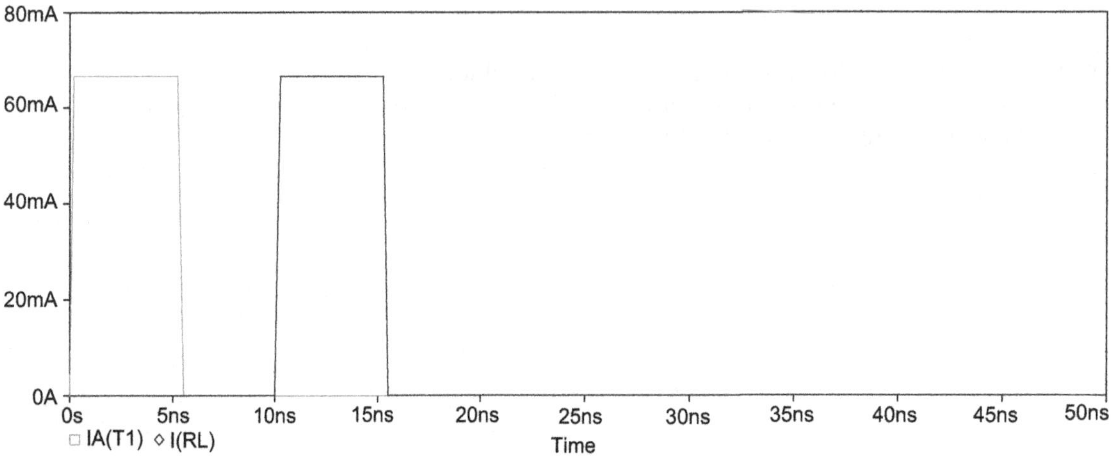

FIGURE 17.11
Matched transmission line current waveforms.

As the load impedance is equal to the impedance of the transmission line, the line is said to be matched. There are no voltage or reflections.

RL REPLACED WITH A SHORT CIRCUIT

7. Delete the current markers and replace the load resistance with a small value resistance of 1 μΩ to represent a short circuit. Place voltage markers on the incident node and the top of the short circuit resistor as shown in Figure 17.12 and rerun the simulation.

FIGURE 17.12
Short circuit load.

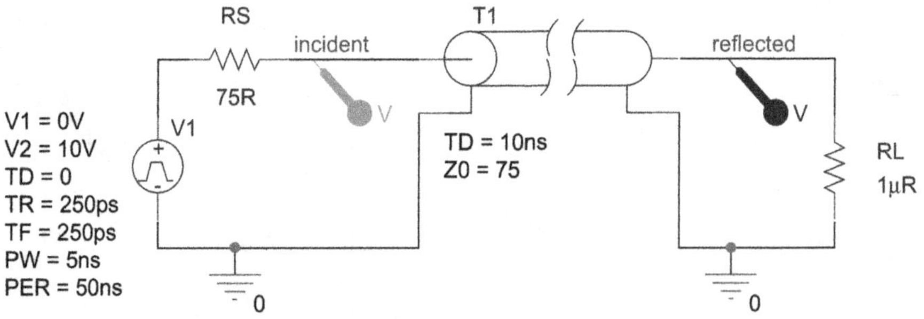

You should see the transmission line response in Figure 17.13. With a short circuit the load voltage will be 0 V and the incident voltage wave will be reflected but 180° out of phase with the incident wave.

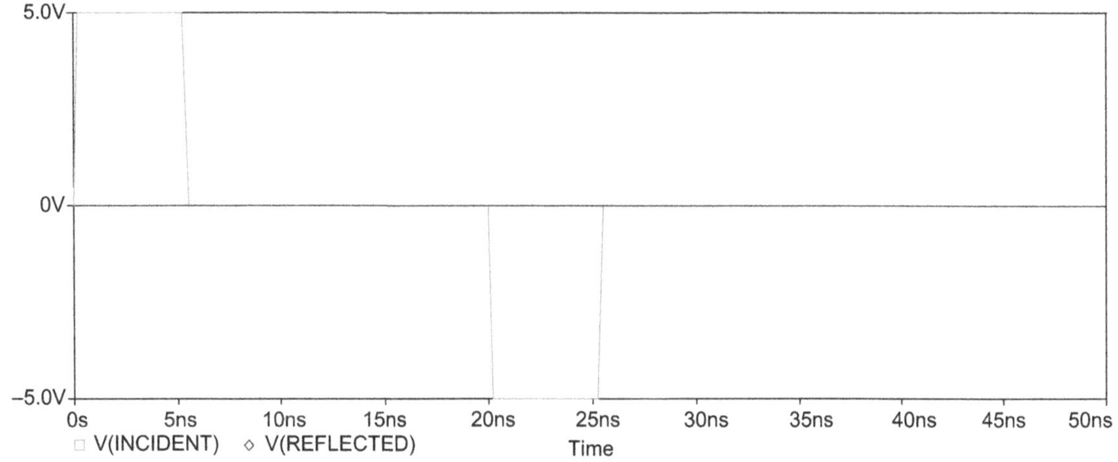

FIGURE 17.13
Short circuit transmission line voltage waveforms.

8. In Capture, delete the voltage markers and place current markers on the input pin (incident) to T1 and the top pin of the RL resistor (Figure 17.14).

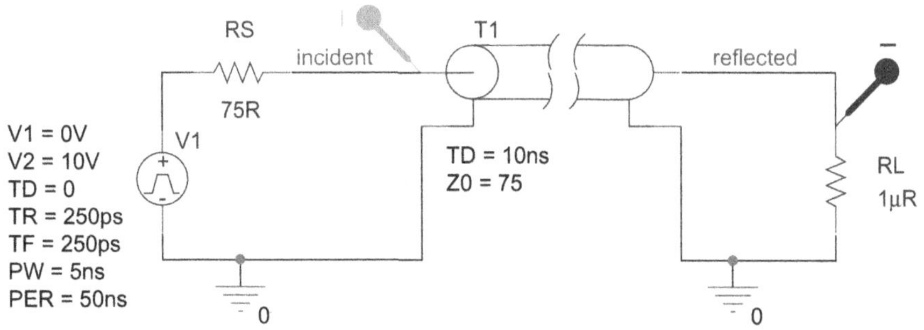

FIGURE 17.14
Placing current markers.

There is no need to rerun the simulation as the waveform display in Probe will automatically be updated. You should see that the incident current wave is reflected with the same amplitude such that the reflected wave is double the magnitude of the incident current wave (Figure 17.15).

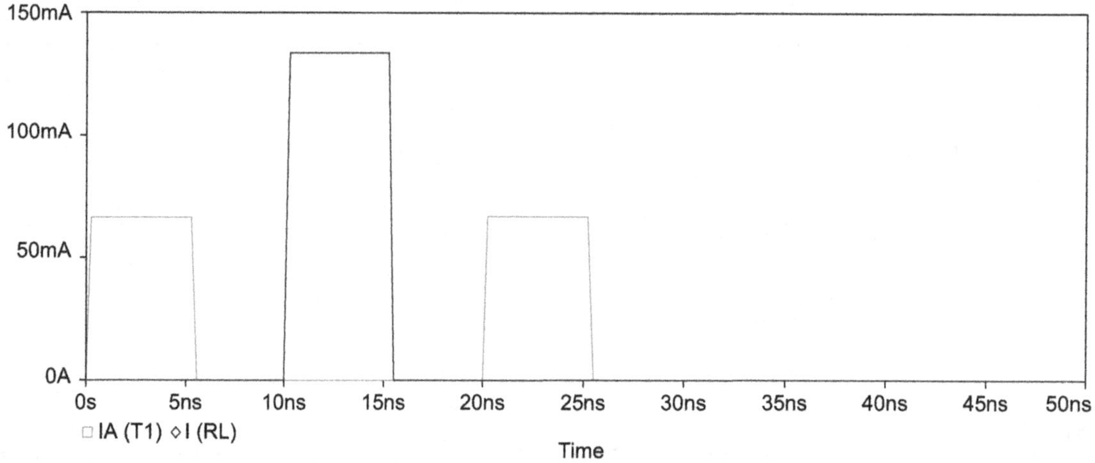

FIGURE 17.15
Short circuit transmission line current waveforms.

RL REPLACED WITH AN OPEN CIRCUIT

9. Delete the current markers and change the load resistance to a 1 TΩ resistor to represent an open circuit. Place voltage markers on the incident node and the top of the short circuit resistor (Figure 17.16) and rerun the simulation.

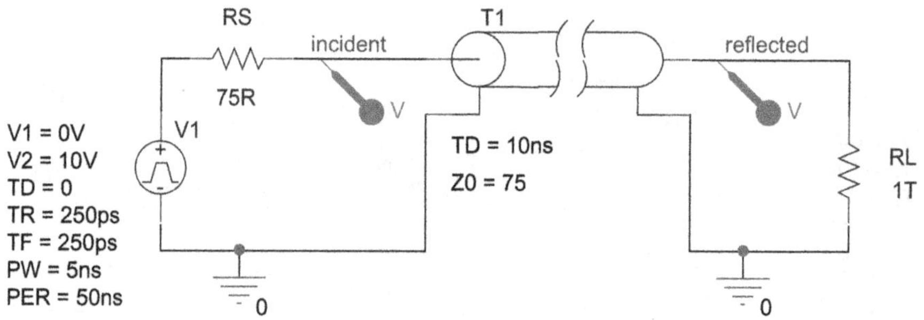

FIGURE 17.16
Placing voltage markers.

You should see the response shown in Figure 17.17. The reflected voltage is equal to the source voltage, so the reflected voltage is double in magnitude to that transmitted.

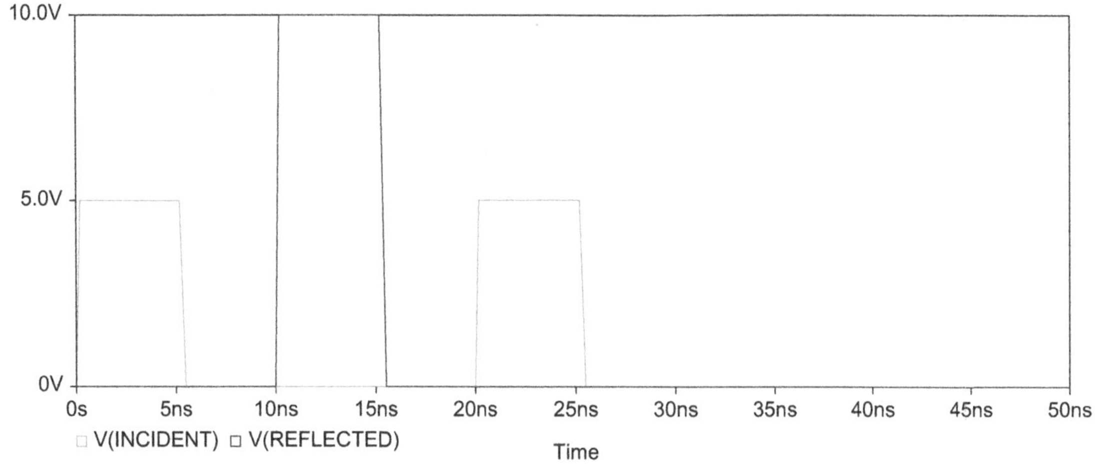

FIGURE 17.17
Open circuit transmission line voltage waveforms.

10. Delete the voltage markers in Capture and place current markers on the input pin (incident) to T1 and the top pin of resistor RL. There is no need to rerun the simulation as the waveform display in Probe will automatically be updated. As the output is an open circuit, no current will flow. The current at the open circuit is reflected back with the same magnitude but is 180° out of phase, as seen in Figure 17.18.

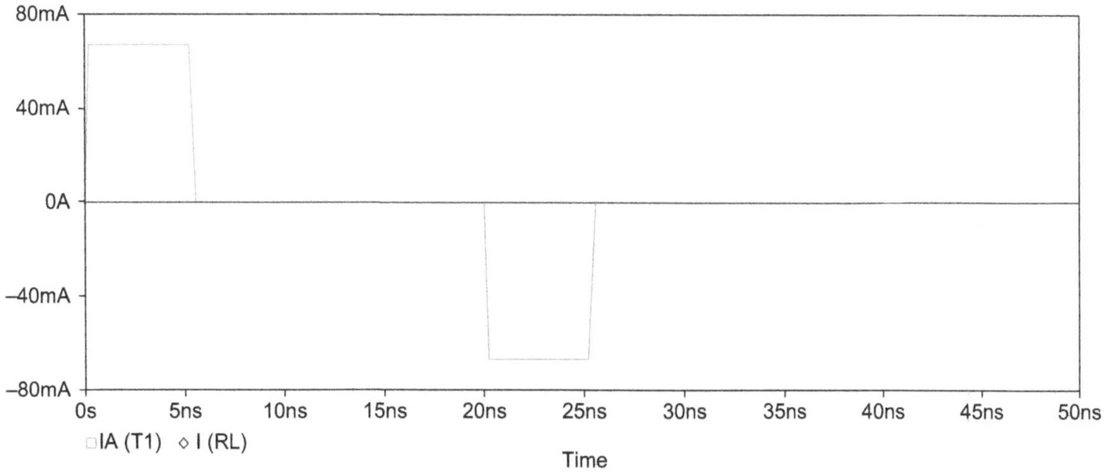

FIGURE 17.18
Open circuit transmission line current waveforms.

Exercise 2

STANDING WAVE RATIO (SWR)

Figure 17.19 shows a lossless transmission line with a short circuit. As shown in Figure 17.13, the incident voltage is reflected with the same amplitude but 180° out of phase. The incident and reflected waves will sum together to produce a standing wave, otherwise known as a stationary wave, as will be demonstrated below.

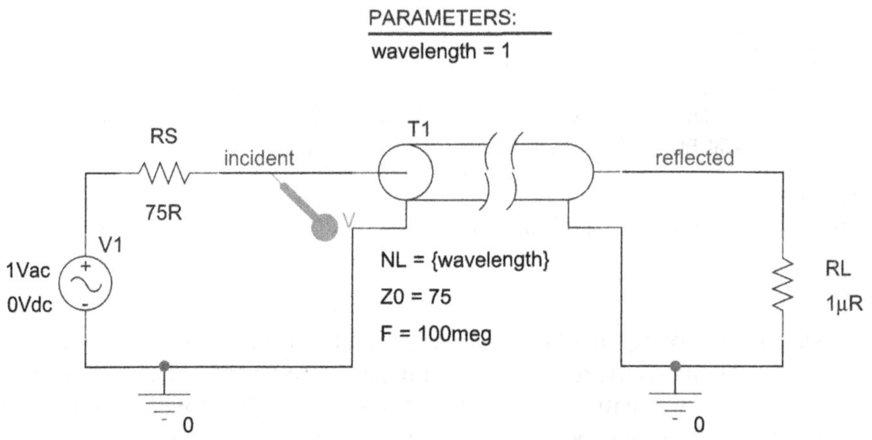

FIGURE 17.19
Demonstrating a standing wave pattern for a short circuit load.

SWR FOR SHORT CIRCUIT

1. Delete the current markers and change the value of RL to 1µR for a short circuit. Delete the voltage pulse, V1, and replace with a VAC source from the source library.
   As mentioned previously, you cannot use TD and NL together, so you can either delete the TD property in the Property Editor or replace the transmission line with a new part.
2. Delete the transmission line T1.
3. Place a T part from the analog library.
4. We are going to vary the value of the transmission line property NL, so we need to parameterize the property value of NL in the Property Editor. Double click on the T part to open the Property Editor. Highlight the NL property value box which has shaded lines and enter {wavelength} (Figure 17.20). As soon as you start typing, the shaded lines in the NL value box will disappear. The {} brackets represent a placeholder for a variable parameter. Do not close the Property Editor.

FIGURE 17.20
Adding a parameterized value to the wavelength property.

5. It is a good idea to display new properties. In the **Property Editor**, highlight the wavelength property and select Display (or **rmb > Display**) and select **Name and Value** as shown in Figure 17.21. Do not close the Property Editor.

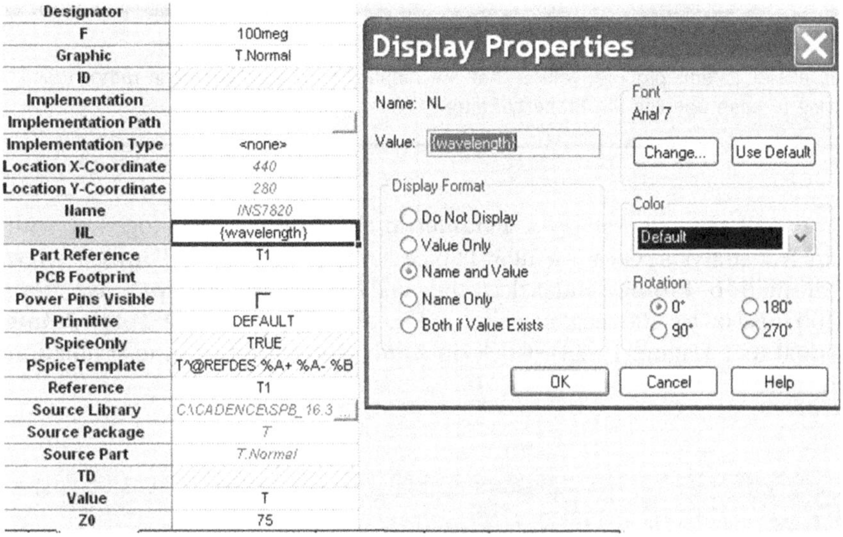

FIGURE 17.21
Displaying name and value of NL property.

6. Set and display, as in Step 5, the name and property values for a frequency, *F*, of 100 megHz and a characteristic impedance, Z0, of 75R, as shown in the circuit diagram Figure 17.19. Close the Property Editor.
7. A default value for the wavelength parameter needs to be defined. Add a **Param** part from the **special** library and double click on the part to open up the Property Editor. Select **New Row** (or New Column) and enter the property **Name** as wavelength and property **Value** as 1 shown in Figure 17.22.

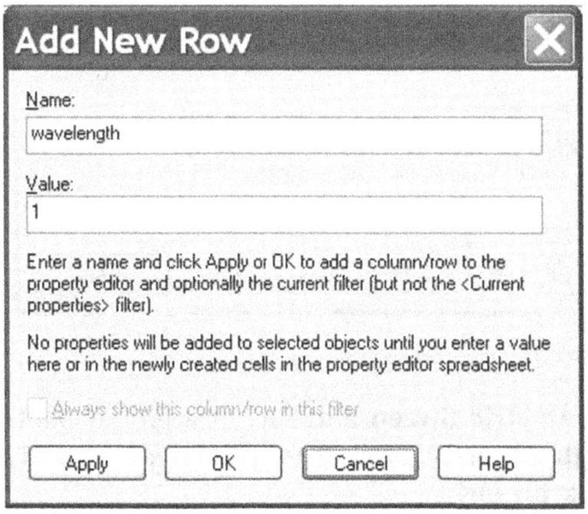

FIGURE 17.22
Adding a new wavelength property with a default value of 1.

8. Display the name and value of the wavelength property and close the Property Editor.
9. Your schematic should now appear as shown in Figure 17.19.

NOTE

It is easier to edit property values that are displayed on the schematic rather than having to keep opening the Property Editor.

10. You will need to set up a **Parametric** simulation sweep together with an **AC analysis**. Create a new PSpice simulation profile, **PSpice > New Simulation Profile**, and select the analysis for **AC Sweep/Noise** from 100 megHz to 200 megHz using a **Linear** sweep with the **Total Points** equal to 1 (Figure 17.23). Click on **Apply** but do not exit the simulation profile.

FIGURE 17.23
AC sweep settings.

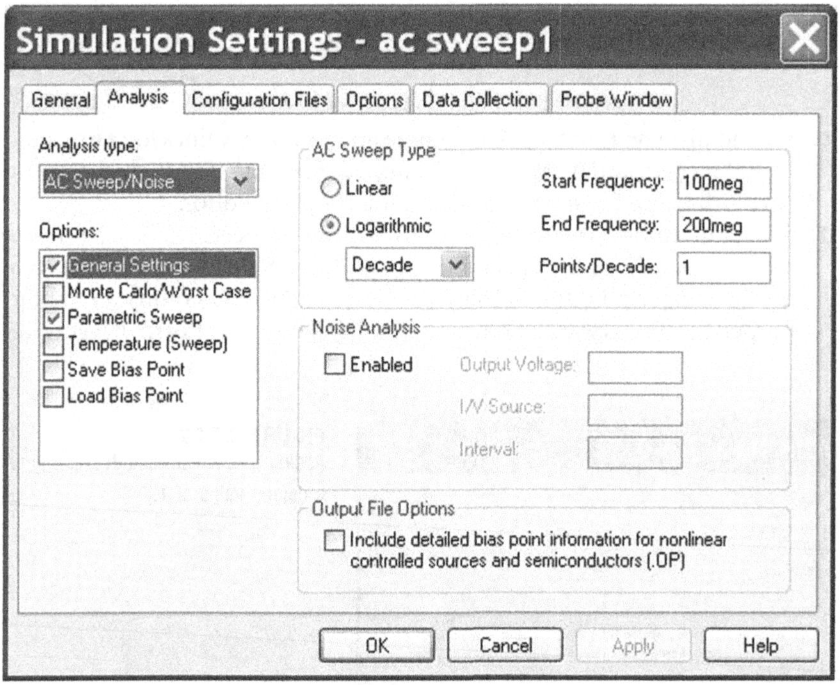

In the **Options** box, select **Parametric Sweep** and set up a global parametric sweep of the **wavelength** property from 0 to 1 in steps of 0.01, as shown in Figure 17.24. Click on OK.

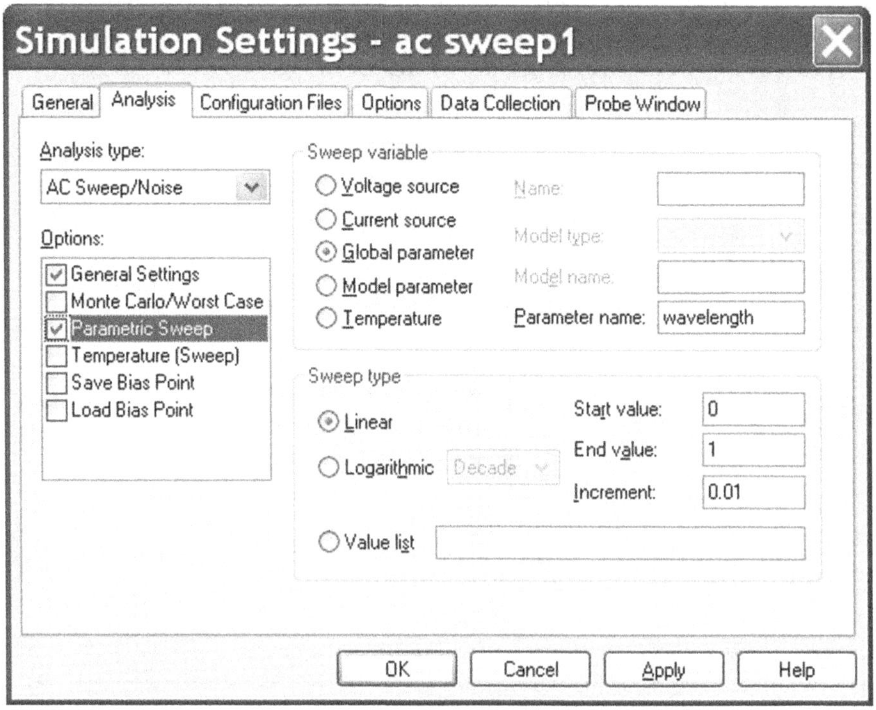

FIGURE 17.24 Parametric sweep settings.

11. Place a voltage marker on the incident node and run the simulation. In the **Available Sections** window, make sure all sections are highlighted and click on OK. You should see the standing wave pattern in Figure 17.25.

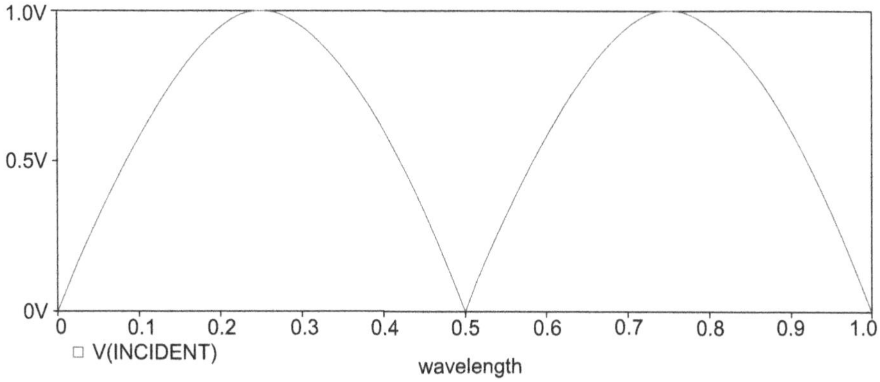

FIGURE 17.25 Standing wave pattern for a short circuit load.

SWR FOR OPEN CIRCUIT

For an open circuit, the voltage is reflected back with the same amplitude but is 180° out of phase.

12. Modify the value of the load resistor in Figure 17.26 to 1T to represent an open circuit and simulate the circuit using the same simulation profile. You should see the standing wave as shown in Figure 17.27.

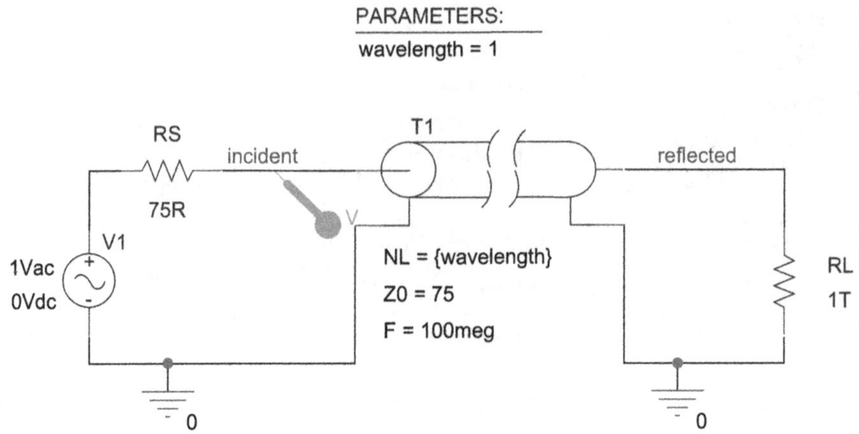

FIGURE 17.26
Demonstrating a standing wave pattern for an open circuit load.

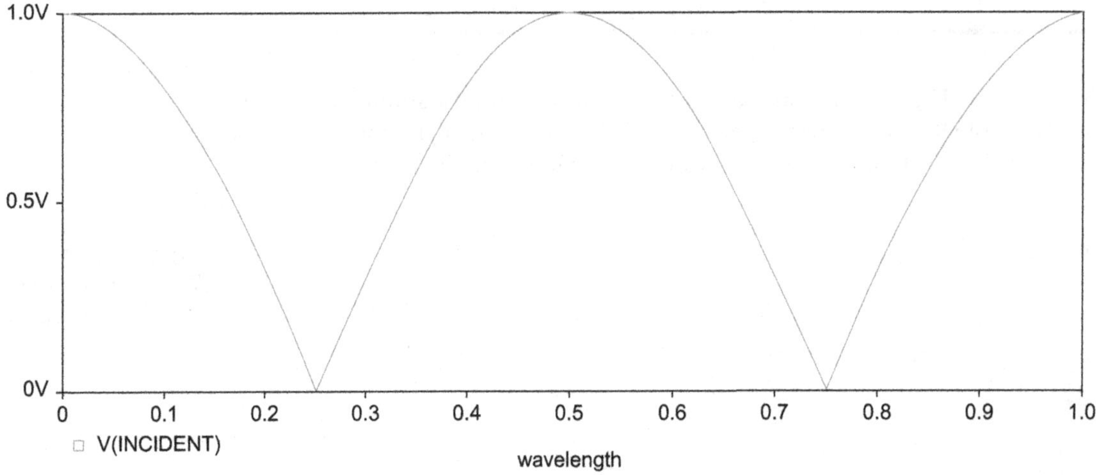

FIGURE 17.27
Standing wave pattern for an open circuit load.

CHAPTER 18

Digital Simulation

Chapter Outline

PSpice uses the same simulation engine for both analog and digital parts. Digital transistor–transistor logic (TTL) and complementary metal oxide semiconductor (CMOS) parts are modeled as subcircuits and include the common digital functions such as gates, registers, flip-flops and inverters. Within each subcircuit, a digital primitive makes up the gate function (AND, OR, etc.) and defines the timing and interface specification for the gate function. Other digital devices include delay lines, AtoD, digital to analog (DtoA), RAM, ROM and programmable logic arrays.

18.1 DIGITAL DEVICE MODELS

A model definition for a two-input CMOS NAND gate is shown below:

```
* CD4011B  CMOS NAND GATE QUAD 2 INPUTS
*
* The CMOS Integrated Circuits Data Book, 1983, RCA Solid State
* tvh 09/29/89    Update interface and model names
*
.subckt CD4011B  A B J
+     optional: VDD=$G_CD4000_VDD VSS=$G_CD4000_VSS
+     params: MNTYMXDLY=0 IO_LEVEL=0
U1 nand(2) VDD VSS
+     A B   J
+  D_CD4011B IO_4000B MNTYMXDLY={MNTYMXDLY} IO_LEVEL={IO_LEVEL}
.ends
```

Analog Design and Simulation using OrCAD Capture and PSpice. DOI: 10.1016/B978-0-08-097095-0.00018-0

The first five lines are comments giving a description of the part and a reference to the data source. On line 6 is the subcircuit definition of the CD4011B with three pins A, B and J. The global power supply is defined by VDD=$G_CD4000_VDD and VSS=$G_CD4000_VSS. The optional parameters are MNTYMXDLY=0, which defines the minimum, typical and maximum delay, and the IO_LEVEL, which defines one of four analog to digital (AtoD) or DtoA interface subcircuits if the digital device is connected to an analog device.

U1 defines a two-input nand(2) primitive which has input terminals VDD, VSS, A, B and J. The '+' signifies a continuation to the next line. The next line (line 11) declares two models, the timing model, D_CD4011B, which defines the timing characteristics such as propagation delay, setup and hold times, and the input/output (I/O) model, IO_4000B, which defines the loading and driving characteristics for the gate. Subcircuits always end with a '.ends' statement, as in line 12. The model D_CD4011B can be found in the CD4000.lib and the model IO_4000B in the dig_io.lib. More detailed information can be found in the PSpice reference manual.

18.2 DIGITAL CIRCUITS

Digital gates by default do not show their power supply pins because this would require a relatively large number of wires to connect all the gates to the power supply, which would overcomplicate the circuit. Instead, TTL and CMOS devices are connected to global power supply nodes which are not displayed and by default are set to 5 V. Different power supplies can be set to accommodate the 3–18 V voltage supply range for CMOS devices. This will not affect the input thresholds and output drives for CMOS devices but the propagation delays will still be defined for a 5 V power supply. For accurate propagation delays, the timing models will have to be modified.

To set digital logic levels on integrated circuit (IC) pins, it is recommended to use digital HI and LO symbols from the **Place > Power** menu and to use digital pull-up resistors from the **dig_misc** library to tie a pin high or low via a resistor. **No Connect** symbols, from the **Place** menu, can be used to identify unconnected pins. Figure 18.1 shows the respective Capture symbols and parts.

In Figure 18.2, a digital clock signal is applied to the input of an 8-bit binary counter (U1A and U1B). In order to enable the counter, the CLR input is tied low

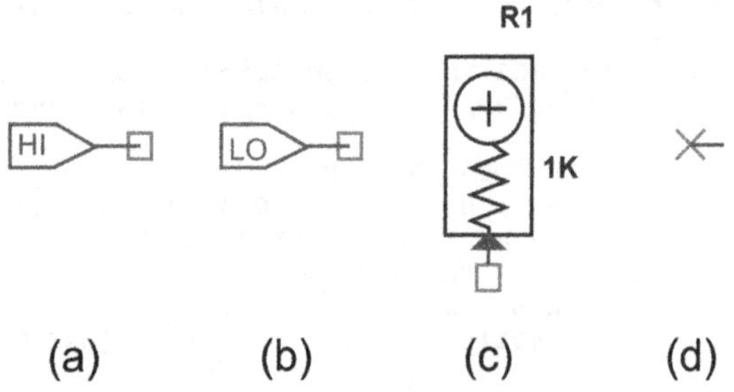

FIGURE 18.1
(a) Digital HI; (b) digital LO; (c) digital pull-up; (d) no connect.

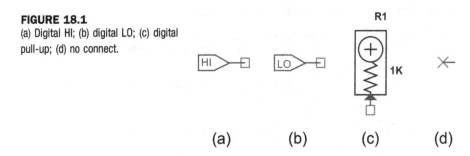

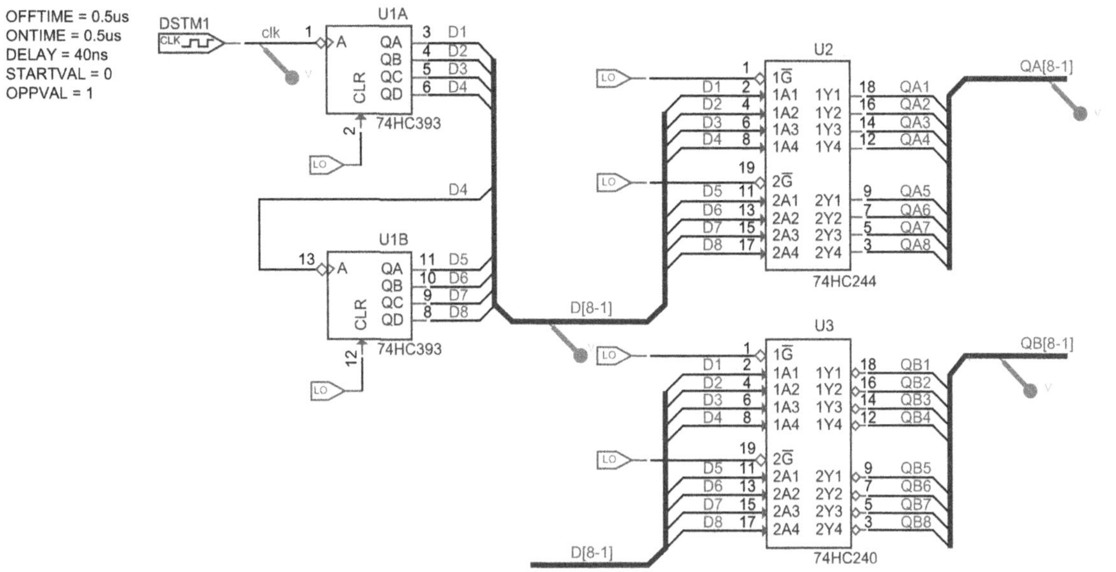

FIGURE 18.2
Connecting the 4-bit counter outputs to an 8-bit bus.

by using a digital LO symbol. Each counter output is connected to an 8-bit bus using bus entry points, via **Place > Bus Entry**, selecting the icon  or pressing E on the keyboard.

NOTE

From version 16.3, connecting pins can automatically be drawn to a bus. Draw a bus and then select **Place > Auto Wire > Connect to Bus**. Click on a connecting pin and then click on the bus (you will be prompted to enter the name of the net). The bus entry point and wire will automatically be drawn.

Each wire connected to a bus entry point has been labeled D1, D2, etc., and the bus itself has a net name of D[8-1], the order of which is msb-lsb. The bus on the data inputs to U3 is also named as D[8-1] and will therefore be connected to the 8-bit bus. The bus can also be labeled as D[7-0] or D[7..0], according to your preference. Only signals of the same type can be grouped together on a bus; mixed busses cannot be defined in Capture. However, in Probe, signals of different types can be collected together and displayed as a bus waveform. Markers can be placed on a bus as well as wires.

18.3 DIGITAL SIMULATION PROFILE

Options for digital simulation can be found in the simulation profile under the **Options** tab and by selecting **Category: > Gate-level Simulation** as shown in

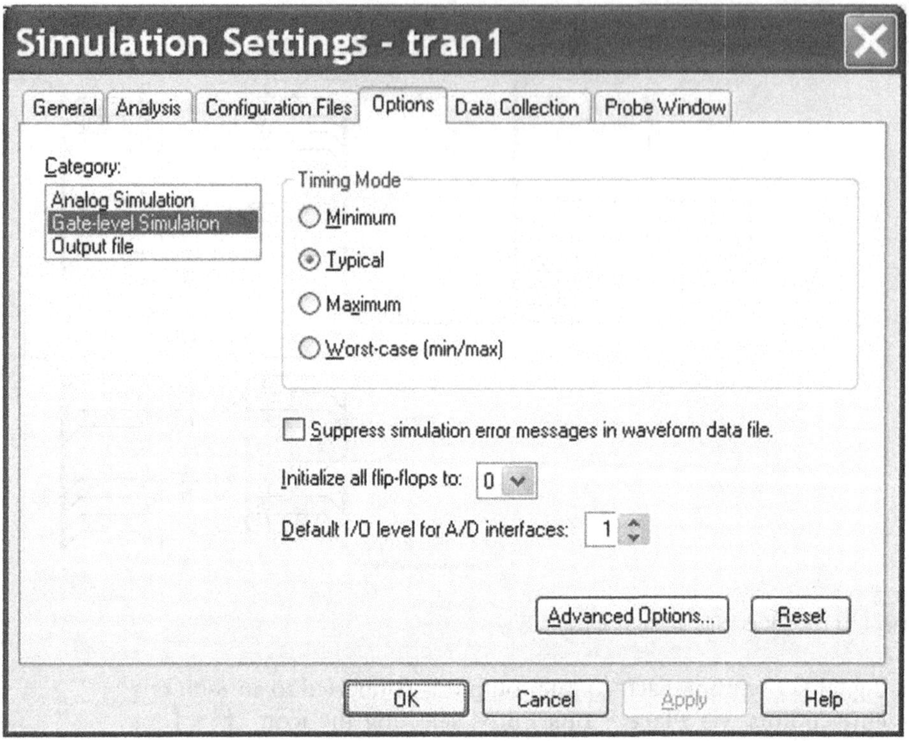

FIGURE 18.3
Digital simulation options.

Figure 18.3. The **Timing Mode** option lets you select the minimum, typical, maximum or worst case timing characteristics for the digital devices. There are four I/O AtoD and DtoA interfaces that you can select and, most importantly of all, you can initialize all flip-flops to either X for do not care, logic 0 or logic 1. There is also the option to suppress simulation error messages, as PSpice will report any digital timing hazards or timing violations.

18.4 DISPLAYING DIGITAL SIGNALS

Digital signals are shown as either high or low logic levels. However, for regions of ambiguity where the transition time is not precisely known, the rising and falling transitions are shown in yellow (Figure 18.4). Unknown states are shown as two red lines and high-impedance states are shown as three blue lines.

NOTE

One common mistake is not to initialize registers (flip-flops) in a circuit, which will result in the two red lines appearing, representing an unknown state. Make sure that you initialize the flip-flops as shown in Figure 18.3.

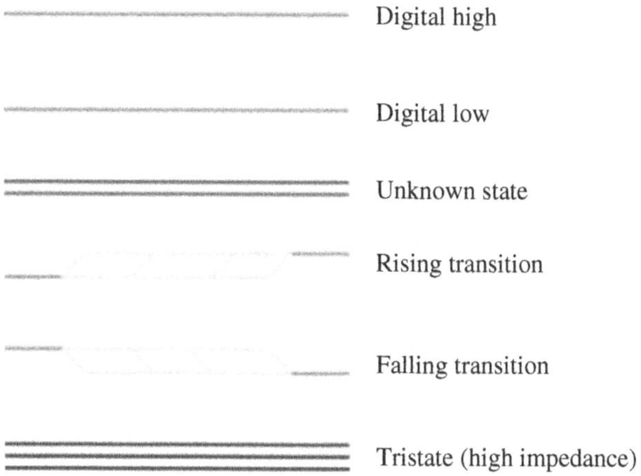

FIGURE 18.4
Digital signals displayed in Probe.

You can group digital signals together and display them as a bus in the Probe window. The bus name can be created in the **Trace Expression** field in the **Trace > Add Trace** window. Up to 32 digital signals can be listed with the order msb to lsb, with a radix of hexadecimal (default), decimal, octal or binary. For example,

{D4 D3 D2 D1};myBus;d	will display D4 to D1 (msb-lsb) labeled as myBus with decimal numbers
{WR RD CE};control;b	will display bus control in binary.

Figure 18.5 shows a QB[8:1] bus shown by default in hexadecimal. The Dbus has been created from a collection of digital signals and is shown in hexadecimal, decimal and binary.

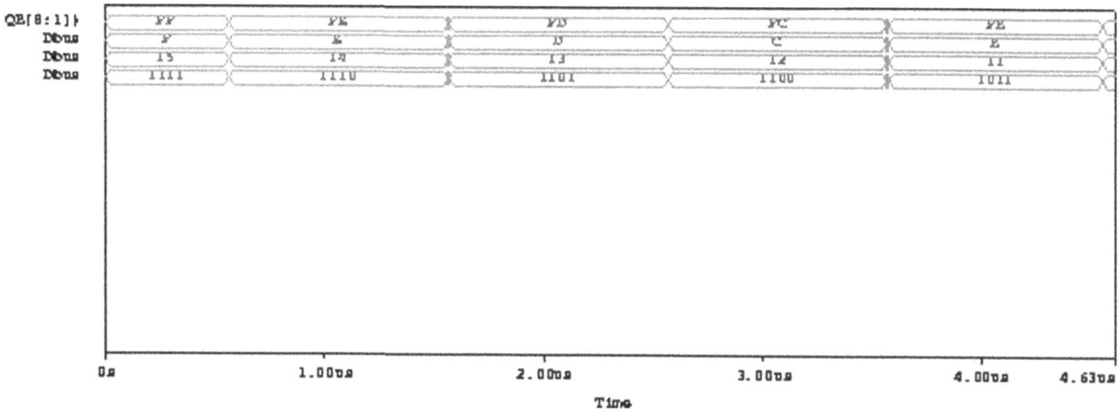

FIGURE 18.5
Bus signals displayed in hexadecimal, decimal and binary.

18.5 EXERCISES

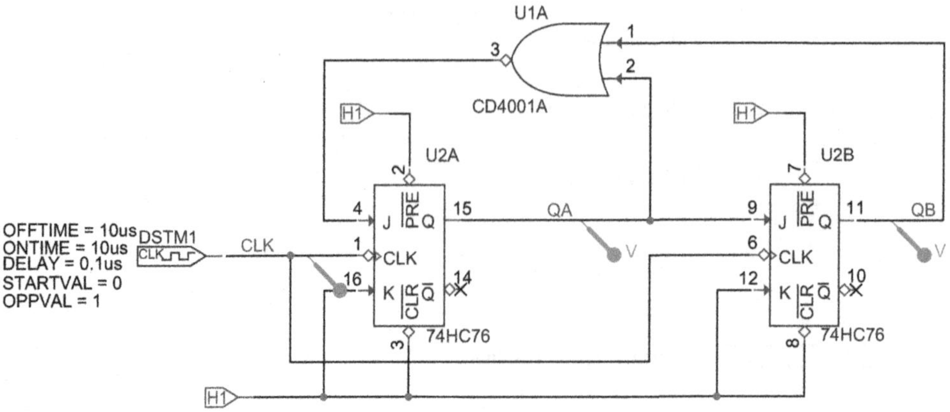

FIGURE 18.6
Modulus 3 synchronous counter.

Exercise 1

You will verify the output sequence of a modulus 3 synchronous counter.

1. Create a project called **Mod 3 Counter**. Rename SCHEMATIC1 to **counter** and draw the modulus 3 synchronous counter in Figure 18.6. The digital flip-flops and the OR gate are from the CD4000 library. The digital HI and LO symbols are from the **Place** menu, or press 'F' on the keyboard. The digital stimulus is **digClock** from the source library.
2. Name the nodes as shown: **Place > Net Alias** or press 'N' on the keyboard.
3. Set up a PSpice simulation profile for 100 μs and select **Options > Category: Gate-level simulation** and set **Initialize all flip-flops** to 0 (Figure 18.7).

FIGURE 18.7
Initializing flip-flops to logic 0.

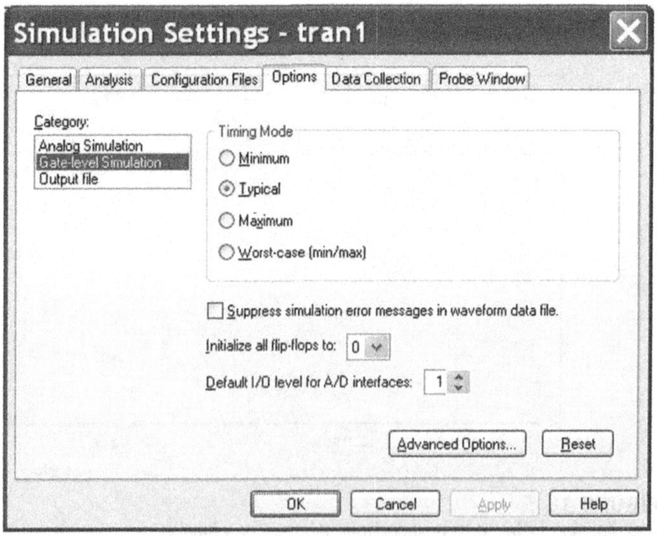

4. Place voltage markers on the CLK, QA and QB nodes.
5. Run the simulation. The trace names will appear in the order that you placed the voltage probes. In Probe, rearrange the trace names such that the CLK is at the top, then QA and QB. This can be done by selectively cutting and pasting. Select CLK and press control-X, which will delete the trace; then press control-V to paste the trace name. Note that both traces QA and QB are initialized to logic 0.
6. Turn on the cursor and move the cursor along the waveforms. The corresponding logic levels should appear on the Y-axis as in Figure 18.8. Note that the output of the flip-flop only changes on the falling edge of the clock signal. The CD4027 flip-flops are negative edge triggered.

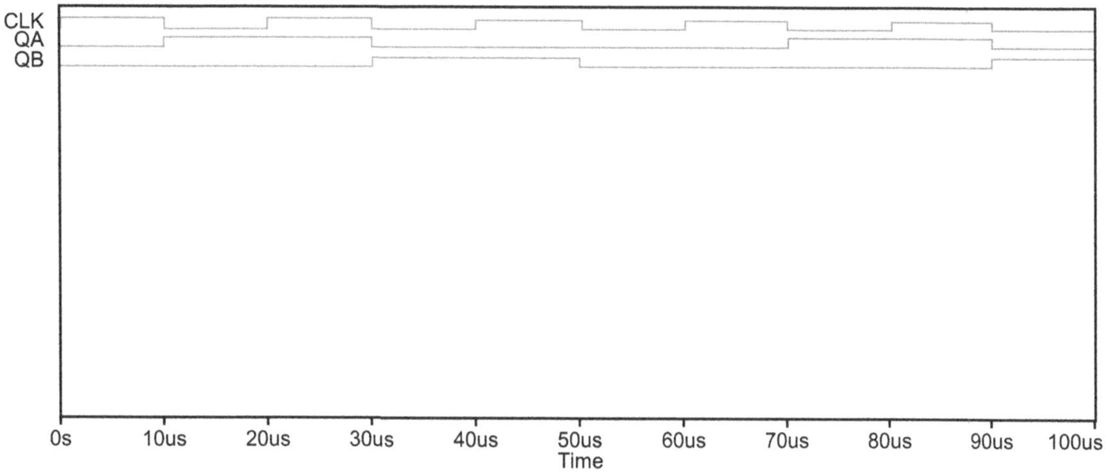

FIGURE 18.8
Digital counter waveforms.

7. You will add a bus to show the binary count. Select **Trace > Add** and in the Trace Expression field enter

```
{QB,QA};count_b;b
```

and click on OK to display the binary count.
8. Select **Trace > Add** and in the Trace Expression field enter

```
{QB,QA};count_d;d
```

and click on OK to display the decimal count.
9. Select **Trace > Add** and in the Trace Expression field enter

```
{QB,QA};count_h;h
```

and click on OK to display the hexadecimal count.
10. Your Probe waveforms should resemble those shown in Figure 18.9, which is that of a modulus 3 counter with a count sequence of 0,1,2,0,1,2,0, etc.

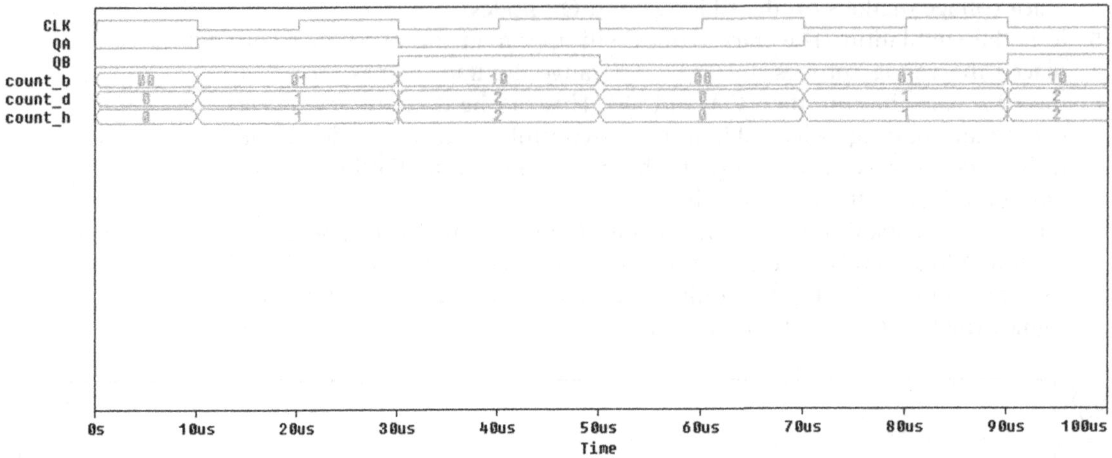

FIGURE 18.9
Counter output digital waveforms.

Exercise 2

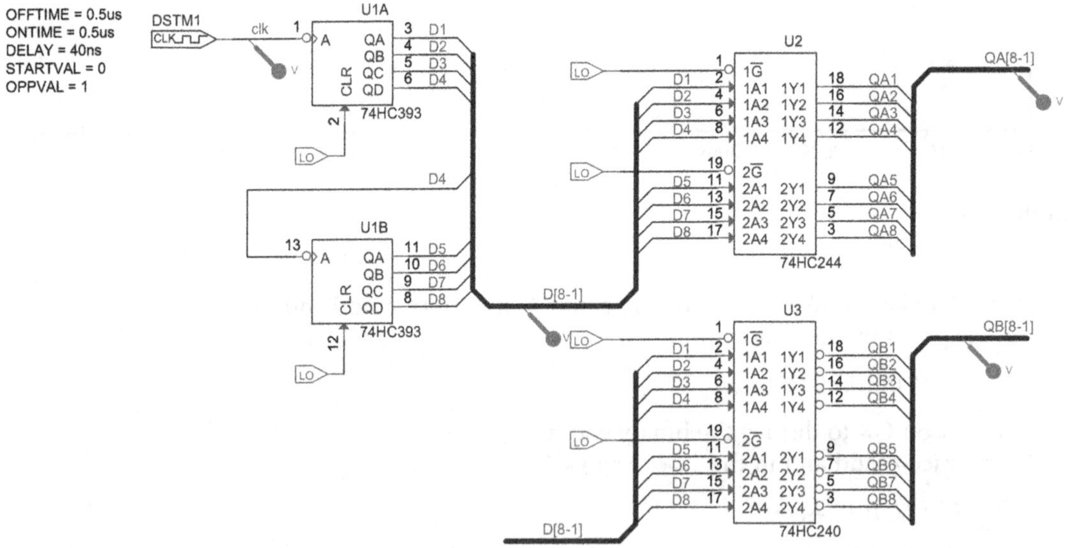

FIGURE 18.10
Connecting signals to a bus.

The circuit in Figure 18.10 is an example of how to connect signals to a bus and how to select the different A and B sections of an IC. A clock signal is divided down by two 4-bit binary counters, U1A and U1B, and then passed to two octal buffers, U2 and U3, via an 8-bit bus. U3 is an inverting octal buffer.

1. The 74HC393 has two identical sections designated A and B, shown in the **Packaging** window in the **Place Part** menu in Figure 18.11. When you select Part B, the pin numbers will change accordingly.

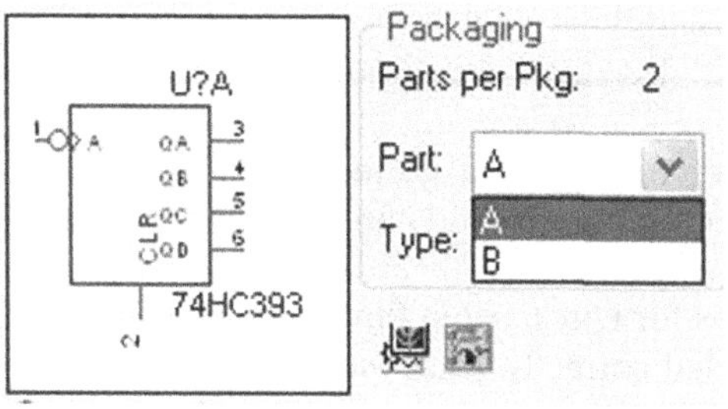

FIGURE 18.11
Selecting different parts (sections) from a package.

2. Place the parts for the circuit in Figure 18.10 but do not connect the wires or bus yet. DSTM1 is a DigClock source from the **source** library and the HC devices can be found in the 74 HC library. The HI and LO symbols are from the **Place > Power** menu.
3. Draw the busses in the circuit. To draw an angled bus, hold down the shift key and left mouse click to define the angle, then draw the bus.
4. Starting with U1A pin 3, place a bus entry on the bus as shown in Figure 18.10. Draw a wire from the bus entry to pin 3 of U1A.
5. Select the wire and place a net alias (press N) labeled D1 on the wire. Press escape or **rmb > End Mode.**
6. Draw a selection box around the wire, net name and bus entry point as shown in Figure 18.12.

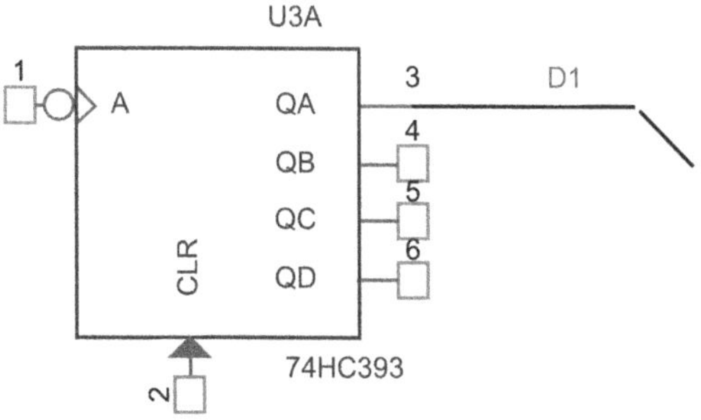

FIGURE 18.12
Selecting the wire, bus entry and net name.

7. Hold the control key down, place the cursor on the wire and drag it down so that it connects to pin 4. The net name automatically increments to D2. With the wire still highlighted, press F4 twice and two more nets, D3 and D4, will appear.

NOTE

In release 16.3 there is now the option to autowire two points: **Select Place > Auto Wire** , select two points and the wire is drawn automatically. Another new feature is the **Place > Auto Wire > Connect to bus.** You click on a connecting pin and then the bus such that the wire and bus entry will be connected automatically. You will also be prompted for the net name.

If you have version 16.3 go to Step 8; otherwise, go to Step 9.

8. Select **Place > Auto Wire > Connect to bus** and wire up the rest of the IC pins as in Figure 18.10.
9. The busses are labeled the same as for wires, using **Place Net >** Alias. Make sure that all three busses are labeled correctly, [msb-lsb].
10. Set up a simulation profile for a transient analysis with a run to time of 10 μs.
    Select **Options > Category: Gate-level simulation** and set **Initialize all flip-flops** to 0 (Figure 18.13).

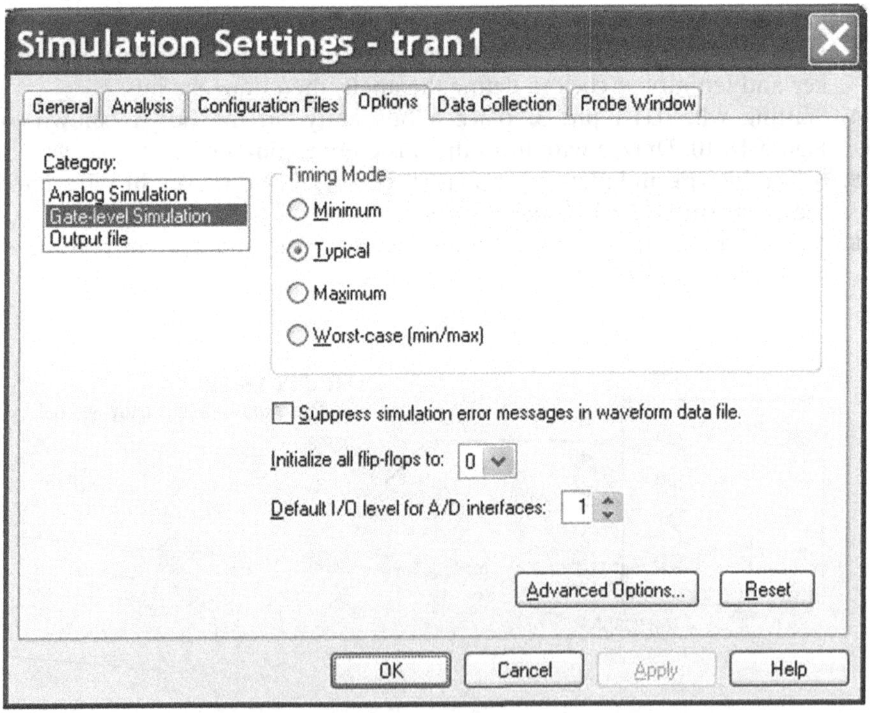

FIGURE 18.13 Initializing flip-flops to logic 0.

11. Place voltage markers on each bus as shown and run the simulation.
12. You should see the count increase for bus D[8-1] and QA[8-1]. As U3 is an inverting buffer, the count for QB[8-1] will start at FF and decrease accordingly. Compare your waveforms with those shown in Figure 18.14.

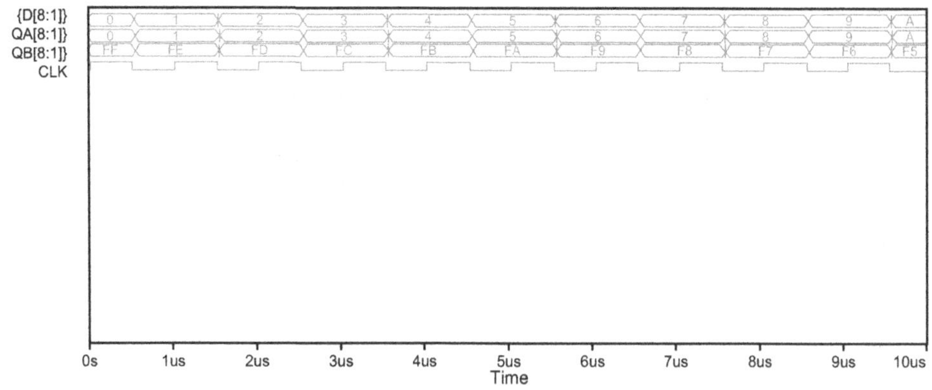

FIGURE 18.14
Digital bus waveforms.

13. Delete all the traces, **Trace > Delete All Traces**.
14. Select **Trace > Add Trace**. On the right-hand side in the Functions or Macros, select the curly brackets {}; the **Trace Expression** box will contain {} with the cursor sitting in the middle waiting for you to select a trace variable name.
15. Select D4, followed by D3, D2 and D1. Add the following text to the expression and click on OK:

```
{D4,D3,D2,D1};myBus;d
```

16. Create a bus called **nibble**, consisting of QA4, QA3, QA2 and QA1, and display the bus in binary:

```
{QA4 QA3 QA2 QA1};nibble;b
```

You should see the waveforms as in Figure 18.15.

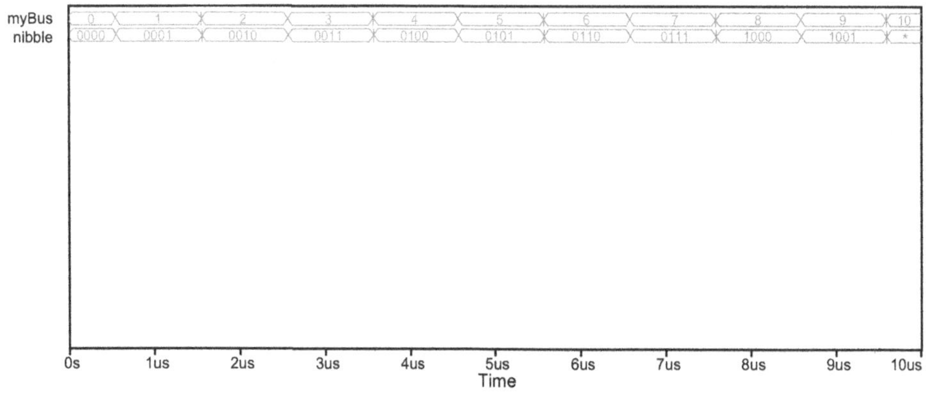

FIGURE 18.15
Custom-made busses, my Bus and nibble.

17. In Capture, disable U2 by setting the enable pins ($\overline{G}$), 1 and 19 high by using a $D_HI symbol from **Place > Power**. Run the simulation.

TIP

Select the $D_HI symbol from the most recently placed part list (Figure 18.16). If the symbol is not in the pull-down menu, type $D_HI in the box and press return.

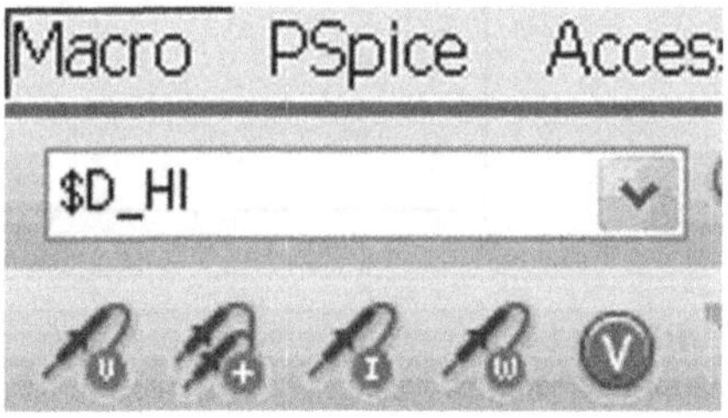

FIGURE 18.16
Most recently placed part.

---

18. Run the simulation.
19. The trace for QA[8-1] is displayed with a Z inside, indicating a high-impedance tristate output (Figure 18.17).

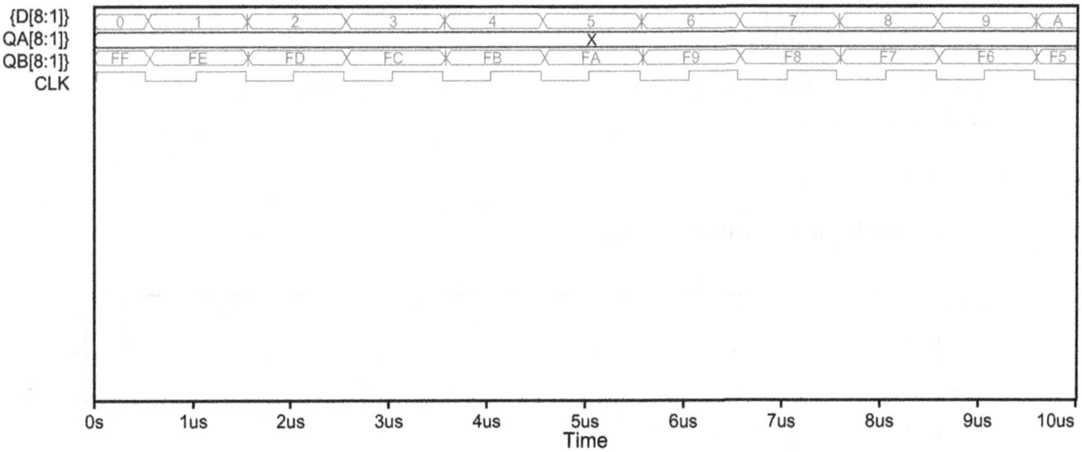

FIGURE 18.17
High impedance shown on U2's outputs.

Exercise 3

PSpice reports and plots timing violations relating to setup times, hold times and minimum pulse width. By decreasing the clock pulse width, we can investigate the reporting of these errors.

1. Change the clock OFFTIME to 0.01 μs and ONTIME to 0.01 μs.
2. Reduce the simulation time from 10 μs to 1 μs.

3. Run the simulation.
4. The Simulation Message window will appear as shown in Figure 18.18.

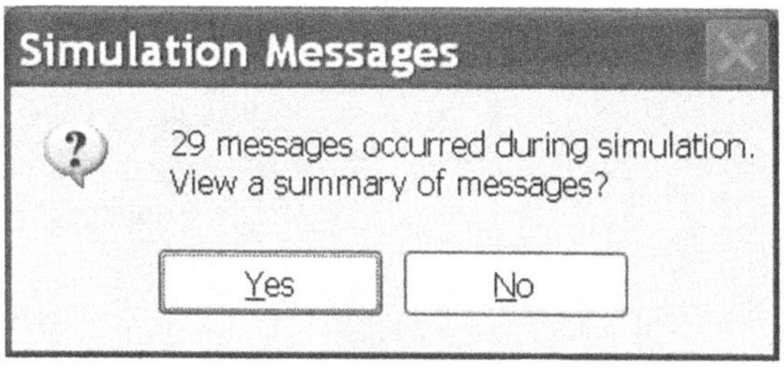

FIGURE 18.18
Number of simulation messages.

5. Click on **Yes** and you will see a list of Warnings (Figure 18.19).

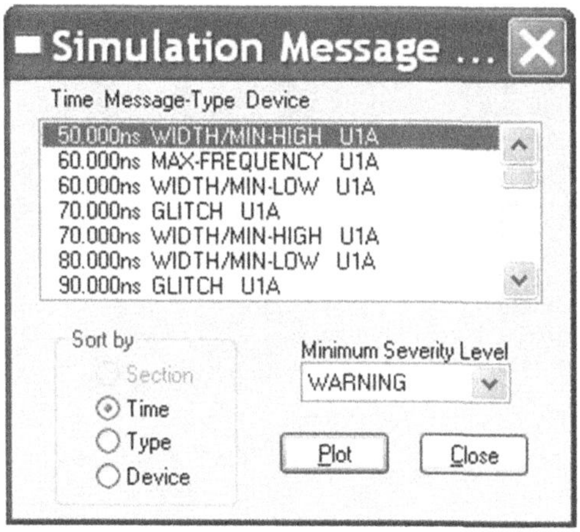

FIGURE 18.19
List of warning messages.

6. The **Minimum Severity Level** pull-down menu lists severity levels for Fatal, Serious, Warning and Info. For Fatal severity, the simulation stops.
   Leave the level on Warning and click on **Plot**.
7. The violation information displayed (Figure 18.20) gives the time at which the violation occurred and with which device. The message gives you not only the measured violation timing value but also the specified timing value. PSpice also plots the occurrence of the violated timing waveforms.
   Selecting each time message will open up a new Probe plot window.

```
WIDTH/MIN-HIGH Violation at time 50ns
  Device: X_U1A.UHC393DLY
  Minimum high WIDTH = 20ns
  NODE: X_U1A.A, measured WIDTH = 10ns
```

FIGURE 18.20
Violation information.

CHAPTER 19
Mixed Simulation

Chapter Outline

PSpice uses the same simulation engine for analog and digital circuits. The simulation results in Probe share the same time axis but are split into separate analog and digital plot windows. Analog and digital components in a circuit are connected together at nodes. In PSpice there are three types of connecting nodes: analog, where all connected parts are analog; digital, where all connected parts are digital; and interface, where there is a mixture of analog and digital parts. Interface nodes are automatically separated into one analog node and one or more digital nodes by inserting analog and digital interface subcircuits, which are either analog to digital (AtoD) or digital to analog (DtoA) interface subcircuits. These subcircuits will also have their own power supply. As this process is automatic and runs behinds the scenes, we do not normally have to worry about the interface subcircuits, although they are available as traces in Probe.

Figure 19.1 shows an analog comparator with an open collector transistor connected to a digital gate. The pull-up resistor is connected to the digital power supply and the output ground for the comparator is connected to digital ground. Figure 19.2 shows the digital waveforms being plotted in the upper area of Probe and the analog waveforms plotted in the lower area.

Mixed analog and digital circuits follow the same procedure for placing parts, creating a simulation profile and simulation.

19.1 EXERCISES

Exercise 1

Figure 19.3 shows an AD7224 digital to analog converter (DAC) with an input digital data word of 0111 1111. From the manufacturer's data sheet, the output voltage is given by:

Analog Design and Simulation using OrCAD Capture and PSpice. DOI: 10.1016/B978-0-08-097095-0.00019-2

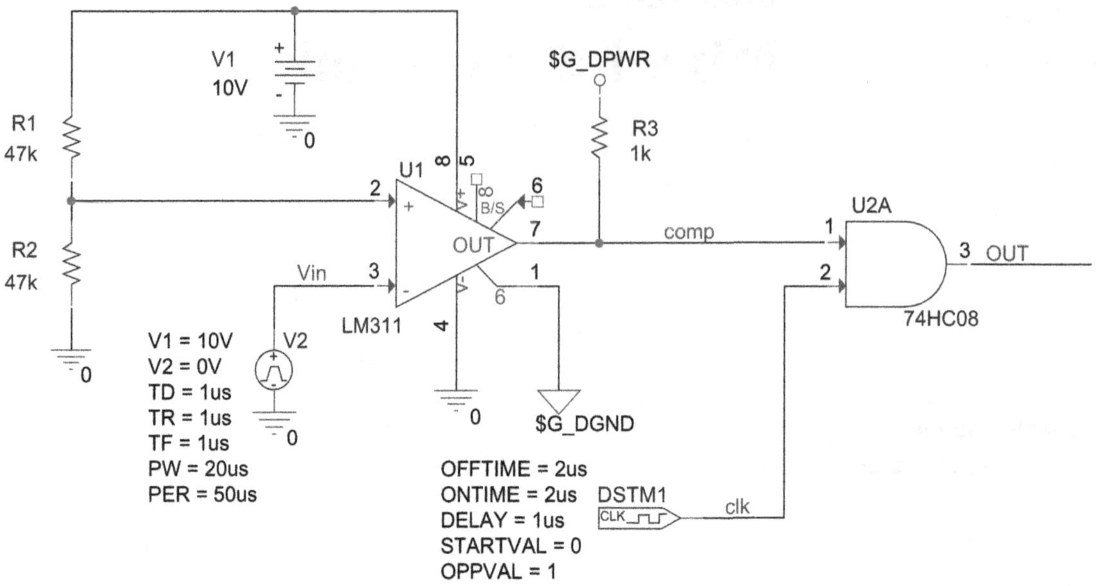

FIGURE 19.1
Analog comparator switching a digital gate.

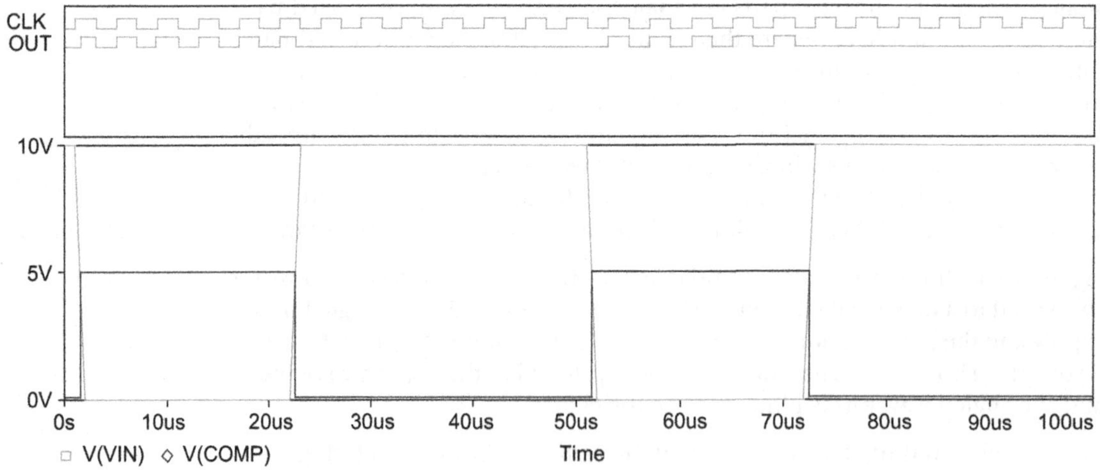

FIGURE 19.2
Analog and digital traces.

$$V_o = V_{REF} \times \frac{127}{256} = 4.96 \text{ V} \qquad (19.1)$$

The timing cycles for the DAC have been set up according to the manufacturer's datasheet.

1. Draw the circuit in Figure 19.3. The AD7224 can be found in the DATACONV library and the DigClock stimuli can be found in the source library.

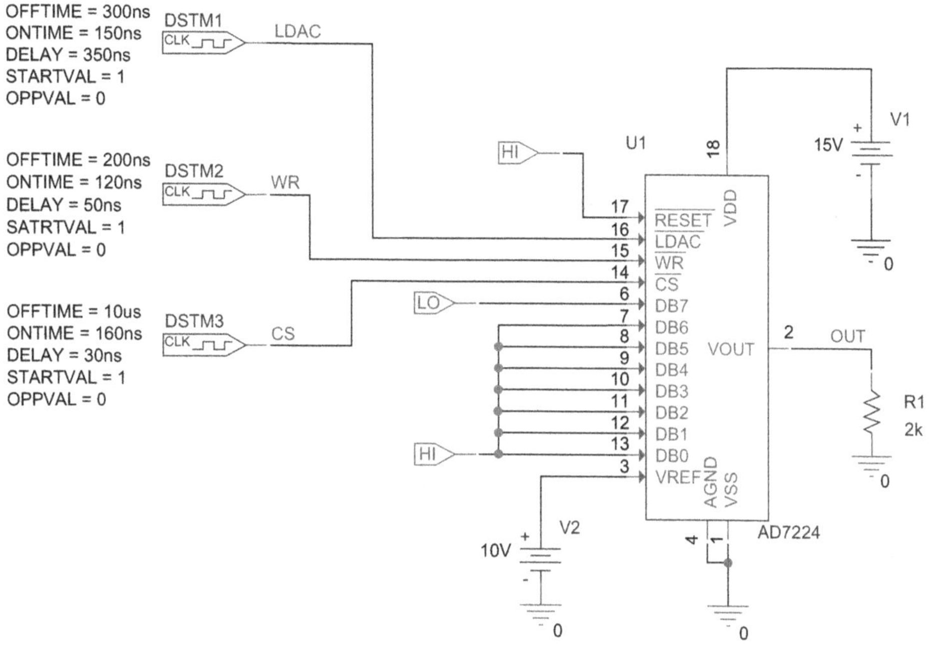

FIGURE 19.3
Digital to analog conversion using the AD7224.

2. Set up a transient analysis with a **run to time** of 5 μs. Select the **Options** tab and select **Category** > **Gate-level Simulation** and set **Initialize all flip-flops** to 0 (Figure 19.4). Close the simulation profile.

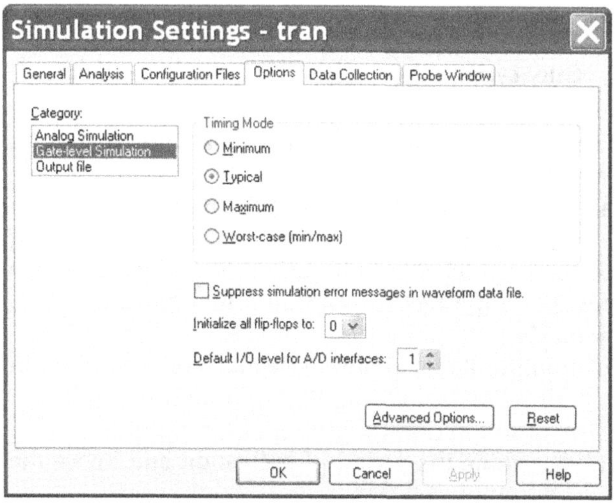

FIGURE 19.4
Initializing flip-flops to logic 0.

3. Place voltage markers on the nets, LDAC, WR, CS and OUT.
4. Run the simulation.
5. In Probe, you will see that the upper plot is for the digital signals and the lower plot is for the analog OUT signal (Figure 19.5).

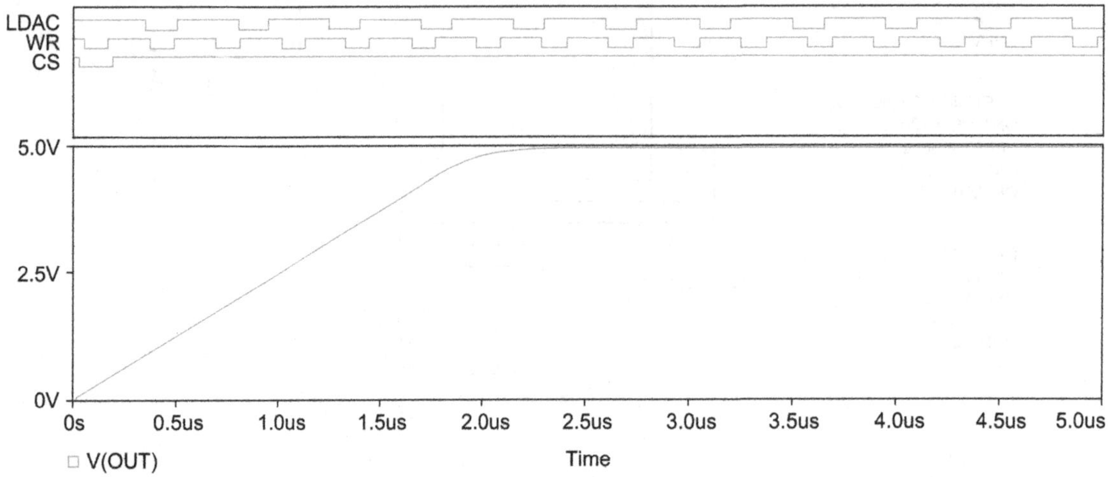

FIGURE 19.5
Analog and digital waveforms.

6. Turn the cursor on and check that the output voltage is 4.96 V as calculated.

Exercise 2

Figure 19.6 shows the ubiquitous NE555 timer, used in countless applications. The timing equations for the NE555 are given as:

$$f = \frac{1.44}{(RA + 2RB)C} \tag{19.2}$$

$$\text{Duty cycle} = \frac{RA + RB}{RA + 2RB} \tag{19.3}$$

Using the components as shown in Figure 19.6 will give a calculated clock frequency of 218 Hz and a duty cycle of 0.67.

1. Create a new project called **Clock Oscillator**. Rename SCHEMATIC1 to **clock** and draw the circuit in Figure 19.6; the 555 can be found in the anl_misc library. There are three versions, 555alt, 555B and 555C, which have the pins arranged differently. Do not forget to place an initial condition, IC1, from the special library on C1.
2. Create a transient simulation profile for 20 ms. Place markers on VC and on OUT.
3. Run the simulation.
4. Display the cursors and determine the period of oscillation and hence the clock frequency.

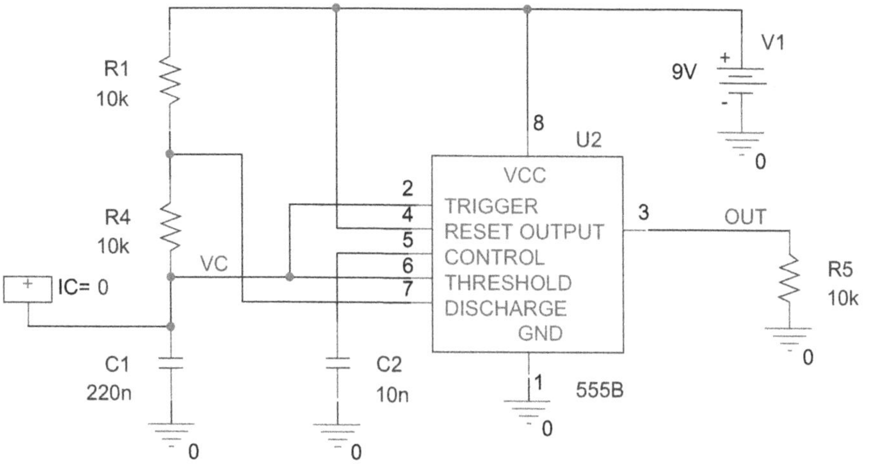

FIGURE 19.6
NE555 clock oscillator.

5. Determine the duty cycle, which is the on time divided by the off time.
6. Confirm your measurements by selecting **Trace > Evaluate Measurement** and Period(1) then selecting V(OUT).

   ```
   Period(V(OUT))
   ```

7. Select **Trace > Evaluate Measurement** and Period_XRange, (1,begin_x, end_x) then select V(OUT), and then enter 5 m and 20 m.

   ```
   Period_XRange(V(out),5m,20m)
   ```

8. Confirm your measurements by selecting **Trace > Evaluate Measurement** and DutyCycle(1), then selecting V(OUT).

   ```
   DutyCycle(V(OUT))
   ```

9. If the results of the measurements are not shown, select **View > Measurement Results**. Your results should be similar to those shown in Figure 19.7.

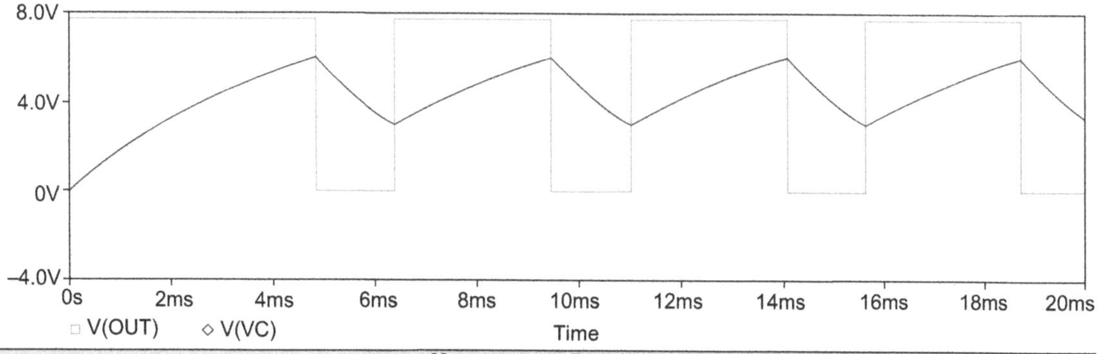

Measurement Results

	Evaluate	Measurement	Value
	☑	Period(v(OUT))	4.62892m
	☑	Period_XRange(V(OUT),5m,20m)	4.62892m
▶	☑	DutyCycle(V(OUT))	665.27676m

FIGURE 19.7
Resultant waveforms and measurement values.

The Cadence\OrCAD software installation includes a good selection of analog, digital and mixed example circuits in the anasim, digsim and mixsim directories. These can be found in the installed directory, for example:

```
<install path>\Cadence\SPB_16.3\tools\pspice\
capture_samples\
```

```
<install path>\Cadence\OrCAD_16.3\tools\pspice\
capture_samples\
```

CHAPTER 20

Creating Hierarchical Designs

Chapter Outline

Capture designs can either be flat, in which signals are connected across pages in the design, or hierarchical, in which the design is partitioned into blocks and signals transverse up and down the hierarchy. Flat designs are represented in the Project Manager as having a single schematic folder with a number of associated pages, whereas hierarchical designs will have more than one schematic folder (Figure 20.1). Each schematic folder in the hierarchy will be represented by a hierarchical block in a schematic. By selecting a hierarchical block, you select the underlying schematic and effectively descend the hierarchy.

For the flat design in Figure 20.1, there is one schematic folder and three pages. For the hierarchical design, there are three schematic folders in the hierarchy each with their own schematic page or pages.

The Project Manager in Figure 20.2a shows two schematic folders, **Top** and **Bottom**. The associated schematics are shown in Figure 20.2b and c, respectively. The **Top** schematic (Figure 20.2b) contains a hierarchical block called **Bottom** which has two hierarchical pins, IN and OUT. To descend the hierarchy, you highlight the

Analog Design and Simulation using OrCAD Capture and PSpice. DOI: 10.1016/B978-0-08-097095-0.00020-9

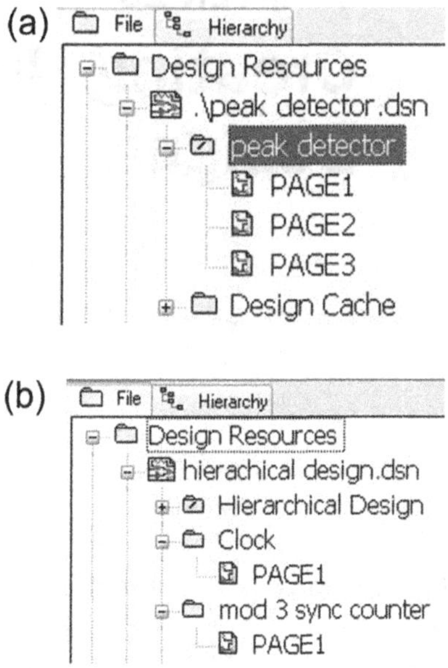

FIGURE 20.1
Project structure: (a) flat design; (b) hierarchical design.

block and **rmb > Descend Hierarchy**, or you can double click on the block and the **Bottom** schematic will be displayed. The connection between the block and the schematic is provided by the hierarchical pins on the Bottom block having the same name as the hierarchical ports in the Bottom schematic, IN and OUT.

In the Project Manager, there is also a Hierarchy tab next to the default **File** tab. By selecting the **Hierarchy** tab, the location of individual parts in the design can be displayed. Figure 20.3 shows that there is a resistor R1, a voltage source V1 and a Hierarchical block HB1 in the Top level schematic and two resistors in the Bottom schematic.

20.1 HIERARCHICAL PORTS AND OFF-PAGE CONNECTORS

As in the case for flat designs, there is normally one folder and one or more pages. In order to connect signals across the pages, off-page connectors are used: **Place > Off-Page Connectors** (Figure 20.4).

Two types are used to indicate the direction of the data flow, i.e. input to output. When a wire connects to an off-page connector, the net name of the wire inherits the name of the connector.

Hierarchical ports connect signals between levels of hierarchy: **Place > Hierarchical Ports** (Figure 20.5). As with off-page connectors, a wire connected to a hierarchical port inherits the name of the port.

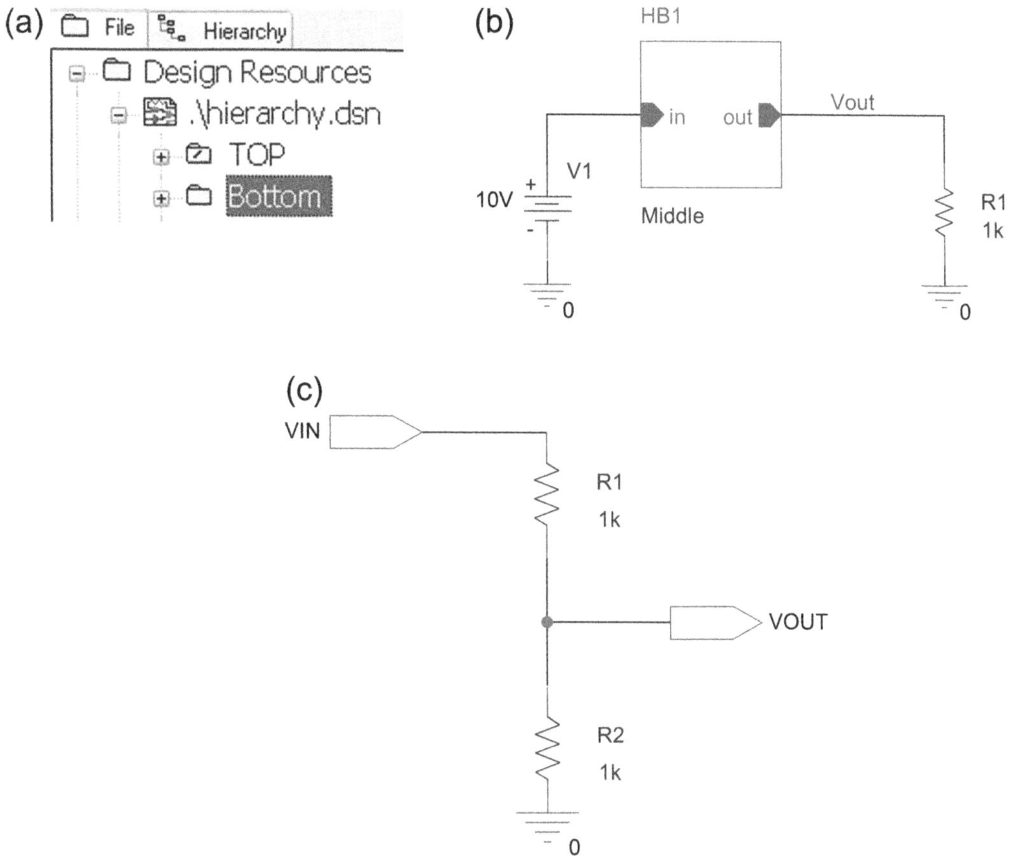

FIGURE 20.2
Hierarchical project: (a) Project Manager; (b) top schematic; (c) bottom schematic.

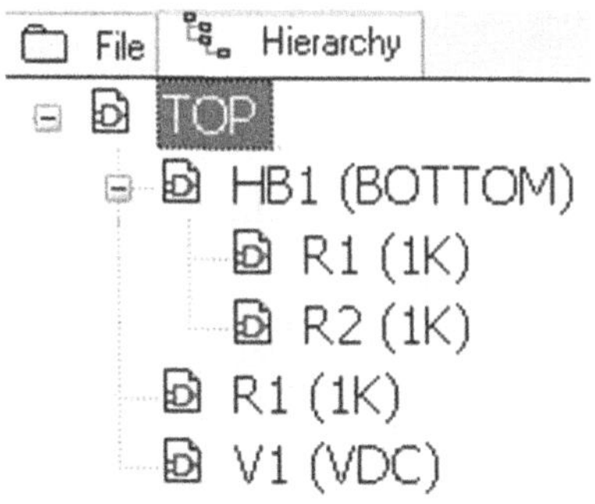

FIGURE 20.3
Design hierarchy showing location of individual components.

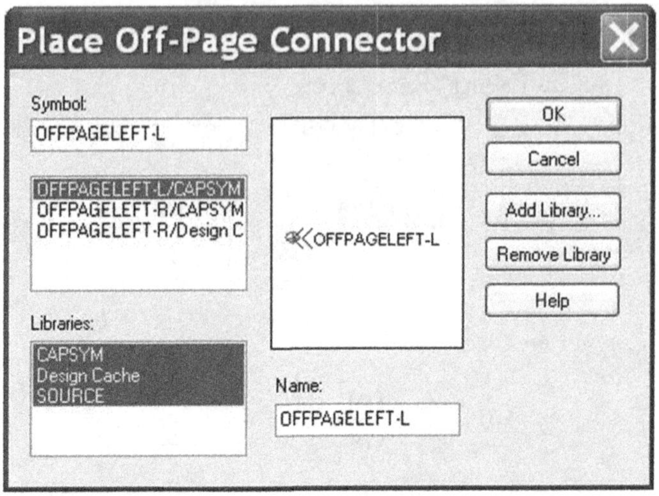

FIGURE 20.4
Off-page connectors.

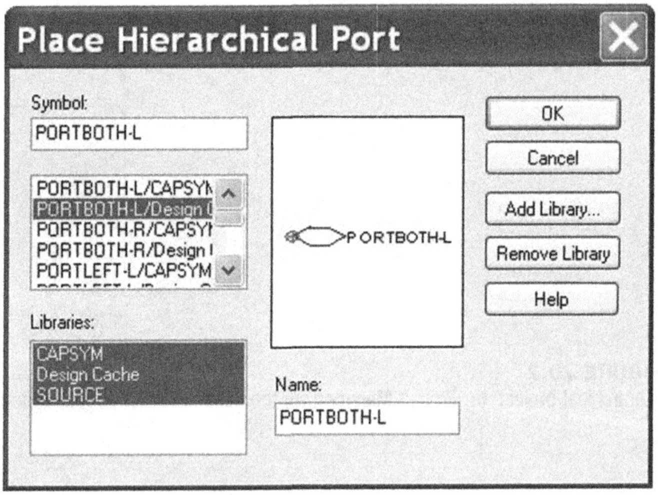

FIGURE 20.5
Place hierarchical ports.

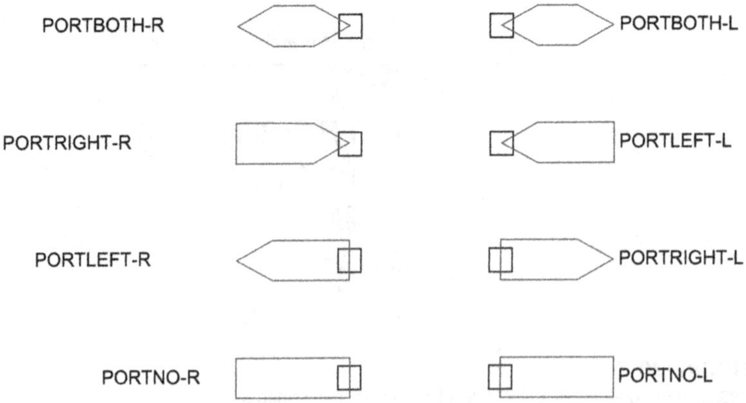

FIGURE 20.6
Available ports.

Different hierarchical ports are available which represent the type of port and direction of data flow. Figure 20.6 shows the types of port available. For example, PORTRIGHT-R is a port that points to the right and has a connection on the right-hand side. Which port you use is entirely your choice.

20.2 HIERARCHICAL BLOCKS AND SYMBOLS

Hierarchical blocks are normally used for top–down designs where the block is drawn on the top-level schematic and associated signal pins are added. Pushing down into the block (descending the hierarchy), the referenced schematic contains the same number of ports as pins with the same associated signal names. The hierarchical blocks cannot be saved to a library as they are drawn 'on the fly' and saved within the schematic file.

Hierarchical symbols are normally used for bottom–up designs where the schematic is drawn first and ports are added to the input and output signals. A symbol is then created with the same number of signal pins and associated names. These hierarchical symbols can be saved to a library for use in other designs.

20.2.1 Hierarchical Blocks

Hierarchical blocks are created 'on the fly' in the schematic: **Place >Hierarchical Block** (Figure 20.7).

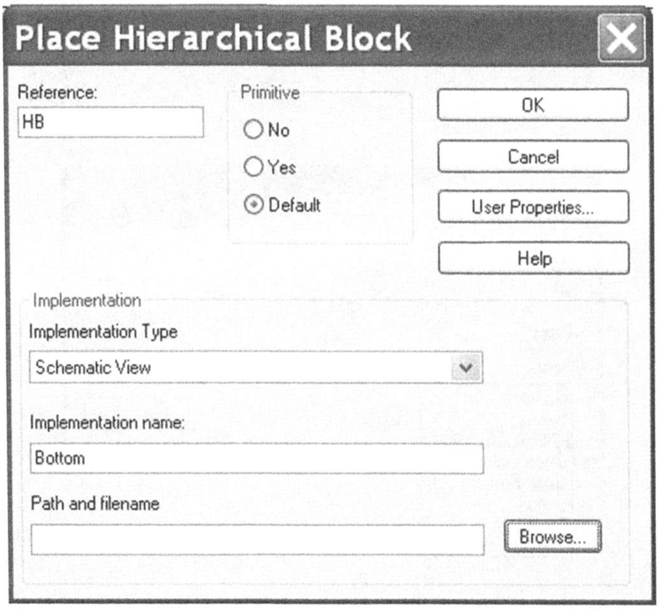

FIGURE 20.7
Creating a hierarchical block.

You define the **Reference** designator, which is your choice, and you then have the option to select the **Implementation Type** and **Implementation Name**. The implementation type defines what the block is referencing and can be any of the types shown in Figure 20.8. Normally for PSpice projects, the **Schematic View** is selected and the **Implementation Name** is the name of the schematic.

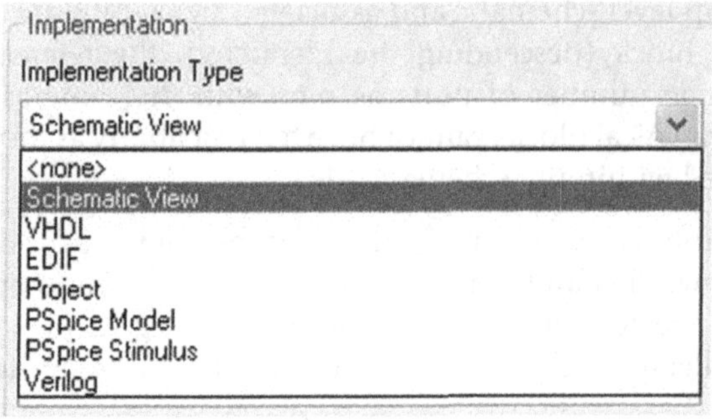

FIGURE 20.8
Implementation types.

You then draw a rectangle for the block and with the block still highlighted select **Place > Hierarchical Pin**. In the **Place Hierarchical Pin** box, you define a name for the pin and its type from the pull-down menu as shown in Figure 20.9. The **Width** option allows you to place a pin representing a bus or a scalar for a single pin. You then place the pin anywhere on the perimeter of the block.

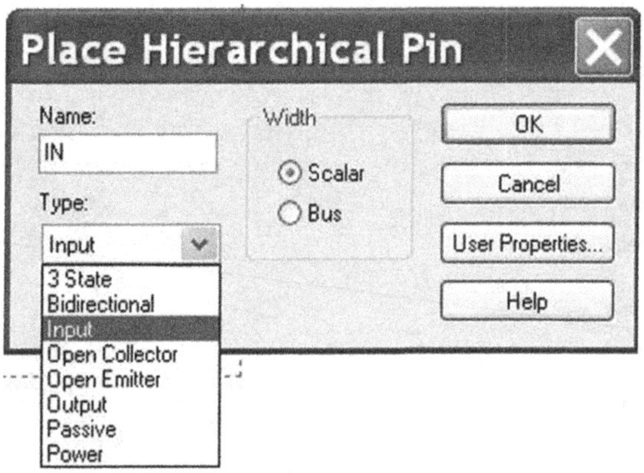

FIGURE 20.9
Defining a hierarchical pin.

20.2.2 Hierarchical Symbols

This is where a circuit is effectively symbolized and a Capture part is generated to represent that circuit. Hierarchical ports are added to the circuit which will appear as pins on the hierarchical symbol. In Capture, a part is generated by selecting **Tools > Generate Part** (Figure 20.10). You select the **Netlist/source file type** as a Capture Schematic/Design file (.dsn), select the name and location of the Part library and select the name of the schematic folder (**Source Schematic name**), and a hierarchical symbol will be generated.

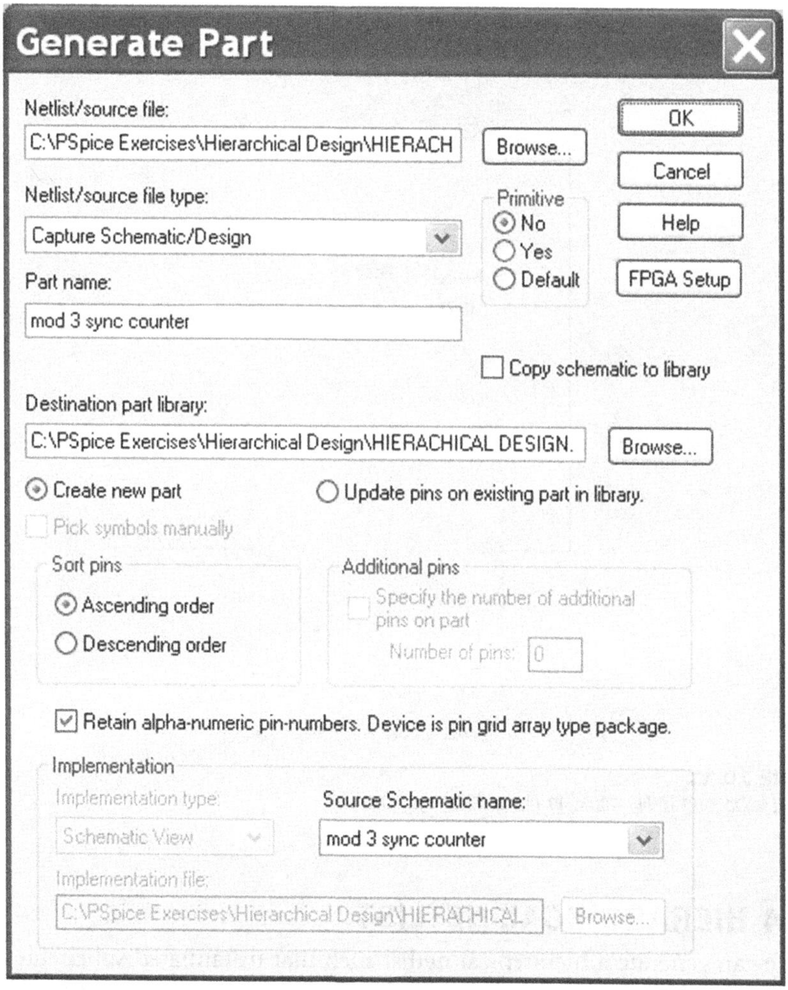

FIGURE 20.10
Generating a capture hierarchical symbol.

20.3 PASSING PARAMETERS

Parameters can be passed between levels of hierarchy using the **Subparam** part from the **Special** library. This allows different parameters to be passed to hierarchical blocks or symbols. For example, you may have a filter block where the gain of the filter is programmed by a single resistor. Using the **Subparam** part, different resistor values can be passed down to each filter to set up different filter gains. Figure 20.11 shows one such implementation where a different value for RVAL sets different filter gains for HB1 and HB2.

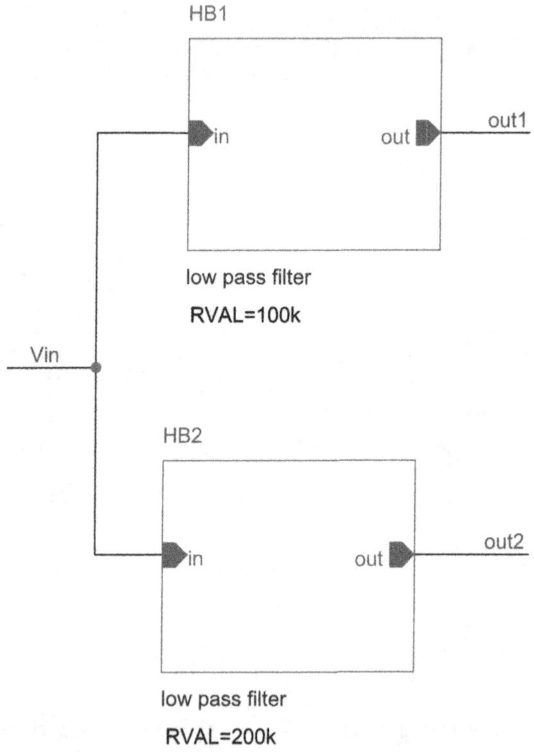

FIGURE 20.11
Passing subparam RVAL values to hierarchical symbols.

20.4 HIERARCHICAL NETLIST

PSpice can generate a hierarchical netlist such that instantiated subcircuit definitions will only appear once in the netlist. In Figure 20.12, there are two subcircuits declared at the top level, X_U1 and X_U2, which reference the subcircuit, osc125Hz. Rather than include the text for both subcircuit definitions in the netlist, only reference calls are made to the subcircuit.

```
* source HIERARCHY
V_V1          N00673 0 12V
V_V2          N02404 0 -12V
X_U1 OUT1 N00673 N02404 osc125Hz PARAMS: RVAL=160k
X_U2 OUT2 N00673 N02404 osc125Hz PARAMS: RVAL=160k

.SUBCKT osc125Hz OUT VCC VSS PARAMS: RVAL=160K
C_C1          N24151 0  0.01u IC=0 TC=0,0
R_R1          N24187 OUT  160k TC=0,0
R_R2          N24151 OUT  160k TC=0,0
X_U1A          N24187 N24151 VCC VSS OUT AD648A
R_R3          0 N24187  910k TC=0,0
.IC          V(N24151 )=0
.ENDS
```

FIGURE 20.12
Hierarchical netlist.

20.5 EXERCISES

Whenever you create a hierarchical design, it is recommended that no power supplies be included in underlying schematics. Power supply ports should be included on the hierarchical symbol and blocks such that the overall power supply connections are made at the top level where they are visible. The ground symbol, though, is global to all designs and so this does not need a port unless you are using separate digital and analog grounds.

Exercise 1

You will create a hierarchical top–down design shown in Figure 20.13, where the top block references an underlying bottom schematic. Hierarchical blocks cannot be saved to a library.

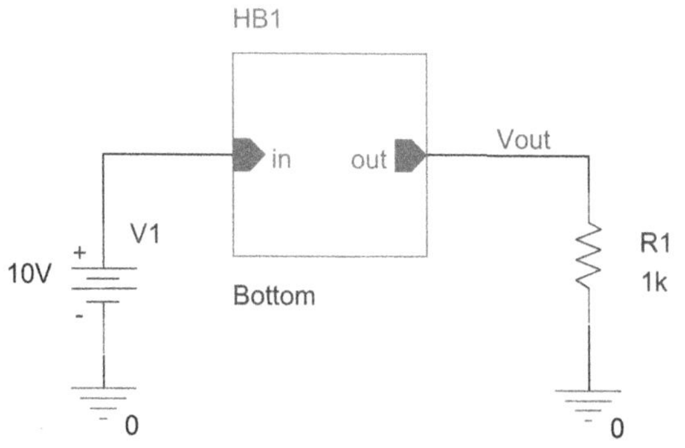

FIGURE 20.13
Block diagram project.

1. Create a new project called Top Down Design.
2. In the Project Manager, rename SCHEMATIC1 to Top.

3. Create a new schematic called bottom, highlight the Top Down Design.dsn file and **rmb > New Schematic** and name it Bottom.
4. In the Project Manager, select the schematic Bottom, **rmb > New Page** and accept the default Page1 name.
5. Your Project Manager should look like that shown in Figure 20.14.

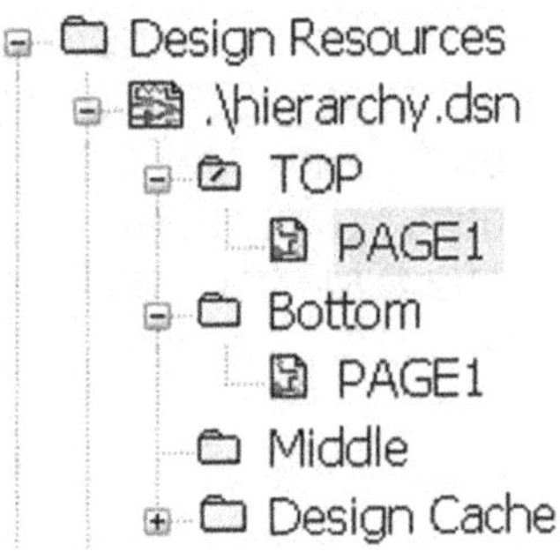

FIGURE 20.14
Hierarchical design folders.

6. Open the schematic for Bottom and draw the circuit in Figure 20.15. For the hierarchical ports, **Place > Hierarchical Port** and place a Hierarchical port (PORTRIGHT-L) on the output node **out** and a PORTRIGHT-R on nodes **in** and **out**. Save and close the schematic page.

FIGURE 20.15
Resistor potential divider.

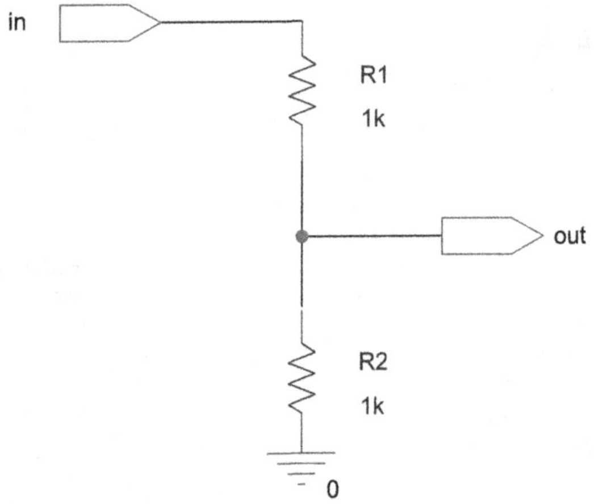

7. Open the schematic for the Top folder.
8. In the Project Manager, select **Place > Hierarchical Block** and enter a value for the reference designator. In **Implementation Type**, select **Schematic View** and enter **Bottom** as the **Implementation name** as shown in Figure 20.16. Leave the **Path and filename** blank as the schematic is part of the project. Click on OK.

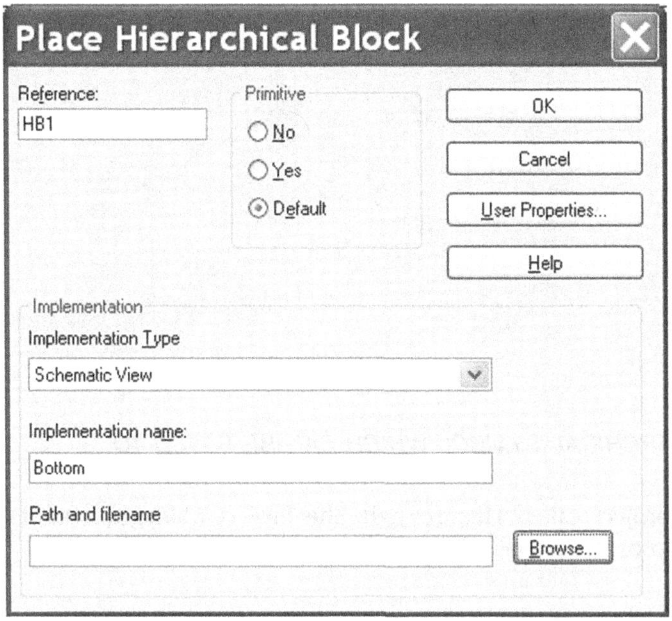

FIGURE 20.16
Defining the hierarchical block and the referenced schematic.

9. When the cursor changes to a cross-hair, mouse click once and then draw a rectangular block. The hierarchical port names on the bottom schematic will appear as hierarchical pins on the block. Move the pins as shown in Figure 20.17.

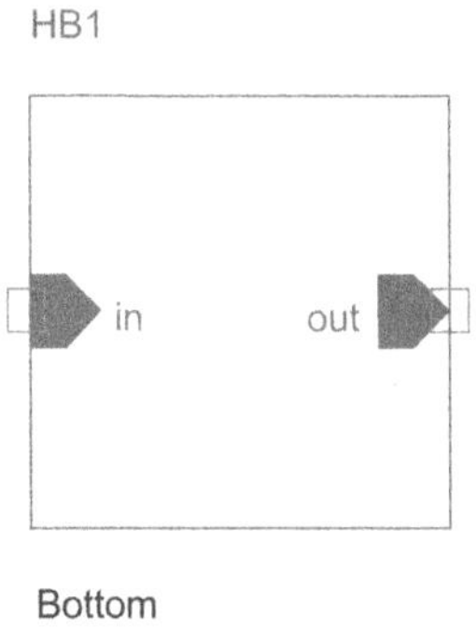

FIGURE 20.17
Hierarchical block with pins added.

10. Select the block and **rmb > Descend Hierarchy** (or double click on the block) and the underlying Bottom schematic will appear.
11. In the schematic, **rmb > Ascend Hierarchy** will take you back up to the top level.
12. Place a V_{DC} voltage source on the input and a resistor on the output as shown in Figure 20.18. Place a voltage marker on the output and run a bias simulation and confirm the output voltage of 5 V.

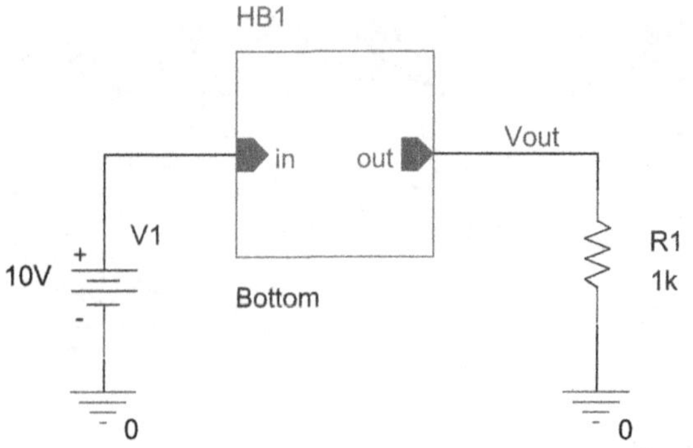

FIGURE 20.18
Testing hierarchical block.

Exercise 2

CREATING A HIERARCHICAL SYMBOL WHICH CAN BE SAVED TO A LIBRARY

1. Create a new project called Hierarchy. In the Project Manager, rename SCHEMATIC1 to osc125Hz (Figure 20.19).

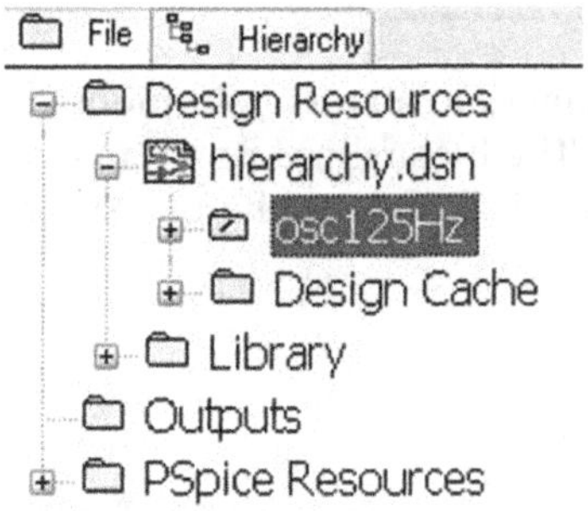

FIGURE 20.19
Rename SCHEMATIC1 folder to osc125Hz.

2. Draw the circuit in Figure 20.20. The AD648 opamp is from the opamp library. If you are using the eval version, use the uA741 opamp from the eval library. For the hierarchical ports, **Place > Hierarchical Port** (Figure 20.21) and place a hierarchical port (PORTRIGHT-L) on the output node **out** and a PORTRIGHT-R for the VCC and VSS power supply connections. Name the power ports VCC and VSS, respectively, and draw a short wire to each port as shown in Figure 20.20.
   The wires connected to the hierarchical ports for VCC and VSS will automatically be named VCC and VSS, respectively, and hence will be connected to the opamp power pins.
   In order for the circuit to oscillate, an initial condition of 0 V needs to be placed on the capacitor. Place an IC1 from the special library on the capacitor C1, as shown in Figure 20.20.

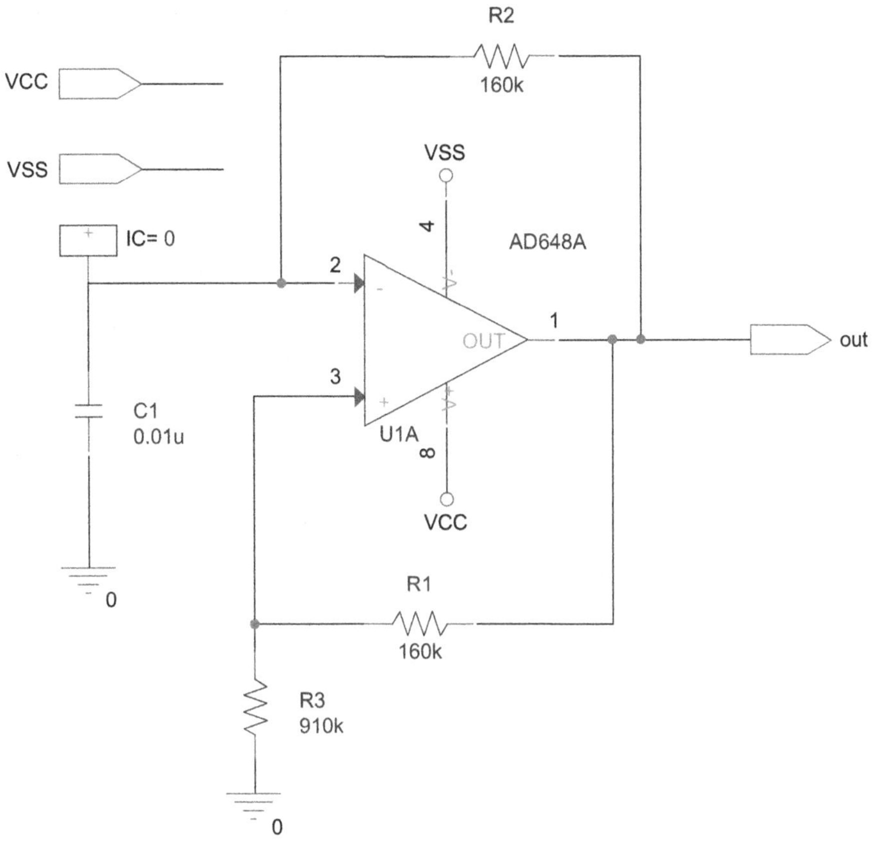

FIGURE 20.20
125Hz oscillator.

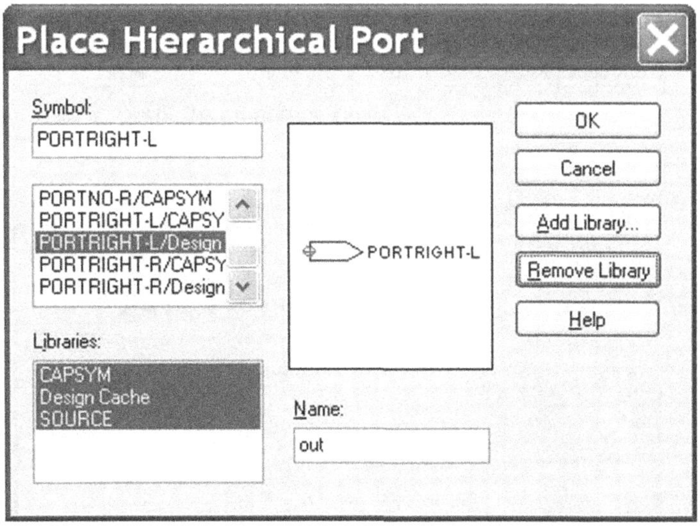

FIGURE 20.21
Hierarchical ports.

TIP

When you first place the hierarchical ports and the net name is some distance away from the port, rotate the port four times and the net name will be placed closer to the port.

3. Save the Project.
4. Create a hierarchical symbol by highlighting the dsn file (Hierarchy.dsn) in the Project Manager and select **Tools > Generate Part**.
5. In the **Generate Part** window, select **Netlist/source file type**: to **Capture Schematic/Design**. Then in the **Netlist/source file**, browse to the **Hierarchy** project folder and select the Hierarchy.dsn file. Do not exit yet.
6. You need to create a library for the new hierarchical part, so in the **Destination Part Library**, call the library Hierarchy and save it in the PSpiceExercises folder or a folder of your choice.
7. Note that the **Source Schematic name** now shows the schematic name osc125Hz rather than SCHEMATIC1. Your **Generate Part** window should look similar to that shown in Figure 20.22.

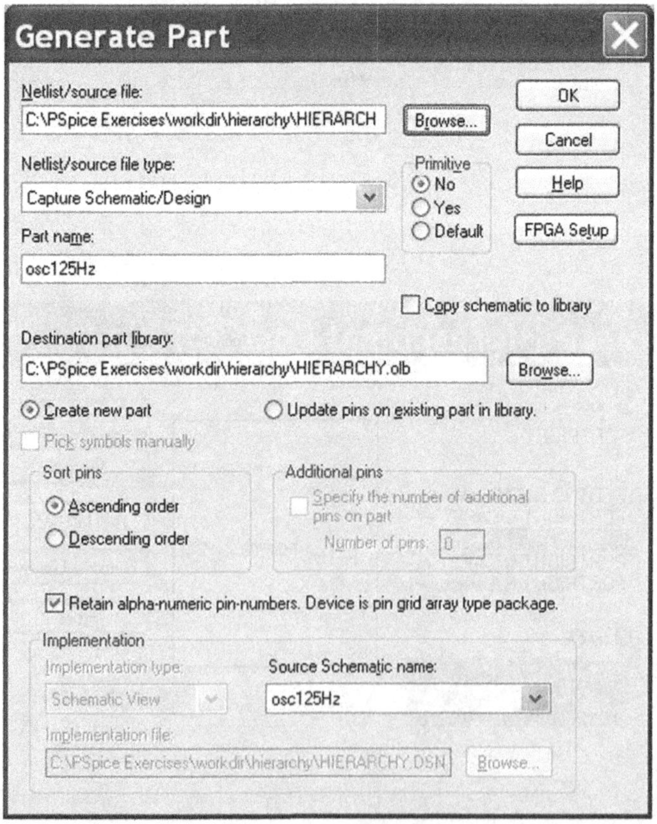

FIGURE 20.22
Generating a hierarchical symbol for the osc125Hz schematic.

8. If you are using version 16.3, the **Split Part Section Input spreadsheet** will open. Click on **Save** and OK.
9. In the Project Manager, you should see the hierarchy.olb library file added to the **Outputs** folder. Move the hierarchy.olb from the **Outputs** folder (or you can cut and paste) to the **Library** folder as shown in Figure 20.23 and expand the library to see the osc125Hz library part.

FIGURE 20.23
The Hierarchy.olb Capture library has been added to the Library folder.

10. Double click on the osc125Hz part in the Hierarchy library, which will open the Part Editor and display the part as shown in Figure 20.24.

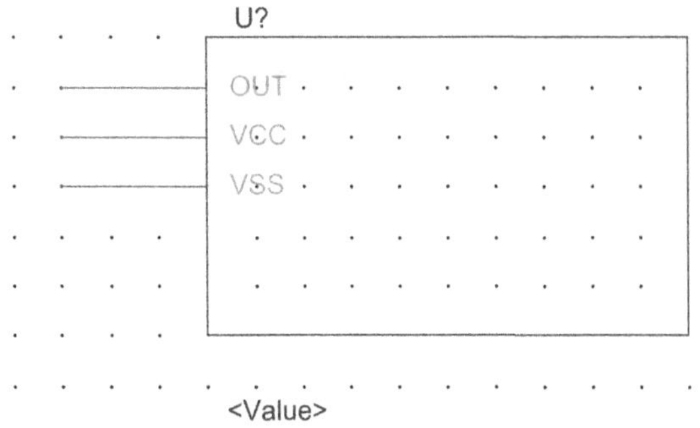

FIGURE 20.24
The generated osc125Hz symbol.

11. Double click on **<Value>** and enter a Value of osc125Hz as shown in Figure 20.25.

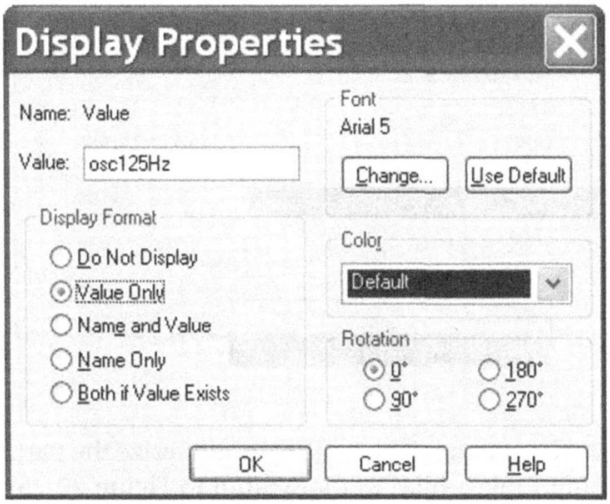

FIGURE 20.25
Entering the displayed value property of osc125Hz.

12. Double click on the pin connected to OUT, and in **Pin Properties**, select the **Shape** as Short, **Type** as Output and pin **Number** as 1 (Figure 20.26) and click on OK.

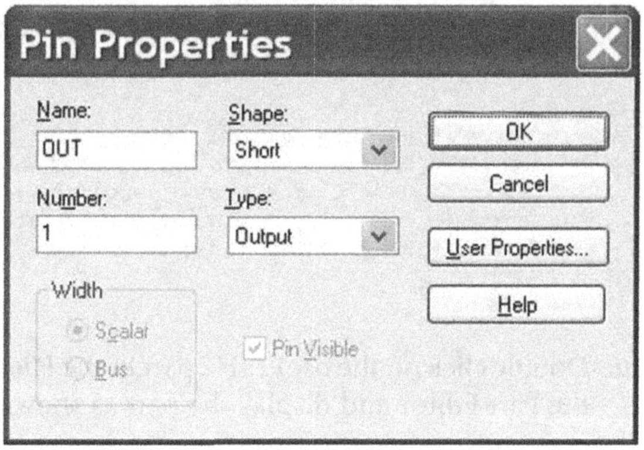

FIGURE 20.26
Pin properties for the osc125Hz.

13. Change the VCC pin shape to **Short** of type **Input** and pin number 2.
14. Change the VSS pin shape to **Short** of type **Input** and pin number 3.
15. Double click anywhere in the Part Editor to display the **User Properties** box (Figure 20.27). Highlight **Pin Numbers Visible** and select True from the pull-down menu. Click on OK.

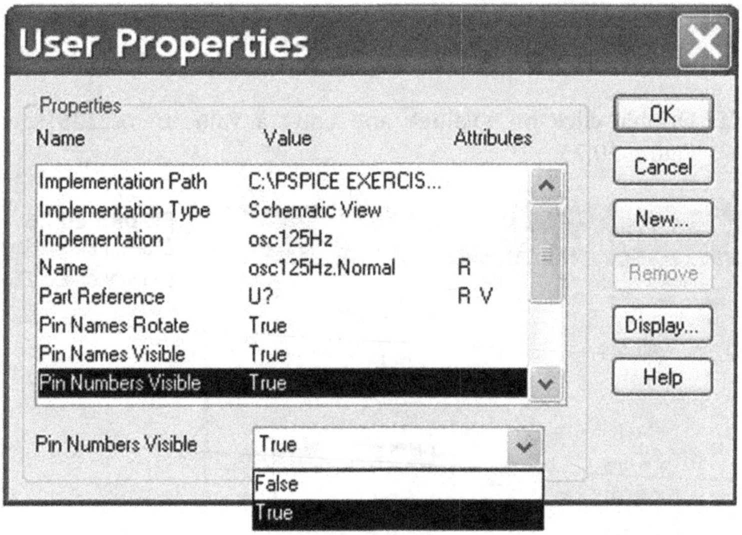

FIGURE 20.27
User properties.

16. Move the output pin from the left- to the right-hand side and resize the part. Your osc125Hz symbol should look similar to the symbol in Figure 20.28.

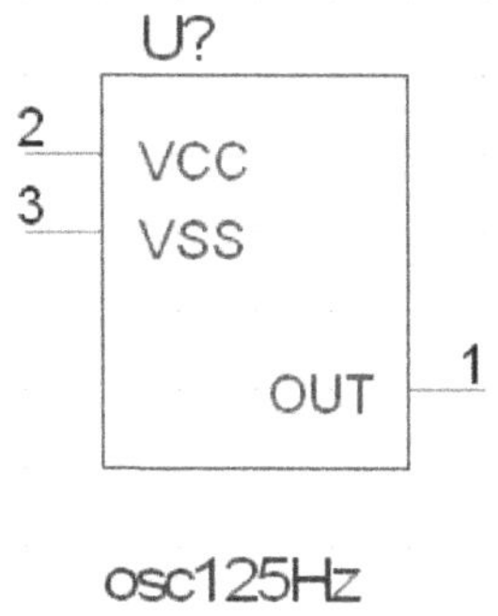

FIGURE 20.28
Modified osc125Hz symbol.

17. Double click anywhere in the Part Editor to display the User Properties window, which displays the generated properties. Click on OK. Close the Part Editor by closing the Hierarcy.olb window (click on the upper right-hand cross in Windows).

Exercise 3

TESTING THE OSCILLATOR IN A HIERARCHY

1. We need to test the osc125Hz hierarchical symbols. Create a new schematic by selecting **Hierarchy.dsn** and **rmb > New Schematic** and name it **Test Osc125Hz**. By default, the new schematic is named SCHEMATIC1. Highlight SCHEMATIC1 and **rmb > Rename** to **Test Osc125Hz**.
2. Highlight **Test Osc125Hz** and **rmb > New Page**, accept the name **PAGE1** and click on OK. Your Project Manager window should look like that in Figure 20.29.

FIGURE 20.29
Project Manager schematic folders.

3. Double click on PAGE1 in Test Osc125Hz and draw the circuit diagram in Figure 20.30. The osc125Hz can be found in the **Hierarchy** library. Make sure V1 is 12 V and V2 is −12 V.

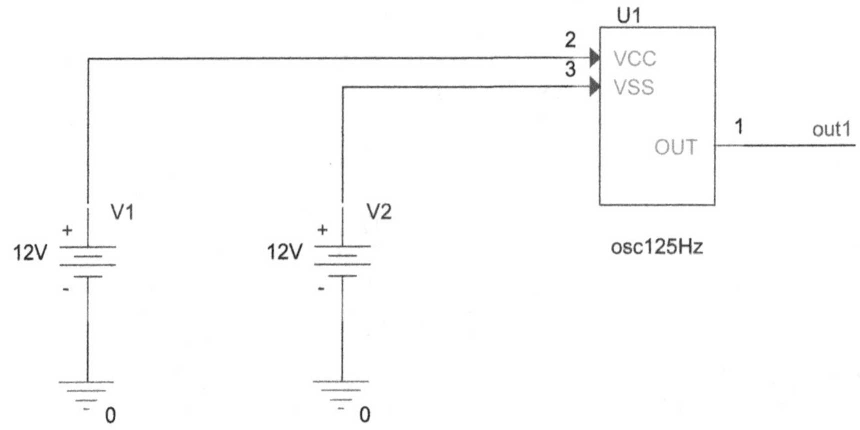

FIGURE 20.30
Test circuit for osc125Hz.

4. Create a simulation profile called transient and run the simulation for 200 ms.
5. Try to place a voltage marker on the net **out1**. You will get a message as shown in Figure 20.31.

FIGURE 20.31
Hierarchy message.

NOTE

The Test Osc125Hz should be at the top level known as the root schematic. You can run a simulation on the Osc125Hz oscillator but not on the Test Osc125Hz as it is not in the hierarchy. This is an ideal way to progressively simulate a circuit from the bottom upwards to test each level of the hierarchy.

6. In the Project Manager, highlight the **Test Osc125Hz** folder and **rmb > Make Root**. You will get another message saying that the design must be saved first.
7. Save the design and highlight the **Test Osc125Hz** and **rmb > Make Root**. This time the **Test Osc125Hz** will be placed at the top of the hierarchy and a slash symbol will appear in the yellow folder (Figure 20.32). If you do not see the response as shown, check the power supplies and make sure that you placed an initial condition on C1.

FIGURE 20.32
Making Test Osc125Hz the root schematic.

8. You will have to create another simulation profile before you place the voltage marker.
9. Run the simulation. You should see the oscillator response as in Figure 20.33.

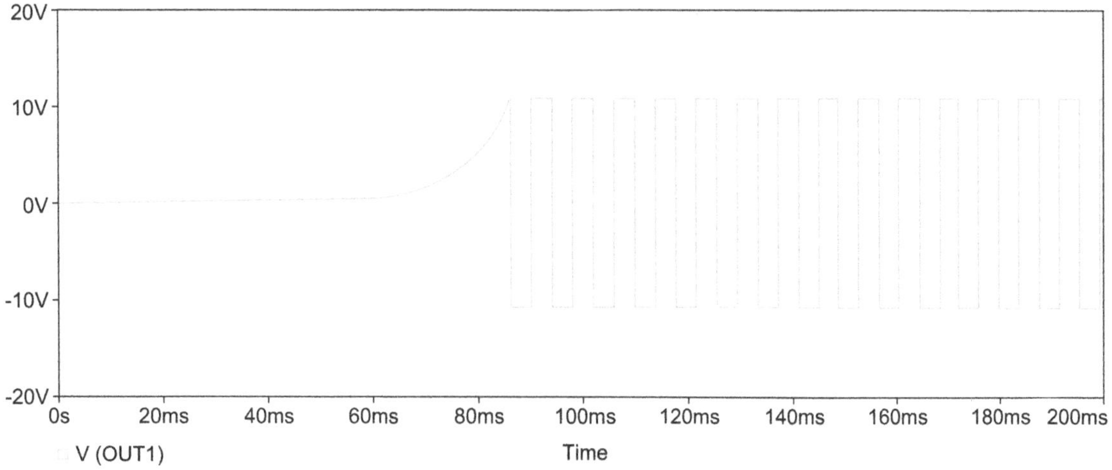

FIGURE 20.33
Output of osc125Hz.

Exercise 4

You will create a subparameter in the osc125Hz circuit which will appear on the hierarchical symbol such that a value can be entered on the symbol and passed down to the schematic.

1. Modify the osc125Hz circuit as shown in Figure 20.34. The **subparam** part can be found in the **special** library.
2. Double click on the subparam parameter and in the Property Editor, add a new row (or column) adding RVAL with a default value of 160 k. Select

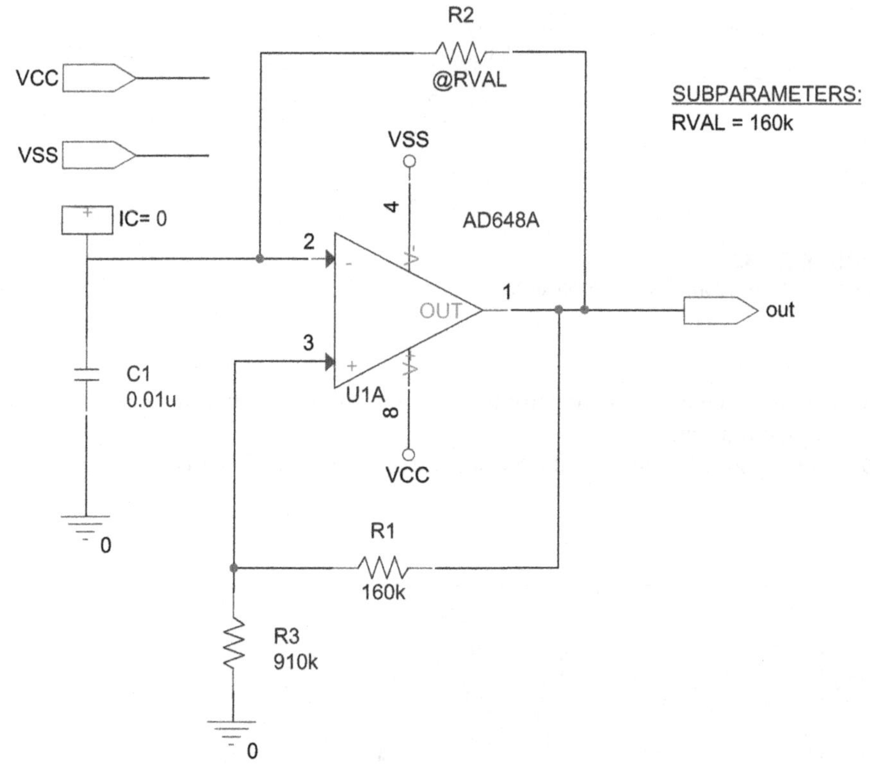

FIGURE 20.34
Making R2 a hierarchical parameter.

RVAL and **rmb > Display** (Figure 20.35). In the **Display Properties** window, select **Name and Value** (Figure 20.36).

3. In the schematic, replace the value of R2 with @RVAL as shown in Figure 20.34.
4. Save the schematic.
5. Generate a Part for the new osc125Hz as in Exercise 2 and modify the pins in the Part Editor as described in Exercise 2.

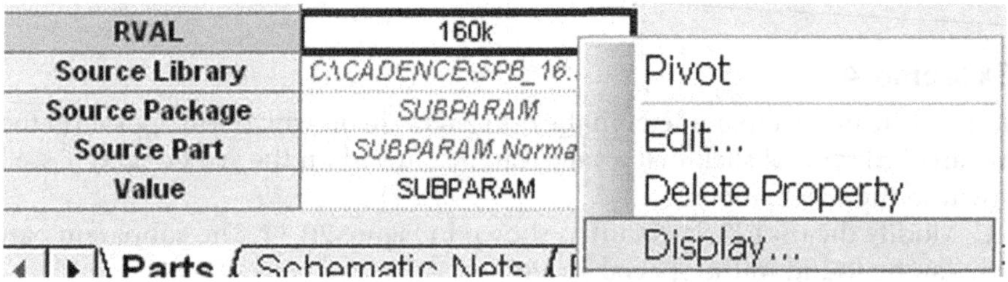
RVAL
160k
Source Library
C:\CADENCE\SPB_16.
Source Package
SUBPARAM
Source Part
SUBPARAM.Norma
Value
SUBPARAM
Pivot
Edit...
Delete Property
Display...
Parts
Schematic Nets

FIGURE 20.35
Creating an RVAL property with a value of 160 k.

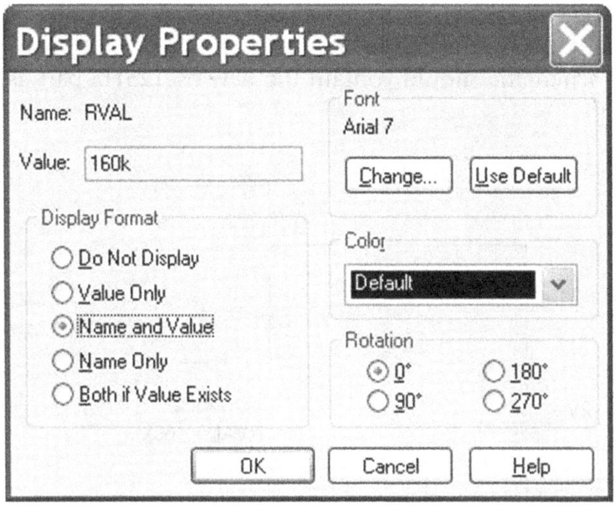

FIGURE 20.36
Displaying name and property of RVAL.

NOTE

The schematic will still contain the previous osc125Hz part, so you will need to delete the old part and replace it with the new part. Alternatively, you can update the part in the Design Cache. The Design Cache is effectively a library which contains all the parts in the schematic. When you delete a part from the schematic, the Design Cache still holds that part until you select **Cleanup Cache**. The Design Cache can be used to update parts and replace parts. From version 16.6, the **Replace Cache** can now be applied to more than one part. Multiple parts can be selected by holding down the Control or Shift key.

---

6. In the Project Manager, expand the Design Cache and select the osc125Hz (Figure 20.37).

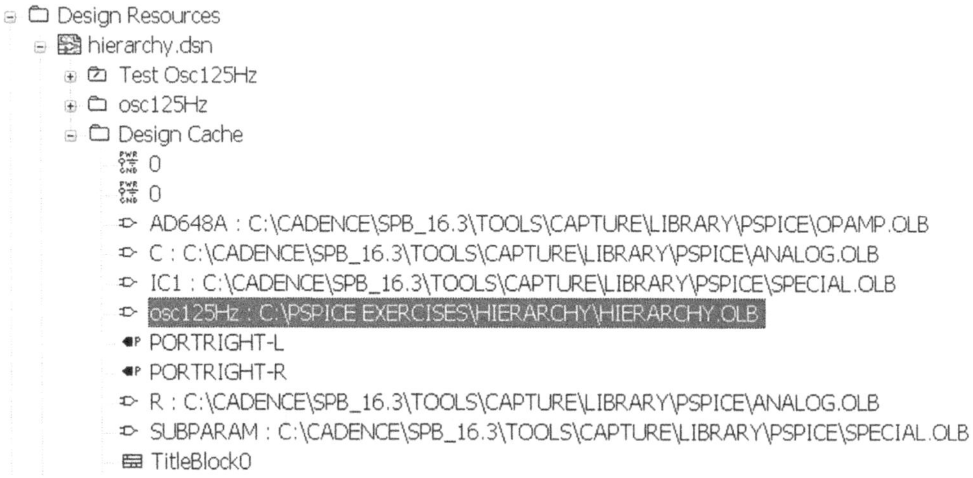

FIGURE 20.37
Design cache showing parts in the schematic.

7. Select **rmb > Update Cache**. Click on YES to Update Cache and click on OK when asked if you want to save the design.
8. Your Test Osc125Hz schematic should contain the new osc125Hz part as shown in Figure 20.38.

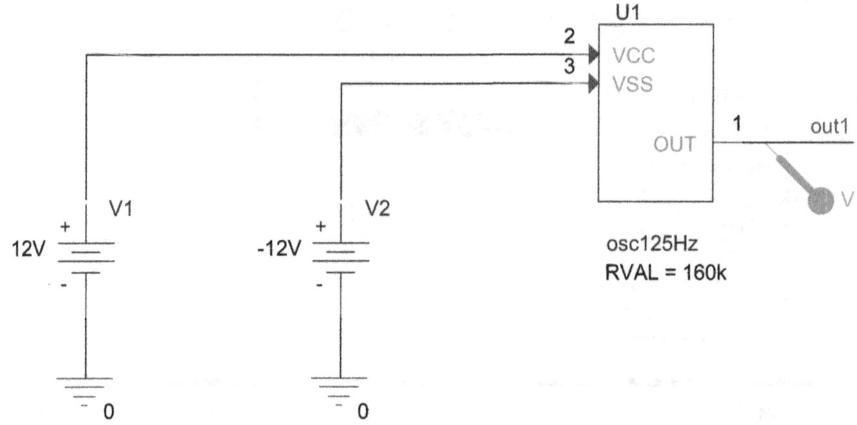

FIGURE 20.38
Test circuit with modified osc125Hz part.

9. Modify the RVAL to 100k and run the simulation. You should see a different oscillator period displayed in Probe, as shown in Figure 20.39.

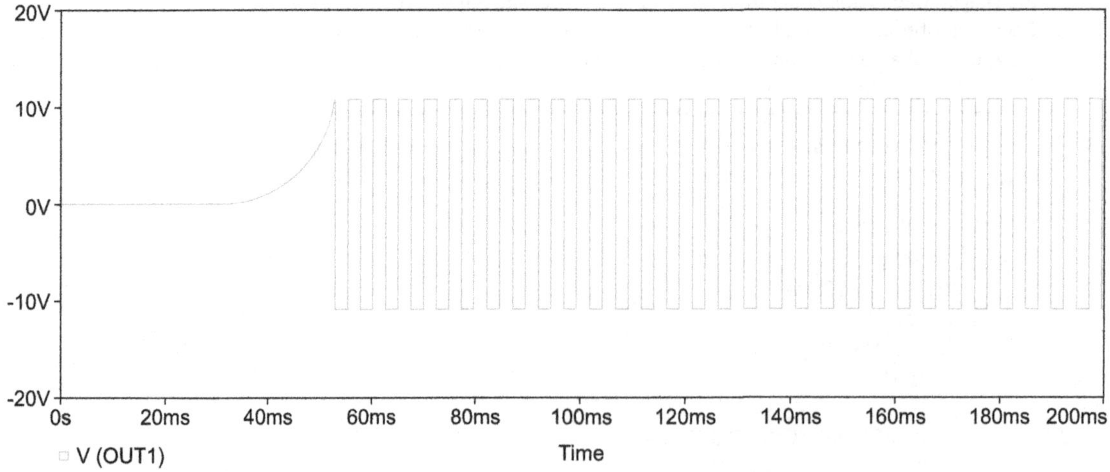

FIGURE 20.39
Using a value of 100k for the oscillator.

Exercise 5

The **Digital Counter** hierarchical design presented here will be used in the Test Bench in Chapter 22.

Figure 20.40 shows a hierarchical design put together based on the 555 clock oscillator in Chapter 19 connected to the mod 3 sync counter in Chapter 18.

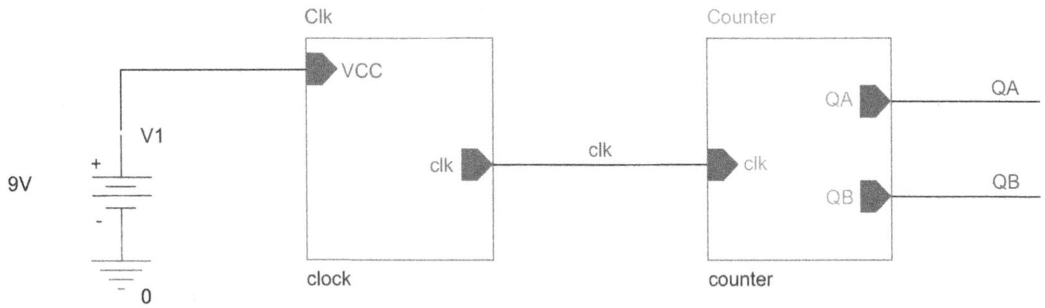

FIGURE 20.40
Hierarchical design of digital counter.

The modified clock and counter circuits that will be used are shown in Figures 20.41 and 20.42, respectively.

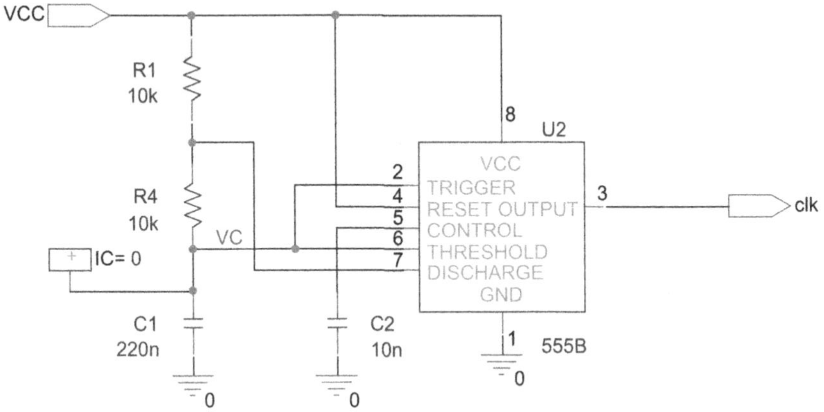

FIGURE 20.41
Clock oscillator.

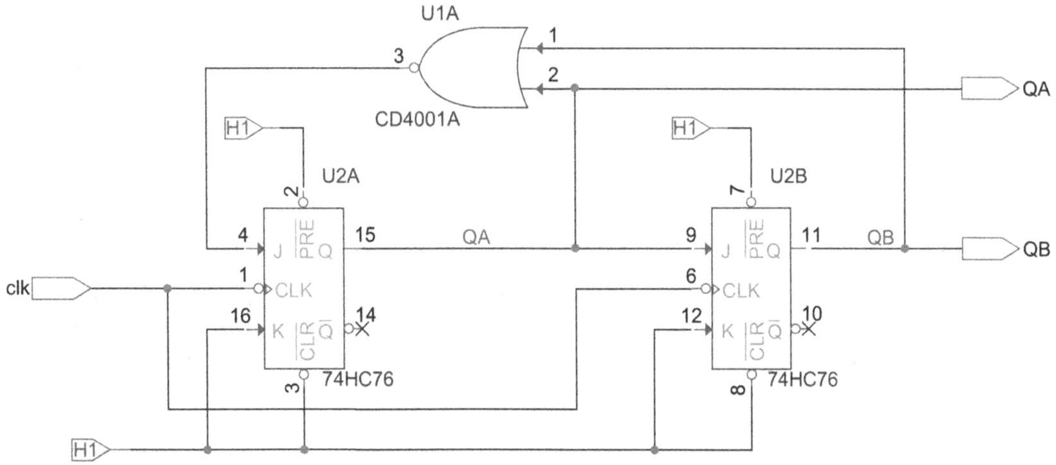

FIGURE 20.42
Modulus 3 counter.

1. Create a new project **Digital Counter** and rename SCHEMATIC1 to **Digital Counter**.
2. Open the project **Clock Oscillator** from Chapter 19, Exercise 2.
3. Place the two Project Managers side by side. You will copy and paste the **clock** schematic folder from the **Clock Oscillator** Project Manager to the **Digital Counter** Project Manager.
   Highlight the **clock** schematic folder and press control-C. Highlight **Digital Counter.dsn** and press control-V. Close the Clock Oscillator project.
   If you did not rename SCHEMATIC1 in the **Clock Oscillator** to clock, select **rmb > Rename**. Close the **Clock Oscillator** project.
4. Open the **Mod 3 Counter** project in Chapter 18, Exercise 1, and as in Step 3, copy the **counter** schematic to the **Digital Counter** Project Manager. Close the **Mod 3 Counter** project. Your Project Manager should appear as in Figure 20.43.

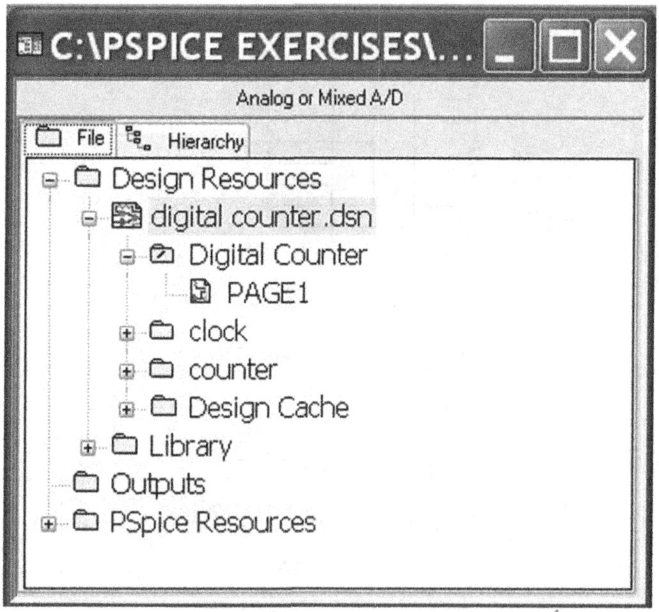

FIGURE 20.43
Clock and counter schematics added to the digital counter project.

You will create a hierarchical design with two hierarchical blocks at the top level referencing the clock and counter circuits as seen in Figures 20.40 and 20.41. Hierarchical ports will be added to the clock and counter circuits such that hierarchical pins will automatically be added when the hierarchical blocks are created.

5. Open the **clock** schematic and delete the 9 V voltage source, V1, and its associated 0 V ground symbol. Delete the load resistor connected to pin 3 of the 555 and delete the net name **out** (if added).

6. Select **Place > Hierarchical Port**. Select the PORTRIGHT-L and name it **clk** (Figure 20.44). Click on OK and add the port to the clock output node. Place a hierarchical port for VCC on the VCC node. It really is your choice as to which type of hierarchical port symbol you use for input and output ports. Your circuit should be similar to that in Figure 20.41.

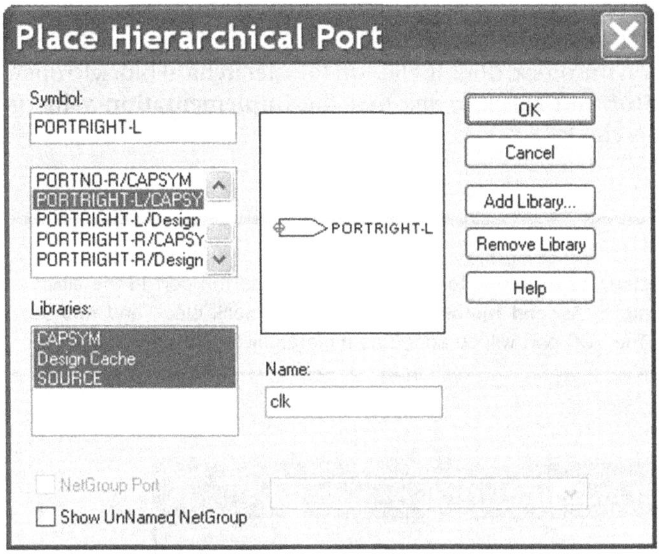

FIGURE 20.44
Adding a hierarchical port and naming it clk.

7. Double click on the **clk** port to open the **Property Editor**. Click on the pull-down menu for the **Type** property value and select **Output** (Figure 20.45). Close the **Property Editor**. The VCC port will be of type **input** by default. Close and save the schematic.

NOTE

By default, hierarchical input ports are placed on the left-hand side and output ports on the right-hand side of hierarchical blocks. You can always change the port types by double clicking on the hierarchical block pins.

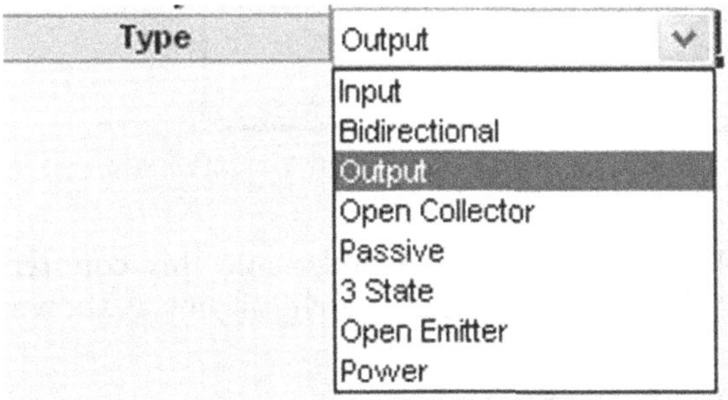

FIGURE 20.45
Hierarchical port types.

8. Open the counter schematic and remove the digital clock source. Place hierarchical ports on the clk, QA and QB wires as shown in Figure 20.42. As in Step 7, change the QA and QB hierarchical ports to type **output**. Close and save the schematic.
9. Open the top-level **Digital Counter** schematic, select **Place > Hierarchical Block** and enter the details as shown in Figure 20.46. Leave the **Path and filename** empty. Click on OK and draw a rectangular block. The VCC and clk pins will appear as seen in Figure 20.40.
   If no pins appear on the block, double click on the hierarchical block to open the **Property Editor** and check to see that the **Implementation** value is named correctly as **clock**.

NOTE

If you have forgotten, for example, to add a VCC port, add the port in the clock schematic and **rmb** > **Ascend Hierarchy**. Highlight the clock block and **rmb** > **Synchronize Up**. The VCC port will be added as a hierarchical pin to the block.

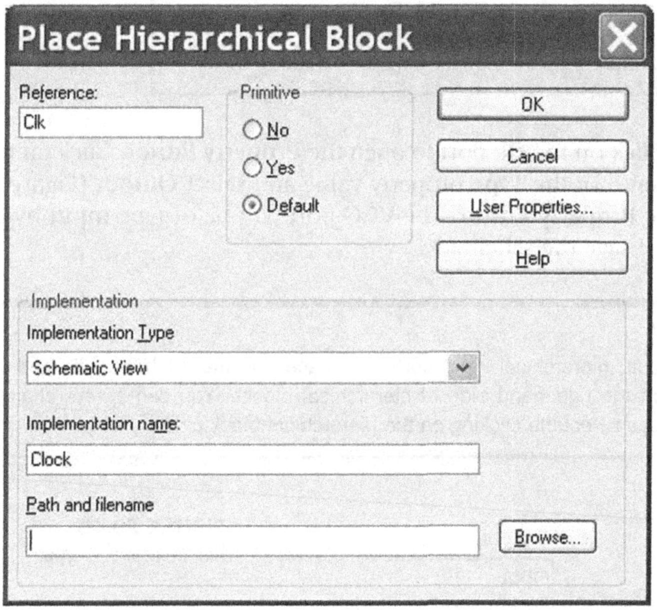

FIGURE 20.46
Creating a hierarchical block.

10. Draw another hierarchical block as in Step 9 and name this **counter** (Figure 20.47). Click on OK and position the hierarchical pins as shown in Figure 20.40.
11. Complete the circuit as in Figure 20.40.

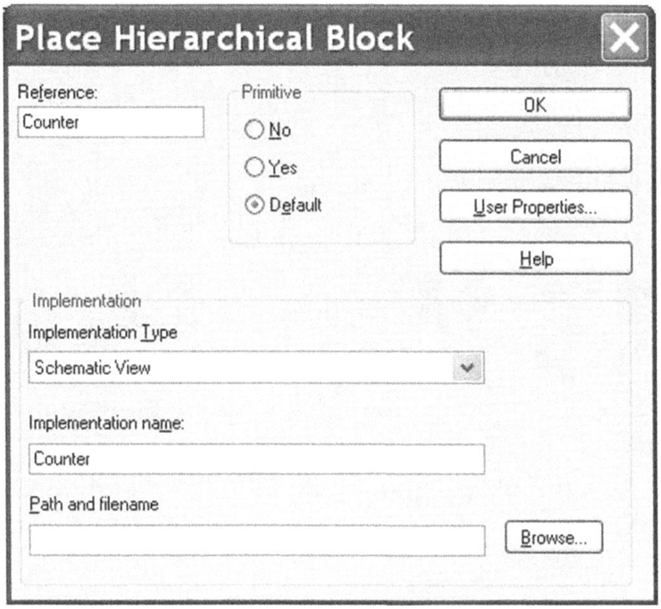

FIGURE 20.47
Creating a hierarchical block for the counter circuit.

CHAPTER 21

Magnetic Parts Editor

Chapter Outline

The Magnetic Parts Editor (MPE) is used for the design of transformers and inductors in switched mode power supply topologies. In particular, the MPE provides a complete transformer design cycle for forward converters, both single and double switch, and a flyback converter operating in discontinuous conduction mode. At the end of the design cycle, the MPE generates a comprehensive data sheet which manufacturers can use for the fabrication of transformers and inductors. The MPE also generates the inductor and transformer PSpice simulation models.

Included with the MPE is a database of commercially available magnetic parts such as wire types, insulation material, bobbins and magnetic cores. You can add to this database by creating your own magnetic parts.

21.1 DESIGN CYCLE

The design cycle consists of a series of design steps which are numbered as you progress through the design. The design steps for a DC–DC converter using the flyback topology will be presented here as an example.

Analog Design and Simulation using OrCAD Capture and PSpice. DOI: 10.1016/B978-0-08-097095-0.00021-0

The specifications are:

- DC input minimum: 50 V
- DC output 12 V with less than 100 mV pk-pk ripple
- DC output 0.5 A with less than 5 mA pk-pk ripple
- switching frequency: 40 kHz
- efficiency: 75%
- maximum duty cycle: 45%

The MPE is started from the PSpice Accessories menu.

21.2 EXERCISES

Exercise 1

Start the MPE from the Start menu. **Start > All Programs > Cadence (OrCAD) <release number> PSpice Accessories > Magnetic Parts Editor**

STEP 1: COMPONENT SELECTION

The first design step in MPE is the selection of one of the components (topologies) to be designed (Figure 21.1), in this case a flyback converter

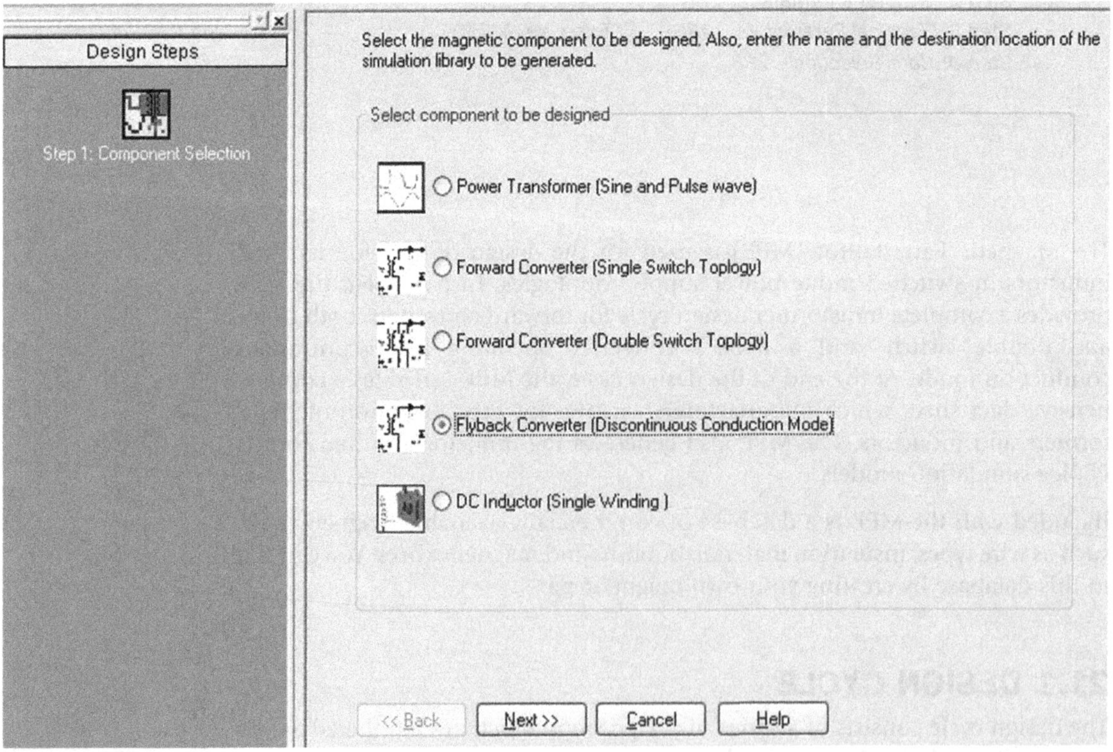

FIGURE 21.1
Topology selection.

(discontinuous conduction mode). Select **File > New** and the Component Selection window will open as shown in Figure 21.1.

On the left-hand side of the window are the completed design steps, which allow you to keep track of which step in the design cycle you are at. You can click on any design step at any time to go back to review the parameters entered. On the right-hand side are the available design components (topologies).

Select **Flyback Converter** and click on **Next >>**.

STEP 2: GENERAL INFORMATION

The second step is to enter the design specifications (Figure 21.2) for the transformer. In this case there will be one secondary; the insulation material will be nylon with a current density of 3 A/mm^2. The efficiency, from the given specifications, is 75%.

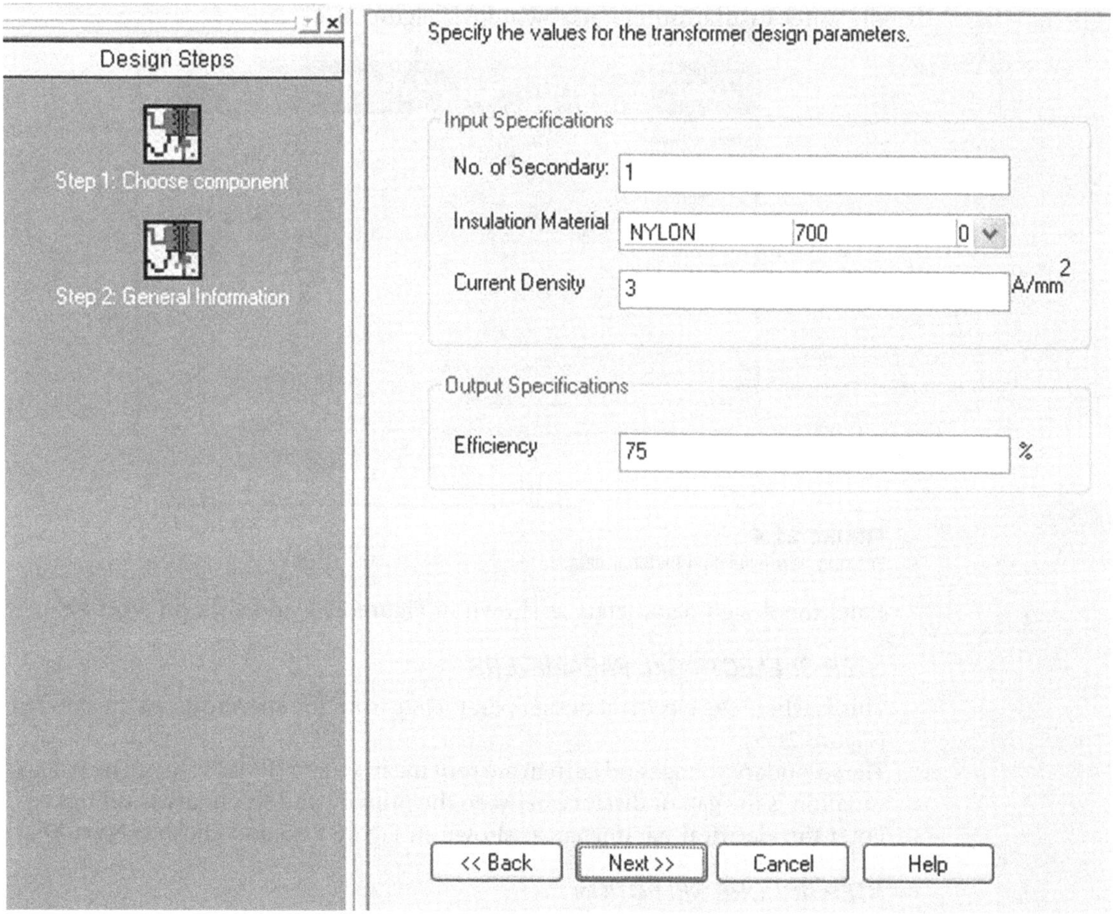

FIGURE 21.2
Transformer design parameters.

The maximum number of secondary windings is nine, but for forward converters only one secondary winding is allowed. You have the option to select the transformer insulation, which by default is nylon. The other provided materials are shown in Figure 21.3.

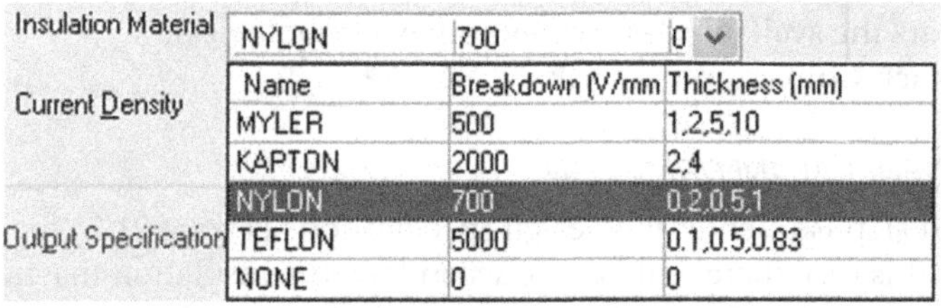

Insulation Material
NYLON
700
0
Current Density
Output Specification
Name
Breakdown (V/mm
Thickness (mm)
MYLER
500
1,2,5,10
KAPTON
2000
2,4
NYLON
700
0.2,0.5,1
TEFLON
5000
0.1,0.5,0.83
NONE
0
0

FIGURE 21.3
Insulation materials available in the installed database.

You can enter your own insulation materials: **Tools > Data Entry > Insulation** in the **Enter insulation material** window (Figure 21.4).

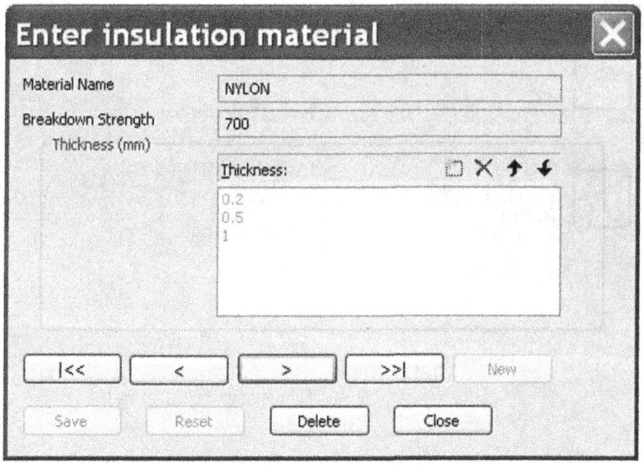

FIGURE 21.4
Entering new insulation material data.

Enter the design parameters as shown in Figure 21.2 and click on **Next >>**.

STEP 3: ELECTRICAL PARAMETERS

This is where the electrical design parameters from the specifications are entered (Figure 21.5).
The secondary voltage and current are root mean square (RMS) values. The voltage isolation is the gap or distance between the primary and secondary windings.
Enter the electrical parameters as shown in Figure 21.5 and click on **Next >>**.

STEP 4: CORE SELECTION

Selection of the magnetic core for the transformer depends on the shape and material. The physical diagram shown for the core in Figure 21.6 is updated when you select another shape such as a toroid, EE or UU.

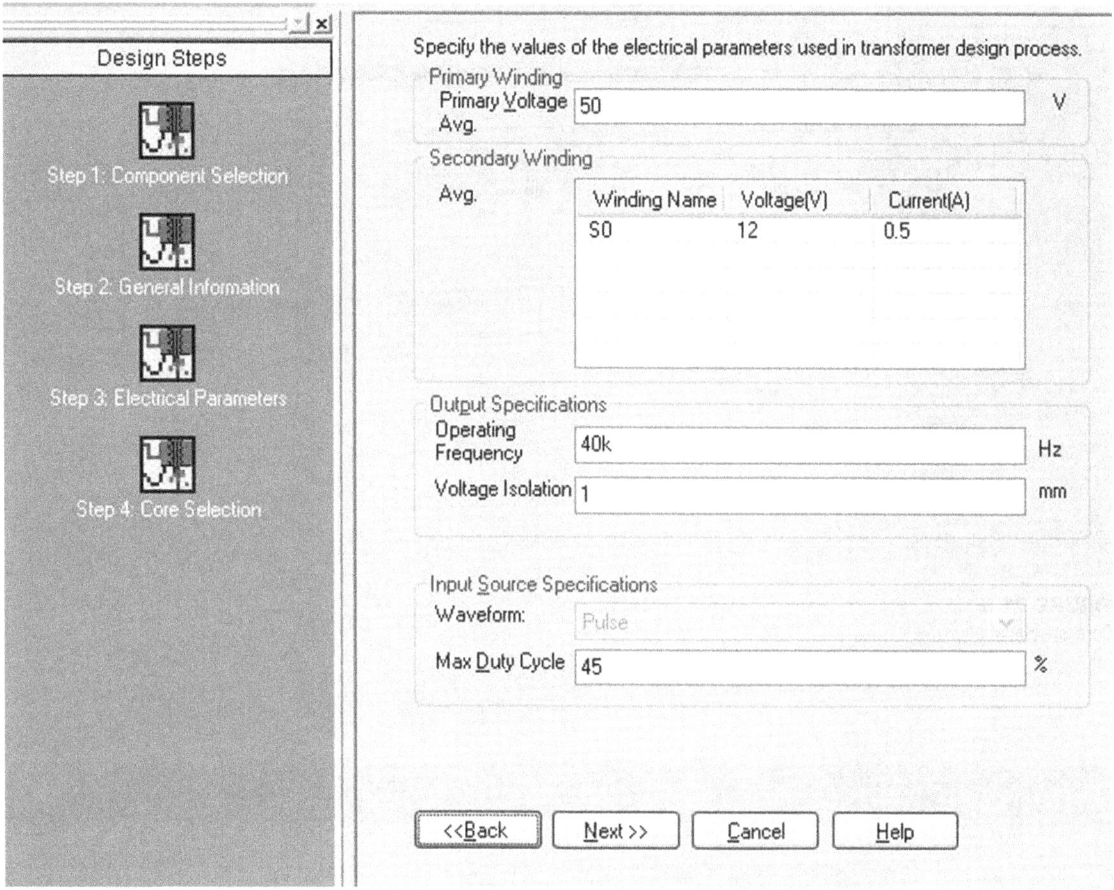

FIGURE 21.5
Specifying the electrical parameters.

Change the Vendor name to Ferroxcube and from the **Family Name** pull-down menu, select Toroid and UU to view the respective geometrical data for the cores. Change the **Family Name** back to EE.

You can enter other manufacturer's core data by selecting **Tools > Data Entry > Core Details > Core** (Figure 21.7).

You can also enter core material data by selecting **Tools > Data Entry > Core Details > Material** (Figure 21.8).

Initially, you will specify a manufacturer's magnetic core material and then let MPE propose a magnetic core using the specified core material. The MPE will determine whether the coil windings will fit on the core.

The MPE core database includes the Ferroxcube range of magnetic cores. For this design, the Ferroxcube EE low-power (10 W) grade 3C81 material will be used as a starting point. These cores are specified for a 67 kHz switching frequency and an output voltage up to 12 V:

Set the **Vendor Name** to Ferroxcube.

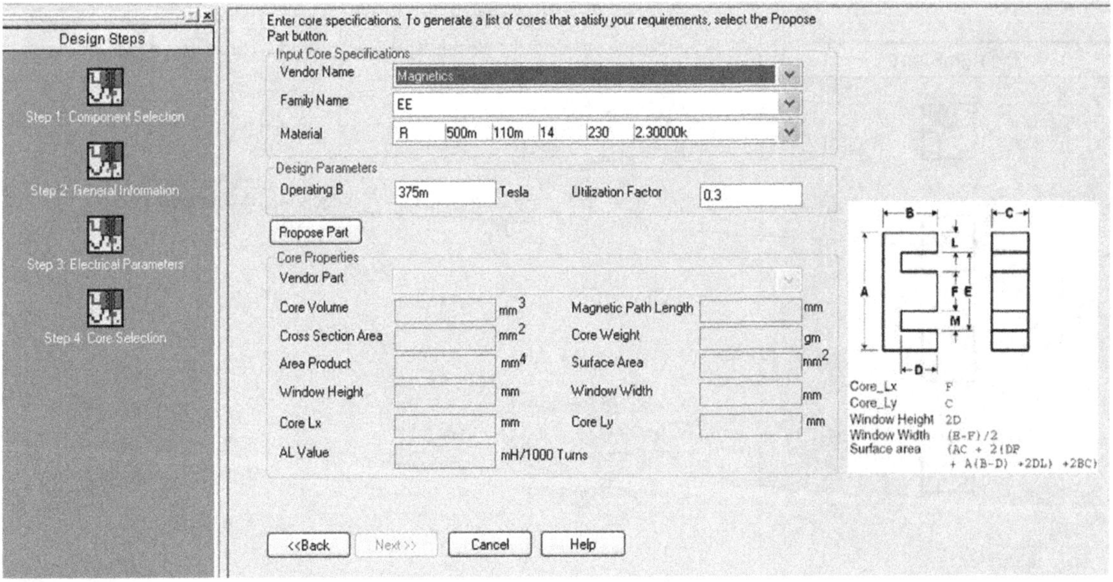

FIGURE 21.6
Core specifications.

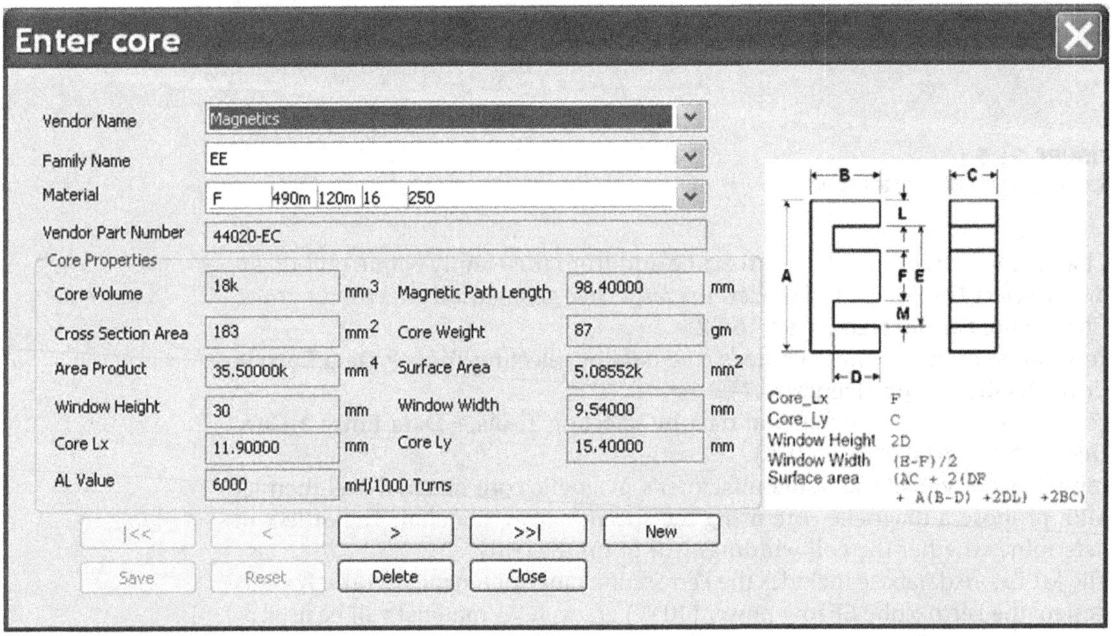

FIGURE 21.7
Creating a magnetic core.

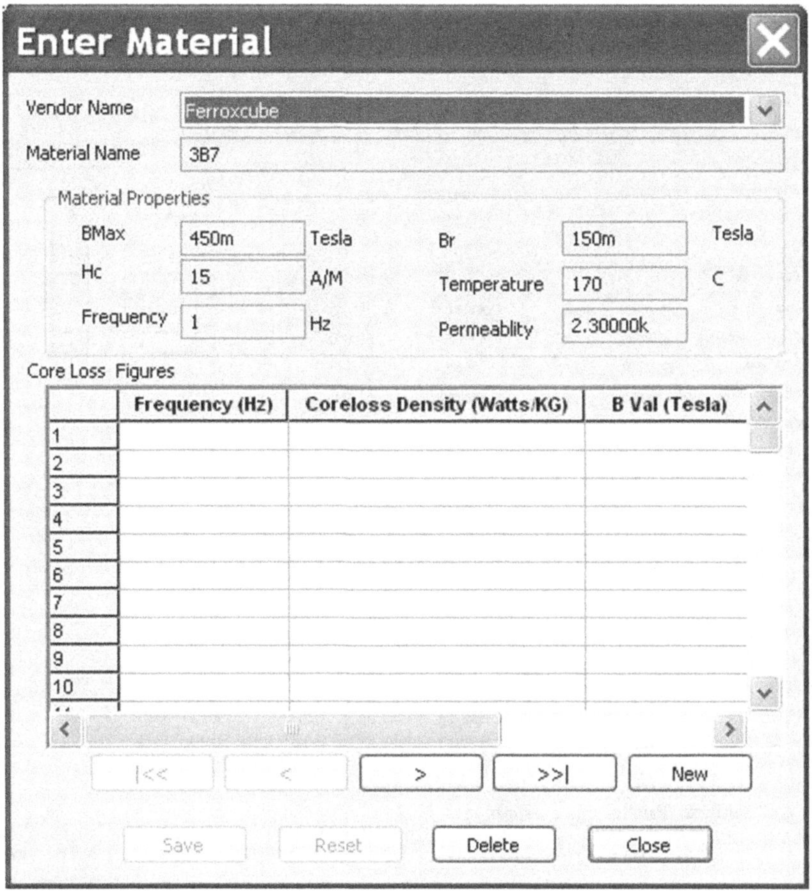

FIGURE 21.8
Creating the magnetic core material.

Set the **Family Name** to EE.
Set the **Material** to 3C81.

Click on **Propose Part**.
The MPE will return a suitable Vendor Part and will show the respective physical dimensions for the core. The pull-down menu for **Vendor Part** contains a list of other suitable cores from the database that will also meet your specification.
In this example, the core E13_6_6 will have been selected, as shown in Figure 21.9.
Click on **Next >>**.

STEP 5: BOBBIN-WINDING SELECTION

This step selects the bobbin which fits on the core and on which the wire is wound. In Figure 21.10, the Bobbin Part No. is shown as NO_NAME, which

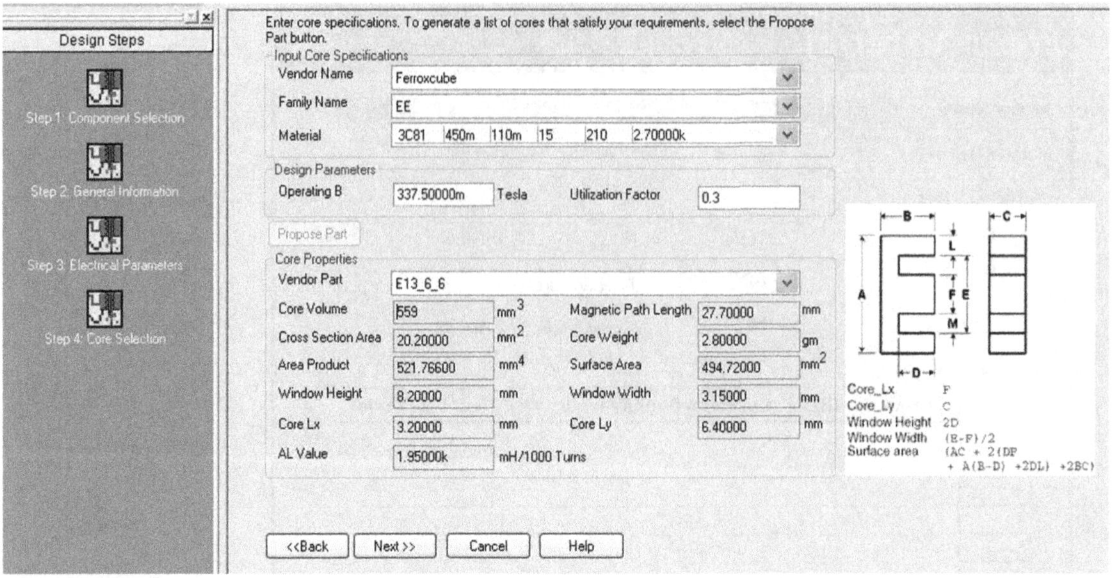

FIGURE 21.9
Bobbin and wire properties.

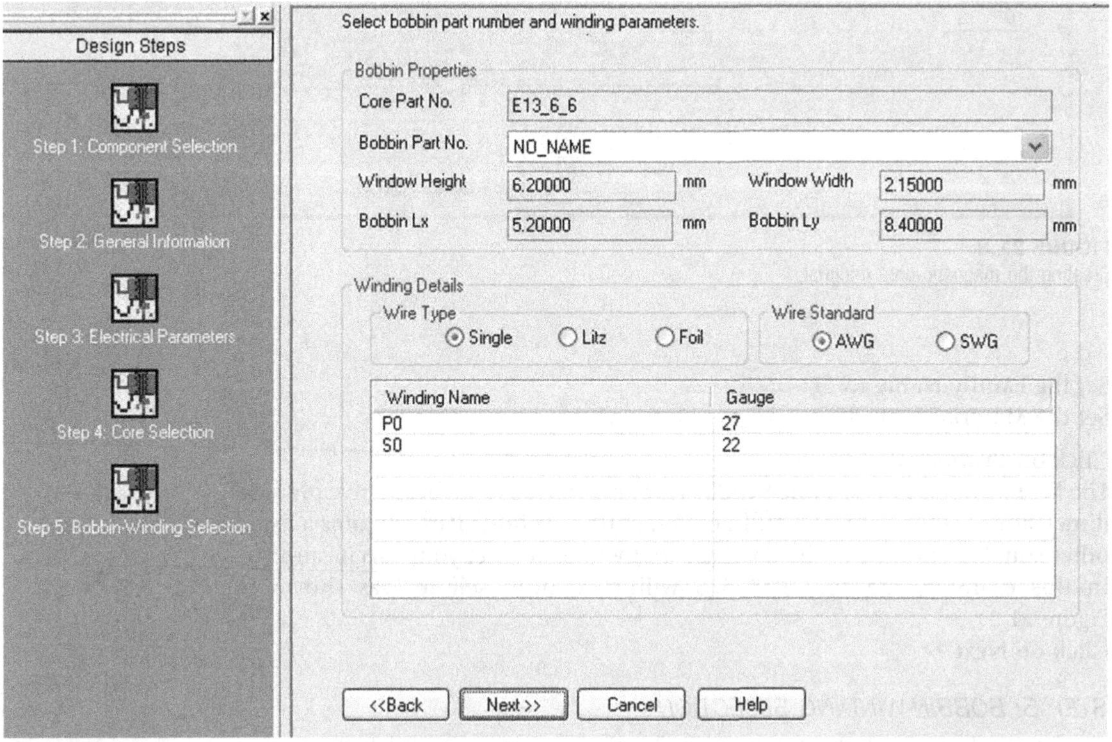

FIGURE 21.10
Bobbin and wire properties.

indicates that there are no bobbins in the MPE database. If no bobbin is specified, then a default bobbin wall thickness of 1 mm is used to calculate the bobbin dimensions based on the core dimensions. To add a bobbin, select **Tools > Data Entry > Core Details > Bobbin** (Figure 21.11).

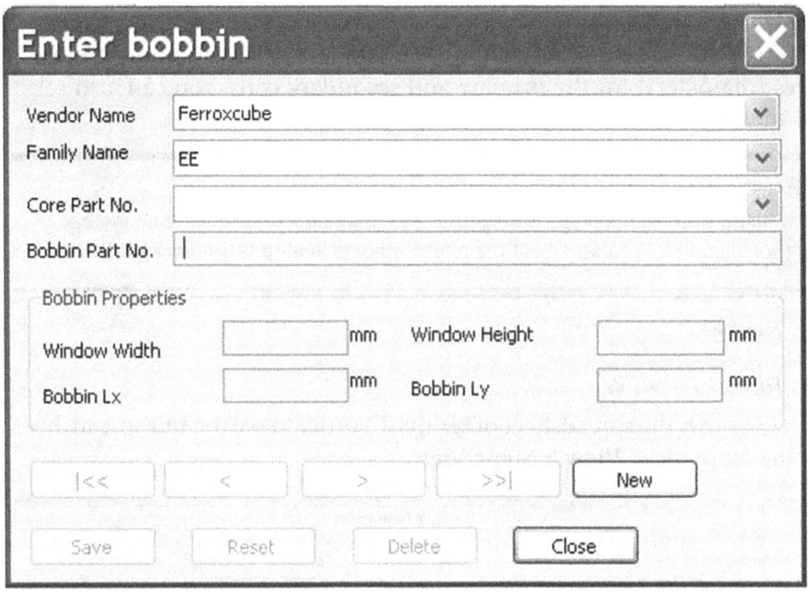

FIGURE 21.11
Creating a new bobbin.

Figure 21.12 shows the bobbin orientation and dimensions given in core vendor datasheets.

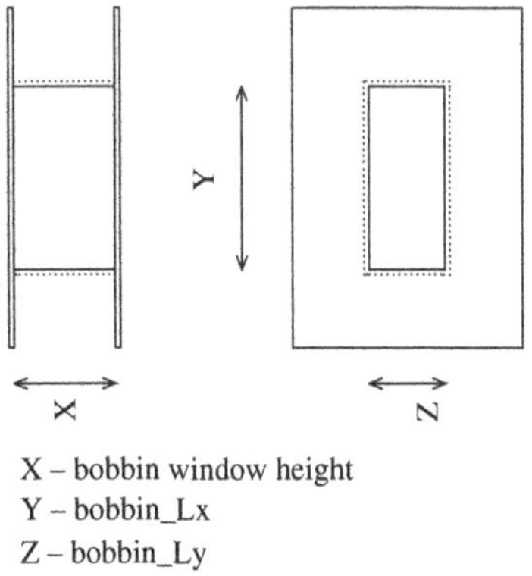

FIGURE 21.12
Bobbin Dimensions.

For the bobbin shown in Figure 21.12, to fit into the core, the calculations are:

bobbin window height = core window height − (2 × bobbin thickness)
bobbin window width = core window width − (1 × bobbin thickness)
bobbin_Lx = core_Lx + (2 × bobbin thickness)
bobbin_Ly = core_Ly + (2 × bobbin thickness)

For this example, the default NO_NAME bobbin will be used, as the core dimensions are not yet finalized.

The Bobbin-Winding Selection window in Figure 21.10 also shows the proposed wire types (diameters) for the primary and secondary coils. You can also select between the different AWG and SWG wire standards.

NOTE

You may see a warning message stating that a LITZ winding should be used instead of a single winding. In this case, select the single winding in step 5 (Figure 21.10).

Click on **Next >>**.

STEP 6: RESULTS VIEW

Figure 21.13 shows the spreadsheet of results. In order to see the full spreadsheet, turn off the Steps view: **View > Steps View.**

Input Parameters		Output Parameters				
Electrical Specifications		**Winding Parameters**			**Calculated Values**	
Primary Voltage (V)	50	Winding Name	P0	S0	Core Loss (Watts)	65.13935m
Secondary Voltage 1 (V)	12	Peak Current (A)	0.7111111111111	1.818181818182	Achieved Efficiency (%)	97.28357
Power (Watts)	6	RMS Current (A)	0.2754121490636	0.7784989441615	Achieved Regulation (%)	
Frequency (Hz.)	40k	No. of Turns	81	24	Window Occupied (%)	102.5065234659
Efficiency (%)	75	Min. Inductance (H)	0.000791015625		Temperature Rise (C)	27.46545
Duty Cycle (%)	45	Wire Gauge'	27	22	Total Buildup (mm)	8.44600
Component Type	Flyback Transformer	Turns/Layer	12	7	Total Copper Loss (Watts)	102.39750m
		No of layer	7	4	Fringing Coefficient	1.22069
Design Status	Error	Inter layer Insulation (mm)'	0.2	0.2	Operating Flux Density (Tesla)'	0.3324890102445
		End Insulation (mm)'	0.2	0.2	AC Flux Density (Tesla)	
Core Details		Winding Buildup (mm)	4.042	3.404		
Vendor Name	Ferroxcube	Winding resistance (Ohm)	0.5668054726953	0.09801702558053		
Part Number	E13_6_6	Copper Loss (Watts)	0.0429932447437	0.0594042579276		
Core Type	EE	Leakage Inductance (H)	0.00021380227247611			
Core Material	3C81	Voltage Drop (V)	0.1561051133361	0.0763061509243		
Bobbin Part Number	NO_NAME					
GAP' (mm)	255.48260m					
Voltage Isolation (mm)'	1	No. Of Strands	1	1		
Maximum Flux Density (Tesla)	337.50000m	Foil Thickness (mm)				
Current Density (A/mm2)'	3	Foil Width (mm)				
Insulation Material'	NYLON					
Wire Type	AWG					

Manufacturer Report | Model View

Original

• INFO: Designing winding layout complete.
• ERROR: The P0 winding could not be fitted in the core. Try design changes to achieve success.

FIGURE 21.13
Spreadsheet of results.

In the spreadsheet, the **Design Status** reports that there is an error (Figure 21.14) and the warning message at the bottom of the results spreadsheet indicates that the P0 winding could not fit into the core.

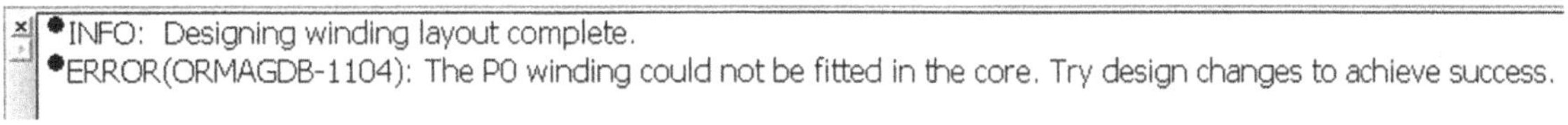

FIGURE 21.14
Design has not been successful.

So either the wire diameter for the secondary needs to be reduced or you need to use a different core material or a larger core. You could also reduce the distance between the primary and secondary windings (Voltage isolation in Step 3).
Go back to **Step 4: Core selection** and select the material as 3C90.
Click on **Propose Part**.
Select the E19_8_9 core, which has a larger core size, as seen in Figure 21.15.
Click on **Next >>** and progress to the **Results View**.

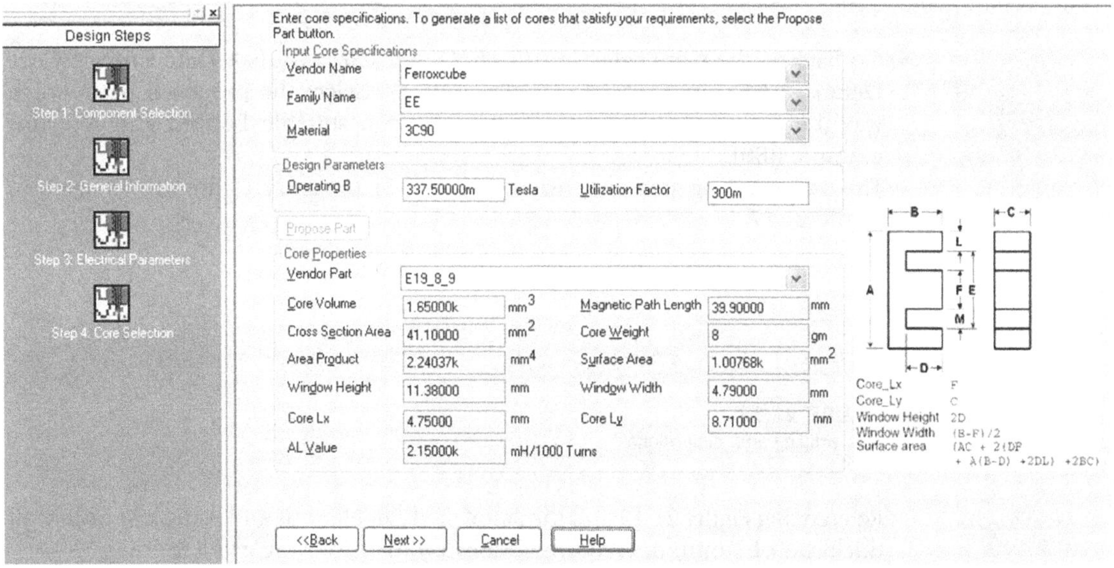

FIGURE 21.15
Changing the core material and core size.

You should see the **Design Status** reporting Success (Figure 21.16).
Now you need to create the bobbin which will fit in the core and run the calculations again.

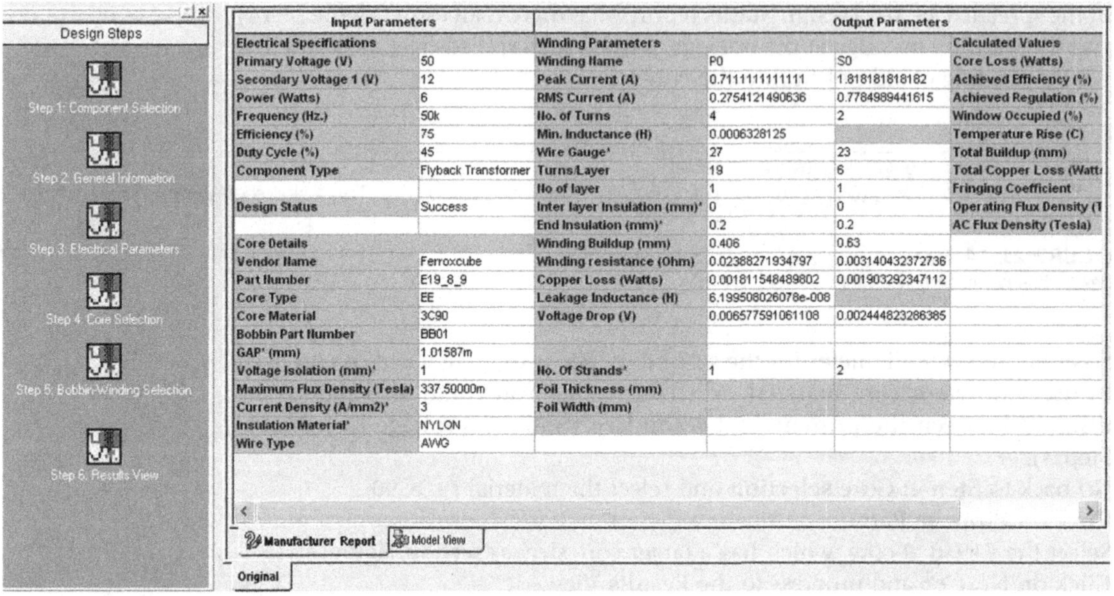

Input Parameters		Output Parameters			
Electrical Specifications		Winding Parameters			Calculated Values
Primary Voltage (V)	50	Winding Name	P0	S0	Core Loss (Watts)
Secondary Voltage 1 (V)	12	Peak Current (A)	0.7111111111111	1.818181818182	Achieved Efficiency (%)
Power (Watts)	6	RMS Current (A)	0.2754121490636	0.7784989441615	Achieved Regulation (%)
Frequency (Hz.)	50k	No. of Turns	4	2	Window Occupied (%)
Efficiency (%)	75	Min. Inductance (H)	0.0006328125		Temperature Rise (C)
Duty Cycle (%)	45	Wire Gauge'	27	23	Total Buildup (mm)
Component Type	Flyback Transformer	Turns/Layer	19	6	Total Copper Loss (Watt
		No of layer	1	1	Fringing Coefficient
Design Status	Success	Inter layer Insulation (mm)'	0	0	Operating Flux Density (T
		End Insulation (mm)'	0.2	0.2	AC Flux Density (Tesla)
Core Details		Winding Buildup (mm)	0.406	0.63	
Vendor Name	Ferroxcube	Winding resistance (Ohm)	0.02388271934797	0.003140432372736	
Part Number	E19_8_9	Copper Loss (Watts)	0.001811548489802	0.001903292347112	
Core Type	EE	Leakage Inductance (H)	6.199508026078e-008		
Core Material	3C90	Voltage Drop (V)	0.006577591061108	0.002444823286385	
Bobbin Part Number	BB01				
GAP' (mm)	1.01587m				
Voltage Isolation (mm)'	1	No. Of Strands'	1	2	
Maximum Flux Density (Tesla)	337.50000m	Foil Thickness (mm)			
Current Density (A/mm2)'	3	Foil Width (mm)			
Insulation Material'	NYLON				
Wire Type	AWG				

FIGURE 21.16
Results view showing success.

Go back to the **Core selection (Step 4)** and select **Tools > Data Entry > Core Details > Bobbin** as shown in Figure 21.17. Select the proposed part, which was the Ferroxcube, EE core 19_8_9. Enter a suitable Bobbin Part No., for example BB01.

The core winding area dimensions given in Figure as 21.15 are shown again in Figure 21.17.

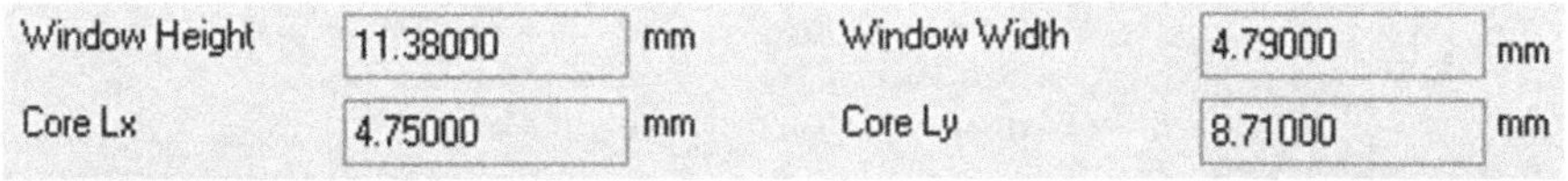

FIGURE 21.17
Core winding area dimensions.

Referencing Figure 21.12 and the associated calculations and using a bobbin wall thickness of 1 mm, the required bobbin dimensions are given as:

bobbin window height = 11.38 − 2 = 9.39 mm
bobbin window width = 4.79 − 1 = 3.79 mm
bobbin_Lx = 4.75 + 2 = 6.75 mm
bobbin_Ly = 8.71 + 2 = 10.71 mm

Go back to **Step 5 (Bobbin-winding selection)** and select **Tools > Data Entry > Core Details Bobbin**. Name the Bobbin BB01, enter the Bobbin data as shown in Figure 21.18 and click on Save.

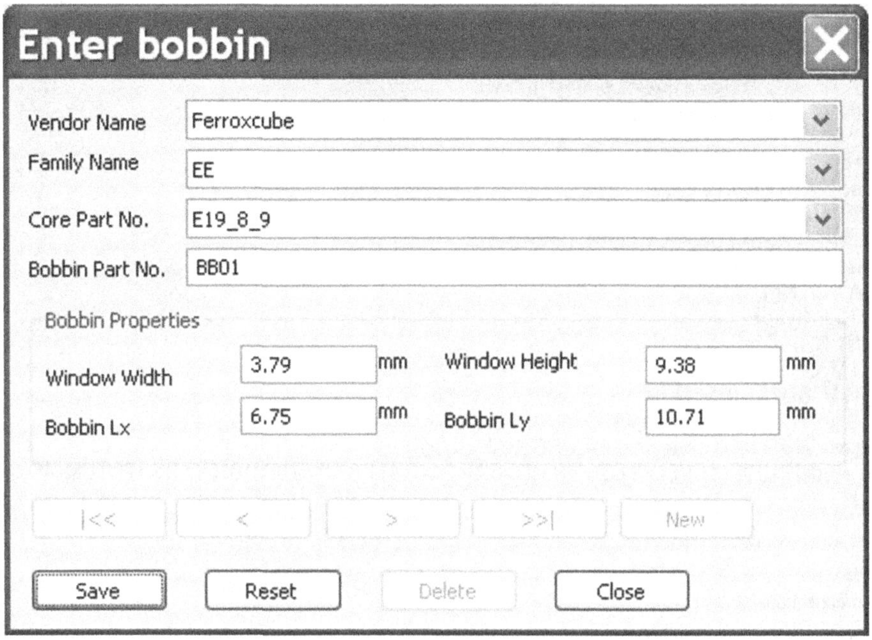

FIGURE 21.18
Creating the bobbin.

When you save the new bobbin, a message will appear informing you that the record was entered successfully into the database. Click on OK.

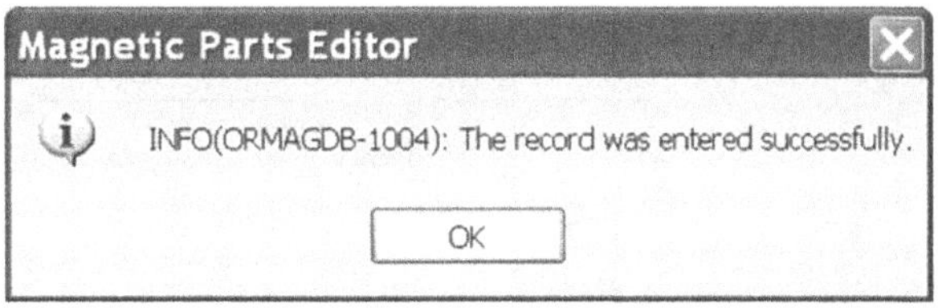

Proceed to **Step 6** and check that the Design Status is still showing **Success**.
Save the design, which will generate a flyback .mgd file.
When you save the design, a PSpice model will be created. In this example the flyback.lib file will be created. By selecting the **Model View** tab at the bottom of the **Results Spreadsheet**, you can view the PSpice transformer model as shown in Figure 21.19.
For the flyback topology, the transformer is modeled as a subcircuit with four terminals, V_IN1, V_IN2, V_OUT11 and V_OUT12. The transformer circuit representation is shown in Figure 21.20.
With the Model View open, click on Save again to save the PSpice model.
Using the Model Editor Wizard in the Model Editor, a Capture part can be generated and associated to a transformer symbol as described in Chapter 16.

```
PSpice Model Name  flyback
PSpice Model .lib  C:\PSpice Exercises\test_flyback          [✓] Default Model

*$
*|Generated by Magnetic Parts Editor on Mon May 30 01:34:56 2011
.subckt flyback V_IN1 V_IN2
+ V_OUT11 V_OUT12
+ PARAMS:  Np=4 RSp=0.0238827 Llp=6.19951e-008
+ Ns1=2 RSs1=0.00314043 Gap = 0.000101587
L_LP NLP V_IN2 {Np}
R_RP NRP NLP {RSp}
L_Leak V_IN1 NRP {Llp}
L_LS1 NLS1 V_OUT12 {Ns1}
R_RS1 NLS1 V_OUT11 {RSs1}
K_K2 L_LP L_LS1  1.0 core_model_K1
.model core_model_K1 AKO:core_model  CORE (GAP={Gap})
.model core_model CORE ( LEVEL=3 OD=3.99 ID=0 AREA=0.411 GAP=0.000101587 Br=1700 Bm=4500 Hc=0.1875 )
.ends flyback
*$
```

FIGURE 21.19
PSpice magnetic core model.

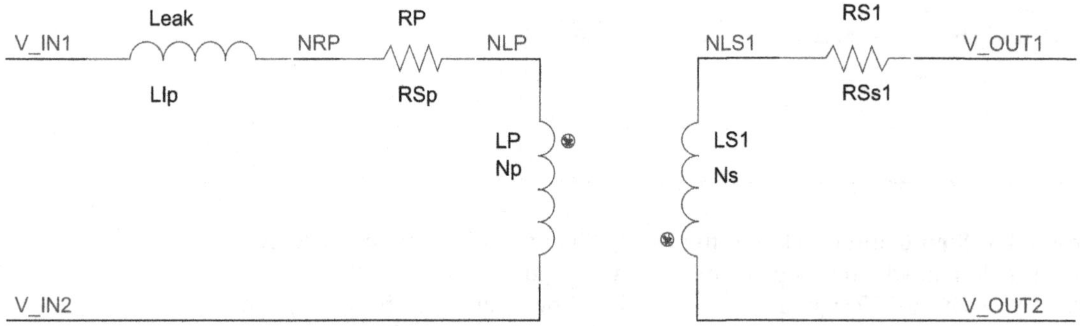

Lip	Leakage inductance referred to primary
RSp	Primary winding resistance
RSs1	Secondary winding resistance
Np	Number of turns in primary winding
Ns	Number of turns in secondary winding

FIGURE 21.20
Flyback transformer model: Llp: leakage inductance referred to primary; RSp: primary winding resistance; RSs1: secondary winding resistance; Np: number of turns in primary winding; Ns: number of turns in secondary winding.

Exercise 2

CREATING A TRANSFORMER MODEL

1. Open the PSpice Model Editor from the **Start** menu
   Start > All Programs > Cadence (or OrCAD) > PSpice > Simulation Accessories > Model Editor
2. In the Model Editor, select, **File > New**.
3. Select **File > Model Import Wizard**.
4. In the Specify Library window (Figure 21.21), browse to the flyback.lib, accept the **Destination Symbol Library** as shown and click on **Next >**.

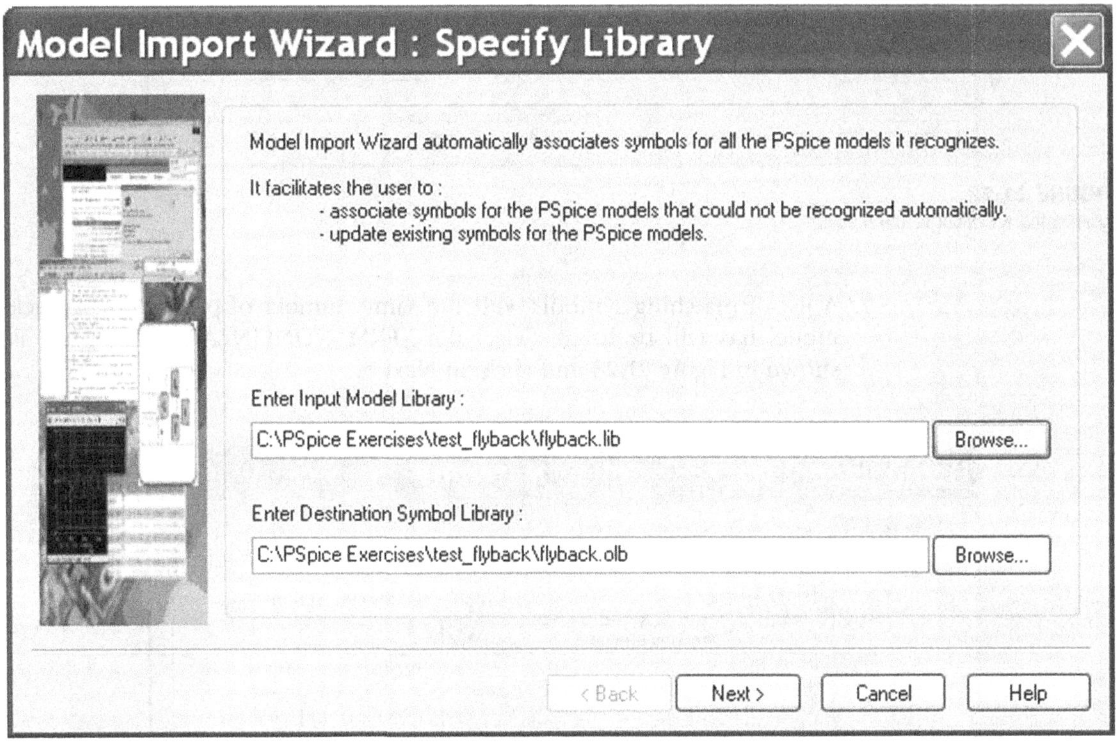

FIGURE 21.21
Enter the flyback.lib PSpice model file.

5. In the **Associate/Replace Symbol** window (Figure 21.22), click on **Associate Symbol**. Select the Flyback model and click on Associate Symbol.
6. In the **Select Matching** window, click on the icon [...] and select the breakout.olb library from the installed folder as **<install path> OrCAD (or Cadence) > version xx.x > Tools > Capture > library > pspice > breakout.olb**

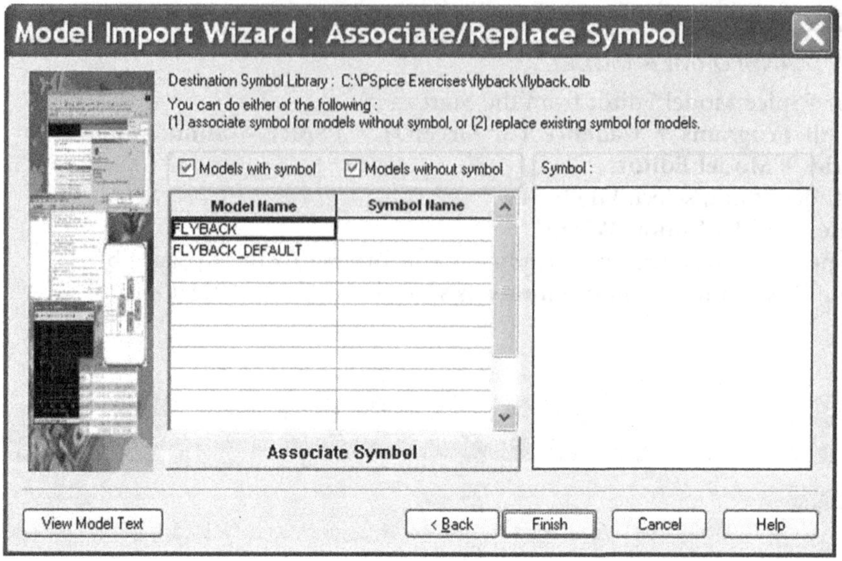

FIGURE 21.22
Associating a symbol to the model.

7. A list of matching symbols with the same number of pins as the flyback model has will be listed. Select the **XFRM_NONLINEAR** transformer as shown in Figure 21.23 and click on **Next >**.

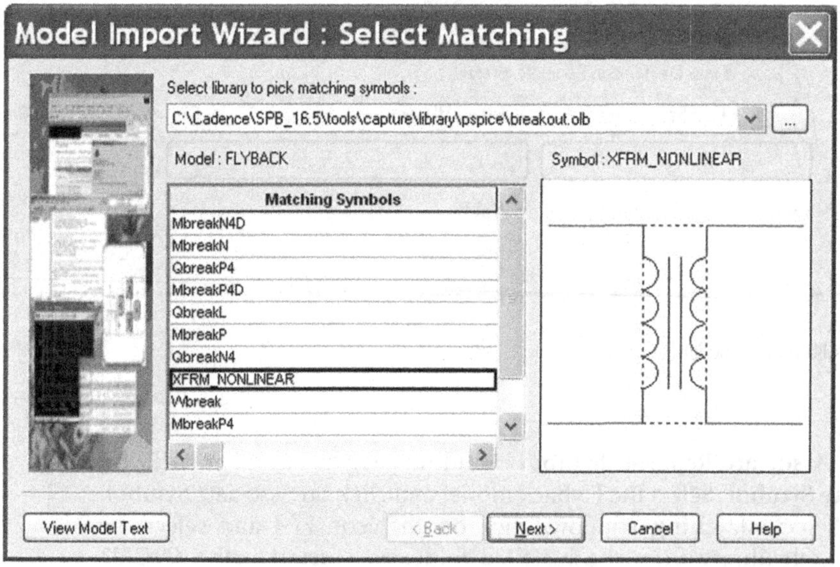

FIGURE 21.23
Select a matching transformer symbol.

8. You have to associate the symbol pins with the model terminals. Select the pins as shown in Figure 21.24 and then click on **Save Symbol**. The pins are in the same order as for the subcircuit model (Figure 21.19).

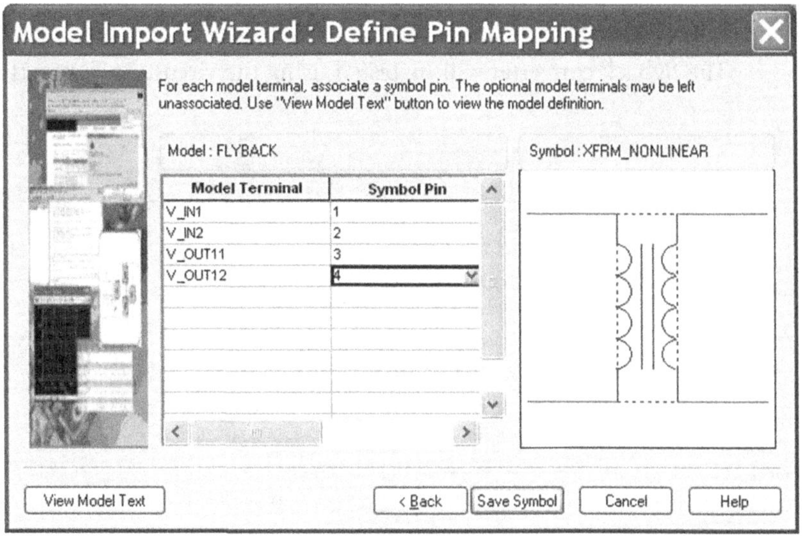

FIGURE 21.24
Associating pins to the symbol.

9. You should see the model name shown with an associated symbol (Figure 21.25).

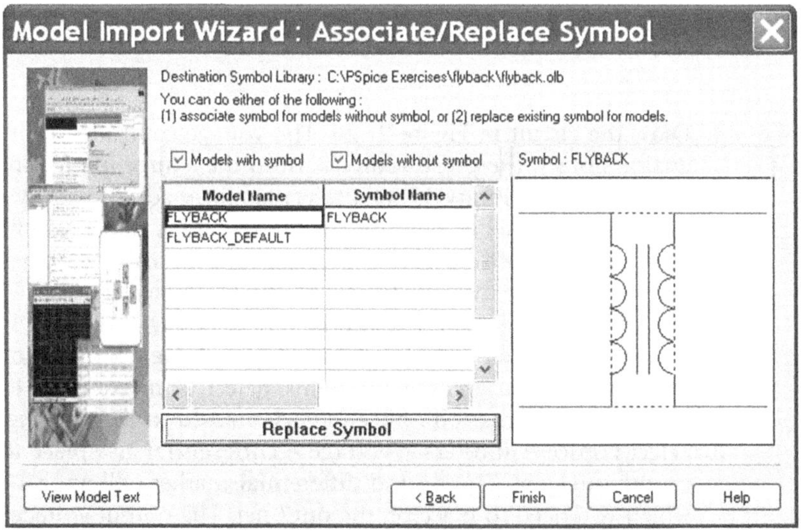

FIGURE 21.25
The Flyback model now has an associated symbol.

10. Click on **Finish** and click on No in the sch2cap window. Check there are no error messages in the Summary Status window and click on OK. The Capture part and model are ready to be used.

Exercise 3

The flyback converter will be tested using the circuit in Figure 21.26.

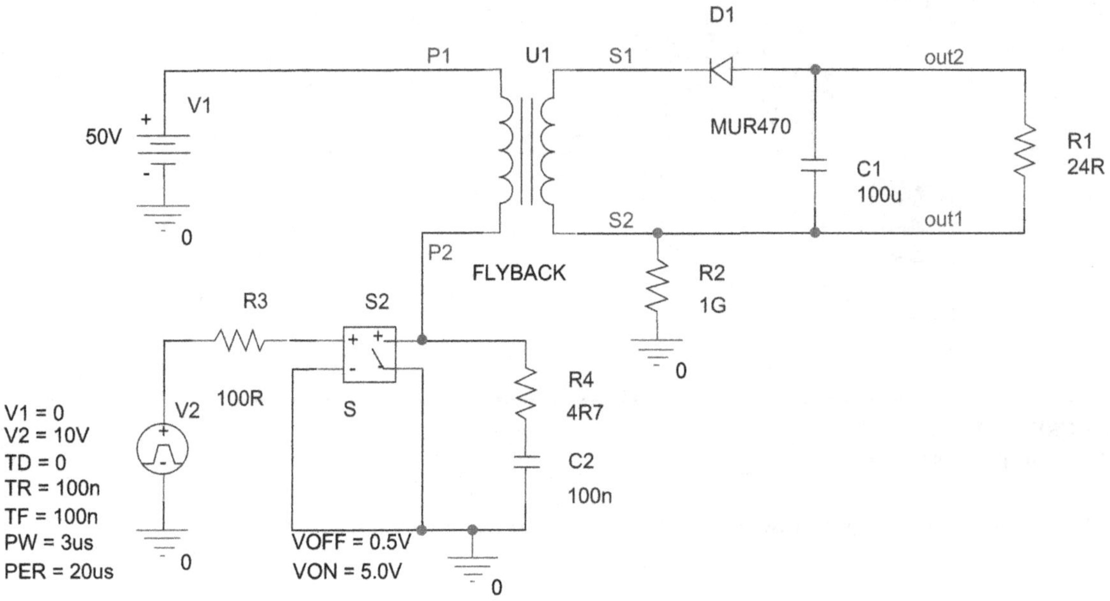

FIGURE 21.26
Flyback converter.

1. Draw the circuit in Figure 21.26. The voltage-controlled switch is from the analog library, the pulse source is from the source library and the diode is from the diode library. In the Project Manager add the newly created library for the transformer by selecting the library folder, **rmb > Add File** and browse to the **Flyback.olb** library. The flyback.olb library will appear in the Place Part menu. Place the transformer as shown in Figure 21.26.
2. You have to make the flyback.lib PSpice library available for the design. Create a PSpice simulation profile, **PSpice > New Simulation Profile**, and set up a transient analysis for 10 ms. Select **Configuration Files > Library**, browse to the flyback.lib file and click on **Add to Design** (Figure 21.27).
3. Select **PSpice > Markers > Voltage > Differential** and place the first marker on the **out1** net. The second differential marker will appear automatically, which you need to place on the **out2** net. The output voltage of the flyback converter is negative; hence the most positive differential marker is placed on **out2**. Run the simulation.

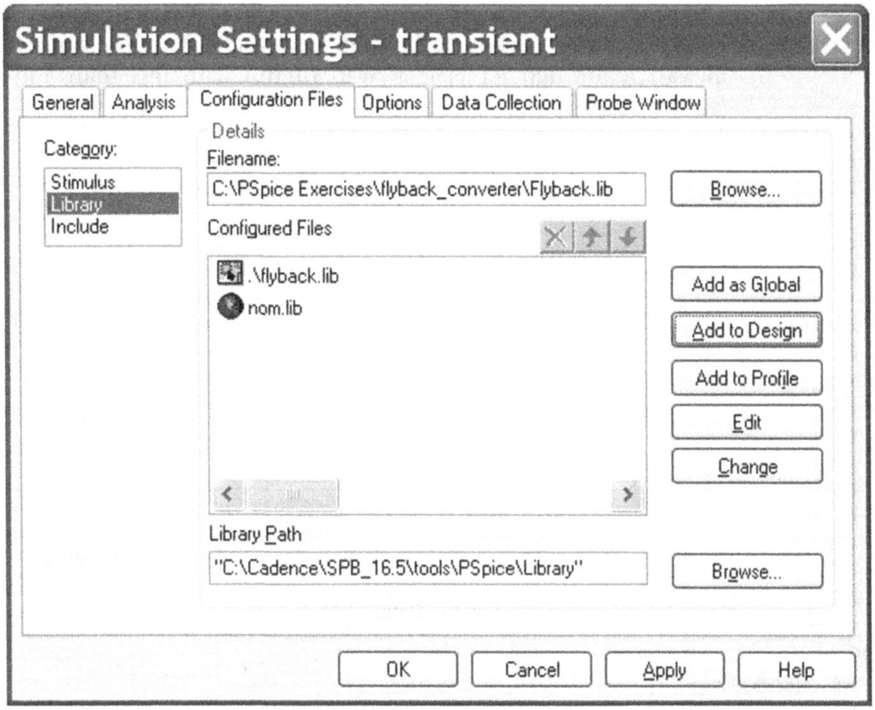

FIGURE 21.27
Adding the flyback.lib model file.

4. You should see that the output voltage is larger than 12 V (Figure 21.28). As the maximum duty cycle was specified as 45%, this can be decreased to

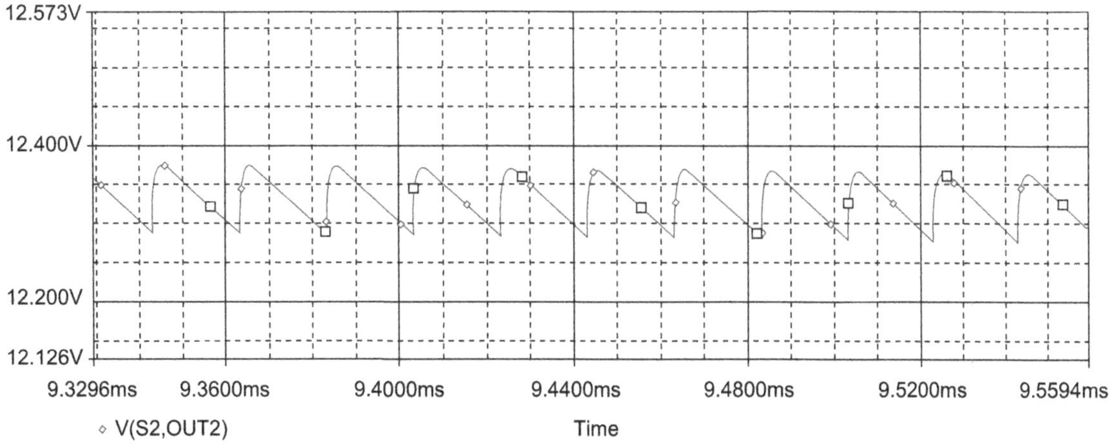

FIGURE 21.28
Output voltage meets specification.

reduce the output voltage. With a duty cycle of 15% (T_{on} is 3 μs) the output voltage reduces to just over 12 V with less than 100 mV of ripple. The current measured through R1 is just over 510 mA with less than 5 mA of ripple (Figure 21.29).

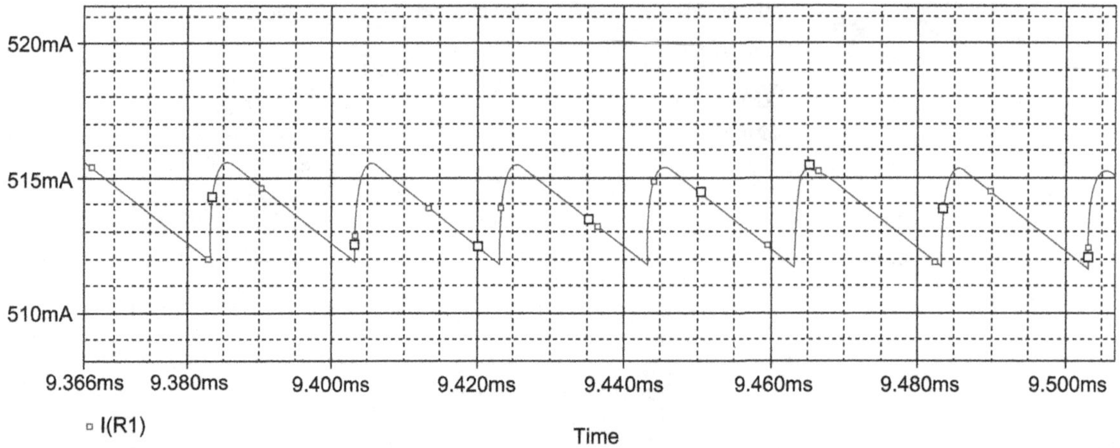

FIGURE 21.29
Output current meets specification.

CHAPTER 22

Test Benches

Chapter Outline

Normally, when you run simulations on a circuit, you add, for example, voltage sources and load resistors to test the circuit. You may even remove components from a circuit in order to run simulations. However, once the simulations are complete, these add-on components will have to be removed and any deleted components restored.

Before version 16.5, you could add a PSpiceOnly property to the parts that are only used for simulation and therefore these parts would not be included, for example, in the printed circuit board (PCB) netlist. From version 16.5, you can use the Partial Design Feature, which uses test benches to allow you to define those components that are used only for simulations. You can also selectively partition designs for different simulation profiles and build up designs using circuits from other projects. Using test benches is very useful when you have a design that has been put together from a collection of circuits from other projects, as it will allow you to test the functionality of each individual circuit as you build up to the complete design.

When you create a test bench, a Test Bench folder, which contains all the design schematics, is added to the bottom of the Project Manager. All the components in all the schematics in the Test Bench folder will be grayed out. You then selectively 'activate' those parts that are required for simulation and add parts

Analog Design and Simulation using OrCAD Capture and PSpice. DOI: 10.1016/B978-0-08-097095-0.00022-2

such as voltage sources and load resistors. Parts can be deselected and selected either from the master design or from the created test benches.

When you create a test bench, another design folder is created in the project folder. The project folder will then contain two folders:

<project name>-PSpiceFiles
<project name>-TBFiles

The Schematic to Schematic (SVS) utility will compare test bench designs to the master design such that the master design can be updated with modified component values.

22.1 SELECTION OF TEST BENCH PARTS

As mentioned previously, there will be two design folders, the master design and the test bench design. You can select those parts required for simulation from either the master design or the test bench design depending on which design you have open, but ultimately you will be simulating the test bench design. For example, Figure 22.1 shows a hierarchical design of a digital counter which contains a clock oscillator and a modulus-3 counter.

Both blocks will be tested separately using two test benches, Test_Clock and Test_Counter. Initially, all schematic parts are grayed out. From the master design you create a test bench by selecting **Tools > Test Bench > Create Test Bench** and name the test bench, which is added to the master design Project Manager. You can create multiple test benches, but only one test bench will be active, denoted by an A in front of the test bench name. You can then select parts from the master design and add to the active test bench by **rmb > Add Part(s) To Active Testbench**. In Figure 22.2, the Test_Counter is the active test bench.

From the master design you can also select and deselect parts using the hierarchy tab in the Project Manager. Figure 22.3 shows the hierarchy tab of the master design Project Manager. In this example, only the parts in the **Clock** hierarchical block will be selected. All the other parts in the design will be grayed out.

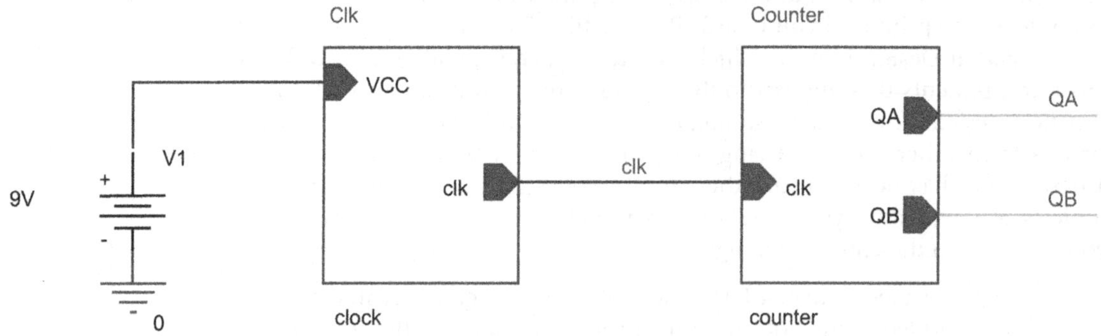

FIGURE 22.1
Hierarchical design of a digital counter.

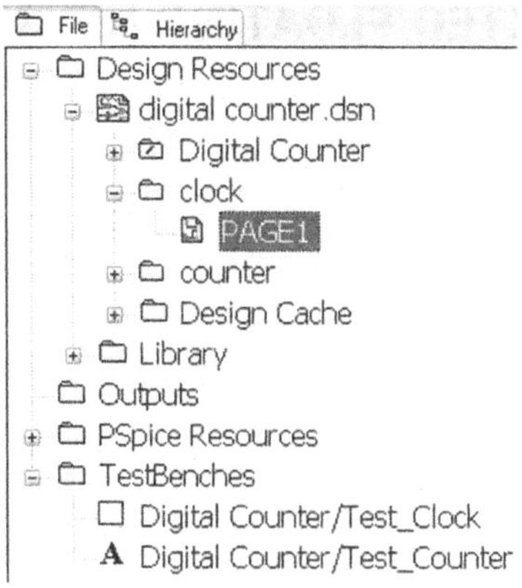

FIGURE 22.2
Master design with two test benches.

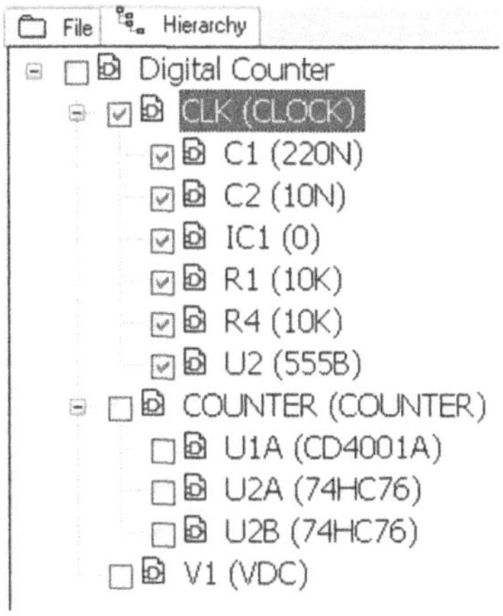

FIGURE 22.3
Hierarchy selection of parts from the master design.

Alternatively, you can add and remove parts from a test bench design. In the above example, if you open the **Test_Clock/Test Bench > clock** schematic, you can select parts and then **rmb > Test Bench >Add Part(s) To Self** or **Remove Part(s) From Self** (Figure 22.4), where **Self** refers to the active **Test_Clock/Test Bench**.

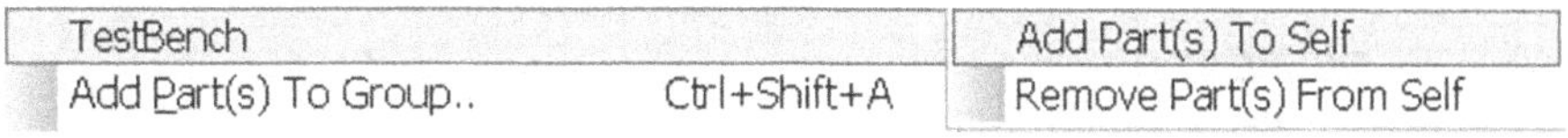

FIGURE 22.4
Adding or removing test bench design parts.

22.2 UNCONNECTED FLOATING NETS

Adding and removing parts in a design can lead to unconnected wires, resulting in floating node errors. As discussed in Chapter 2, all nodes must have a DC path to ground. You can search for floating nets using the **Text to Search Box**, which presents a list of searchable objects (Figure 22.5), one of which is floating nets.

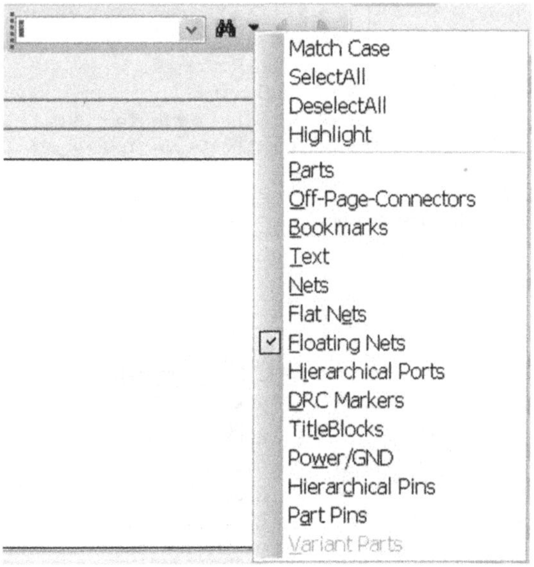

FIGURE 22.5
Searching for floating nets.

Floating nets must be resolved, otherwise the simulation will not proceed. Sometimes, all that is required is to connect a resistor between the floating net and ground to provide the DC path to ground.

In 16.6 two advanced search features have been added as shown in Figure 22.6. **Regular Expressions** and **Property Name=Value**.

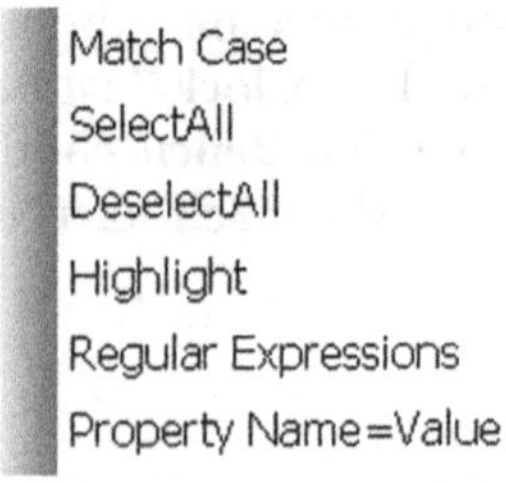

FIGURE 22.6
New search features.

Property Name=Value requires the full Property name whereas for Value, the * wildcard and character ? marks can be used. For example in the Digital Counter, to find all ICs, you select **Property Name=Value** as shown in Figure 22.7.

FIGURE 22.7
Select Property Name=Value.

Then enter, Part Reference=U*. 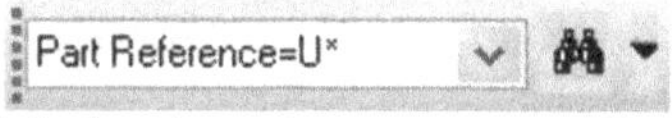

To be more specific in searching only for digital 74 series type 76 JK flip flops ICs, regardless of the technology, ie LS, HC, AC etc, then you can enter; Value=74??76. 

The **Regular Expression** offers more flexibility in providing conditional searching for strings, ie you can specify a range of values or can be selective in using AND or OR (|) functions. For example, if you want to find the first resistors R1,R2 **OR** the first capacitors C1, C2 in the Digital Counter, then you could enter; Part Reference=(C|R)[1-2]. Note that both the **Regular Expression** and **Property Name=Value** are both selected. 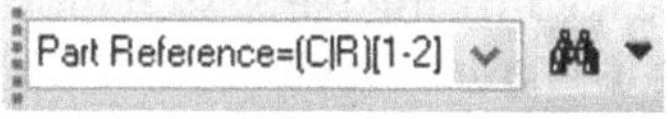

The result of the search finds R1, C1, C2 and IC1, the initial condition part as shown in Figure 22.8.

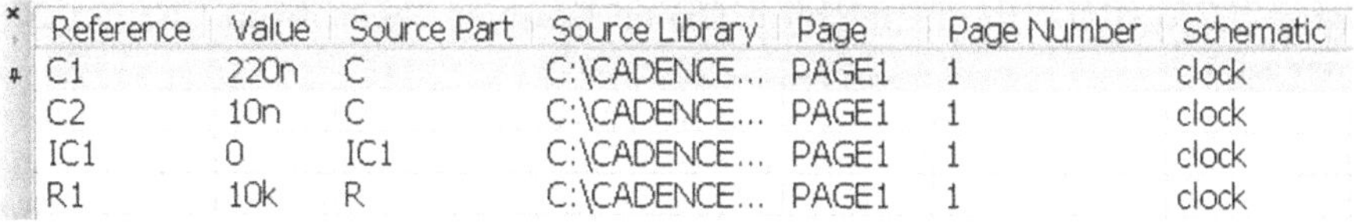

Reference	Value	Source Part	Source Library	Page	Page Number	Schematic
C1	220n	C	C:\CADENCE...	PAGE1	1	clock
C2	10n	C	C:\CADENCE...	PAGE1	1	clock
IC1	0	IC1	C:\CADENCE...	PAGE1	1	clock
R1	10k	R	C:\CADENCE...	PAGE1	1	clock

FIGURE 22.8
Result of Regular Expression search.

22.3 COMPARING AND UPDATING DIFFERENCES BETWEEN THE MASTER DESIGN AND TEST BENCH DESIGNS

Design differences between the master design and test bench designs can be viewed in the SVS utility, which displays the differences using a color-coded system to highlight missing parts (red), unmatched parts (yellow) and matched parts (white). In Figure 22.9, the test bench design is shown on the left-hand side and the master design on the right-hand side. The test bench clock schematic contains an extra capacitor C3, shown in red, and a different value for R4.

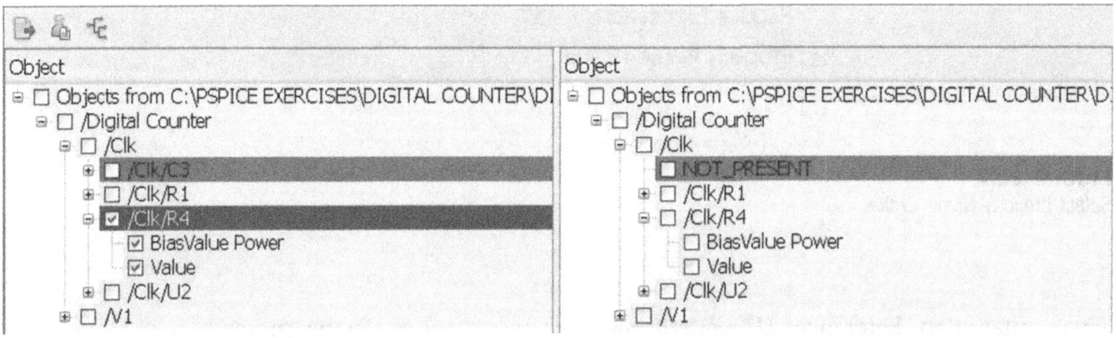

FIGURE 22.9
Comparing test bench design with master design.

You can update the test bench design with modified component values to the master design by selecting the **Accept Left** icon. However, only modified values will be updated to the master design. In the example above, if you check the Value box for R4 in the left-hand panel, the modified value for R4 will be updated from the test bench to the master design. The extra capacitor C3 will not be updated to the master design. In addition, the master design cannot be updated if there are missing parts, so removed parts in test benches will not be removed from the master design.

22.4 EXERCISES

Exercise 1

Figure 22.10 shows the hierarchical Digital Counter design from Chapter 20, Exercise 5.

You will now create a Test_Clock Test Bench to simulate and verify the performance of the clock oscillator only. In the master design, you will only add clock parts to the active test bench by selecting those parts in the Hierarchy tab in the master design Project Manager.

1. Select **Digital Counter.dsn** and from the top toolbar select **Tools > Test Bench > Create Test Bench**. Name the test bench **Test_Clock**, as shown in

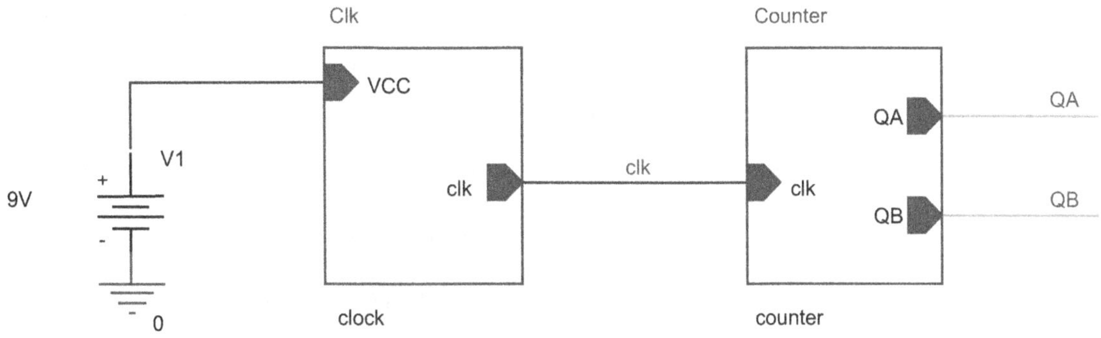

FIGURE 22.10
Hierarchical digital counter design.

Figure 22.11. Click on OK and click on OK again if prompted to save the design.

The **Test_Clock** test bench will be placed at the bottom of the Project Manager in the **TestBenches** folder (Figure 22.12).

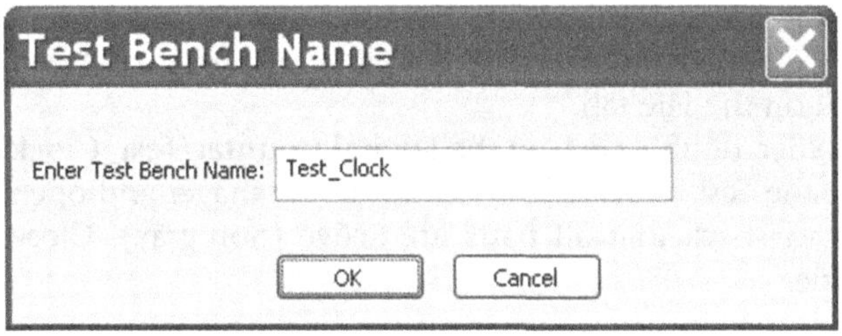

FIGURE 22.11
Creating a Test_Clock test bench.

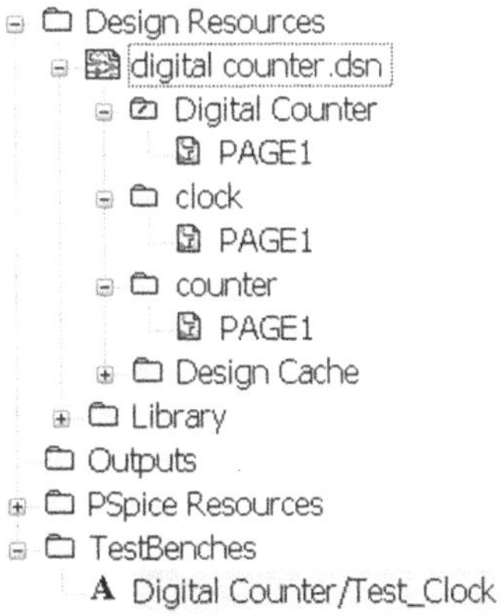

FIGURE 22.12
Test_Clock test bench.

2. In the **TestBenches** folder, double click on **Digital Counter/Test_Clock**, which will open the **Test_Clock.dsn** design.
3. In the **Test_Clock** Project Manager, double click on **Test_Clock.dsn** and open the **counter** Page1 schematic. You will see that all the components are grayed out. Close the schematic page.
4. Still in the **Test_Clock** Project Manager, open the **clock** schematic and you will also see all the components grayed out.
   You will now activate the clock components in the Test_Clock test bench from the master design.
5. Select the master design Project Manager by either selecting the Digital Counter tab above the schematics (Figure 22.13) or selecting **Window > Digital Counter**. It may be easier to place the master and test bench designs side by side.

FIGURE 22.13
Selecting the Project Manager.

6. In the master design (Digital Counter) Project Manager, click on the **Hierarchy** tab to display the hierarchical design. Expand **Digital Counter** and **Clock** if not already expanded and check all the parts for **Clock** as shown in Figure 22.14. Click on the File tab.
7. In the **TestBenches** folder, double click on the **Digital Counter/Test_Clock** test bench to activate the Test_Clock test bench Project Manager and open the **clock** schematic to check that all parts are active (non-gray). Close and save the schematic.

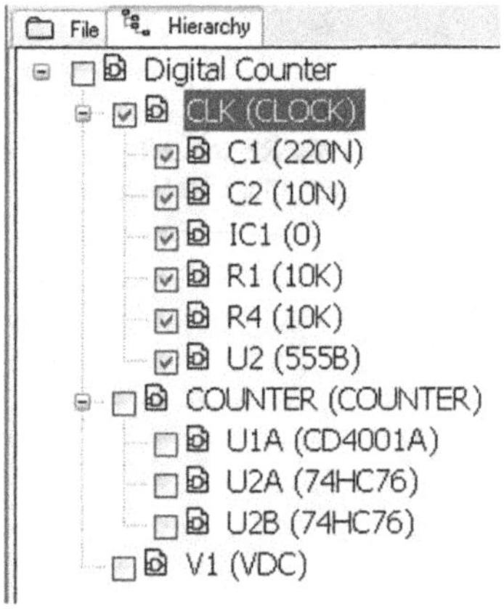

FIGURE 22.14
Selecting the clock oscillator circuit.

8. Still in the Test_Clock Project Manager, open the **Digital Counter** schematic page as shown in Figure 22.15.
   The voltage source V1 is still grayed out. Draw a box around V1, the connecting wires and the 0 V symbol and **rmb > TestBench > Add Part(s) To Self** (Figure 22.16).

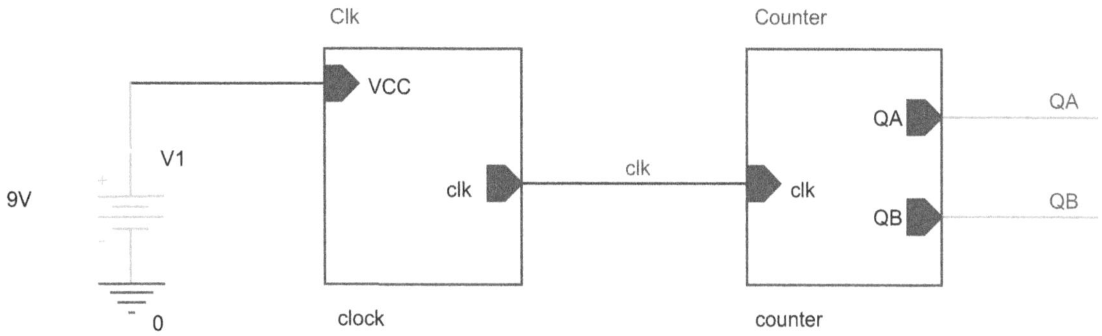

FIGURE 22.15
Test_Clock > Digital Counter schematic.

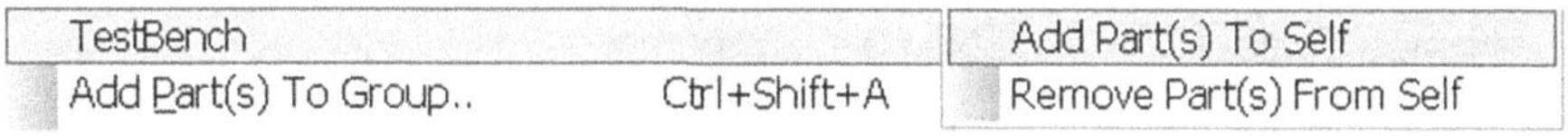

FIGURE 22.16
Adding the voltage source, V1, to the active test bench.

Now that only the clock oscillator circuit is active; you need to search for unconnected nets in the **Test_Clock** test bench. Make sure the Search menu is displayed: select **View > Toolbar > Search**.

9. Highlight the **Test_Clock.dsn** and select, from the top toolbar, the pull-down menu to the right of the Binocular, which is next to the **Text to Search Box** (Figure 22.17).
10. Click on **DeselectAll**. Next, select **Floating Nets** and then click on the Binocular icon.
    The **Find** window at the bottom of the screen, below the schematic, reports a floating net on the clk net (Figure 22.18).

NOTE

If you see a 0 (Global) net reported as floating this is because there is no connecting wire between V1 and the 0 V symbol. Select the 0 V symbol and drag it down so that a length of connecting wire appears, and run the search for floating nets again so that only the clk net is floating.

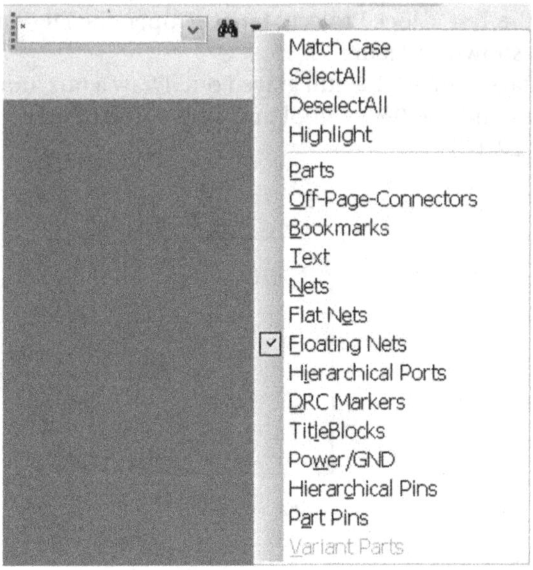

FIGURE 22.17
Search for floating nets.

Object ID
Net Name
Page
Page Number
Schematic
Pin
clk(Wire Alias)
CLK
PAGE1
1
Digital Counter\
Counter.clk,clk.clk
Floating Nets
Find Window

FIGURE 22.18
Reported floating nets.

11. Place a 1k resistor from the **clk** net to ground as shown in Figure 22.19. This provides a DC path to ground for the clk net.
12. Create a PSpice simulation profile for a transient analysis with a run to time of 20 ms. Do not exit the simulation profile.
    You will need to initialize all flip-flops to 0. Select the **Options tab > Gate-level Simulation** and initialize all flips-flops to 0.
13. Place a voltage marker on the **clk** net and run the simulation.
14. You should see the clock output waveform as in Figure 22.20, which has a frequency of 216 Hz.
15. Double click on the clock hierarchical block to open the clock schematic and place another 220 nF in parallel with C1. Change the value of R4 in Figure 20.42 to 6k8 and rerun the simulation. The clock frequency will now be 138 Hz. Close and save the clock schematic.
    Now you will verify the operation of the counter circuit, but this time you will select the active parts from an active Test Bench. The counter parts

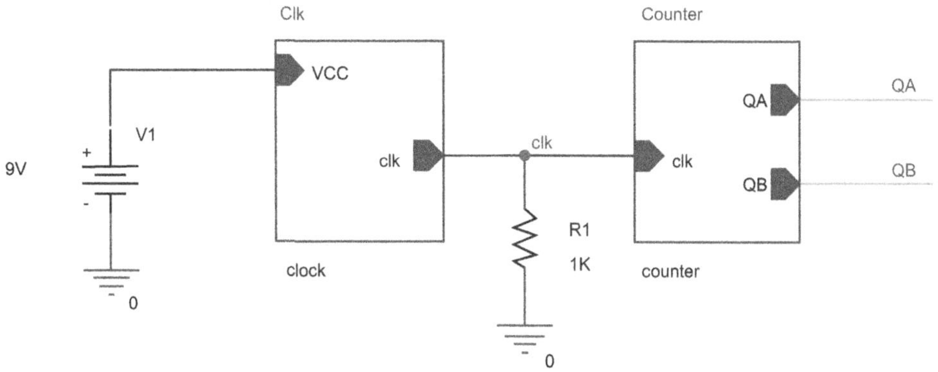

FIGURE 22.19
Adding a 1 k resistor from the clk net to 0 V.

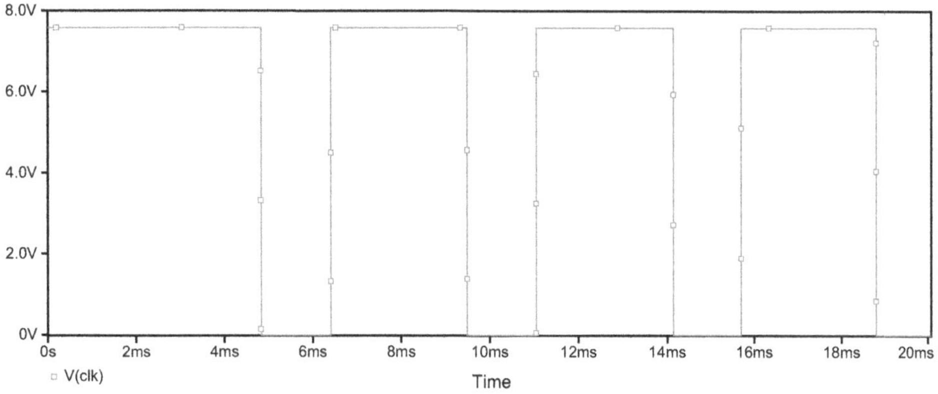

FIGURE 22.20
Clock output waveform.

will be added to the active Test Bench and the clock parts will be removed from the active Test Bench.

16. Open the Project Manager for the Digital Counter (master design).
17. Highlight the **Digital Counter.dsn** and select **Tools > Test Bench > Create Test Bench**. Name the test bench **Test_Counter** (Figure 22.21).

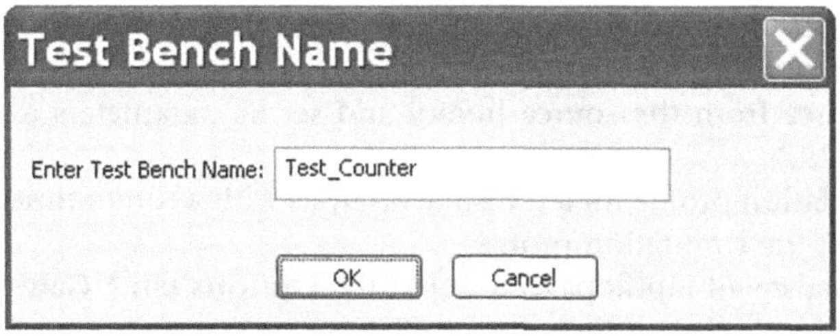

FIGURE 22.21
Creating the Test_Counter test bench.

18. There will be two test benches in the master design counter, as shown in Figure 22.22. The **A** in front of the Test Counter test bench represents the active test bench. You can make a test bench active by **rmb > Make Active**.

FIGURE 22.22
The Test_Counter is the active test bench.

19. Double click on the **Test_Counter** test bench to open the Test_Counter design.
20. Double click on **Test_Counter.dsn** and open the **Digital Counter** schematic Page1.
21. Draw a box around the **clock** hierarchical block, V1 and the 0V symbol and then **rmb > Remove Part(s) From Self** (Figure 22.23).

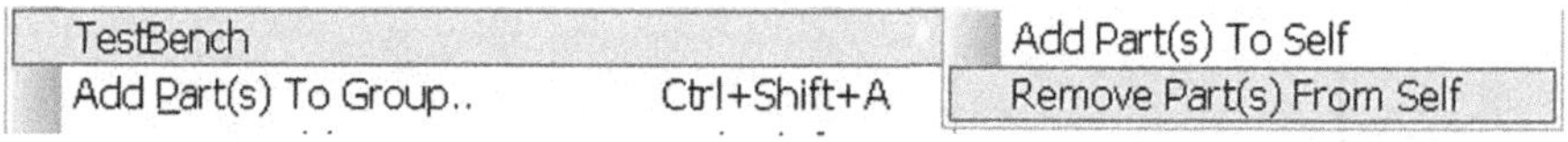

FIGURE 22.23
Removing the clock and associated parts from the active Test_Counter test bench.

22. Draw a box around the **counter** hierarchical block and make sure the clk, QA and QB nets are also selected; **rmb > TestBench > Add Part(s) To Self** (Figure 22.24).

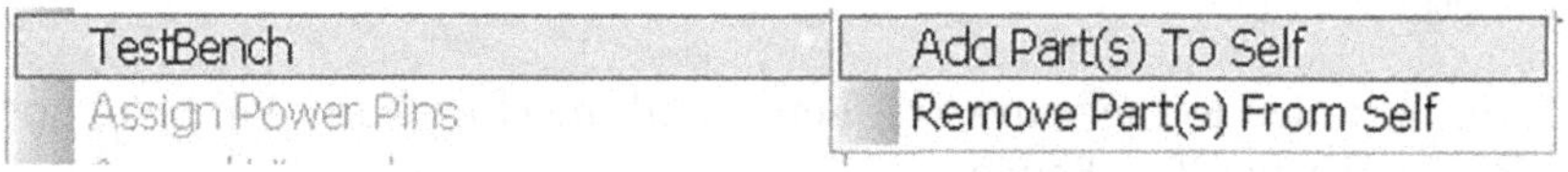

FIGURE 22.24
Adding the counter parts to the Test_Counter test bench.

23. Repeat Step 17 and check for floating nets. The QA and QB should be reported as floating. These are digital output nodes and therefore do not require a DC path to ground. Ignore the warning.
24. Add a DigClock source from the source library and set its parameters as shown in Figure 22.25.
25. Create a PSpice simulation profile for a transient analysis with a run to time of 20 ms. Do not exit the simulation profile.
Do not forget to initialize all flip-flops to 0. Select the **Options tab > Gate-level Simulation** and initialize all flips-flops to 0.

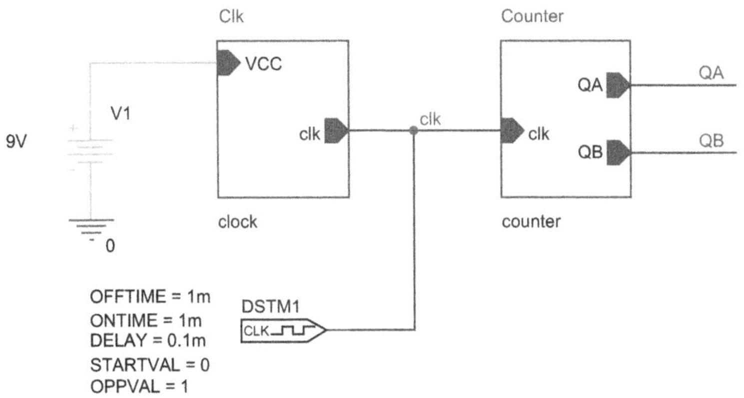

FIGURE 22.25
Testing the operation of the counter circuit.

26. Place voltage markers on QA and QB and run the simulation. You should see the circuit response as shown in Figure 22.26.

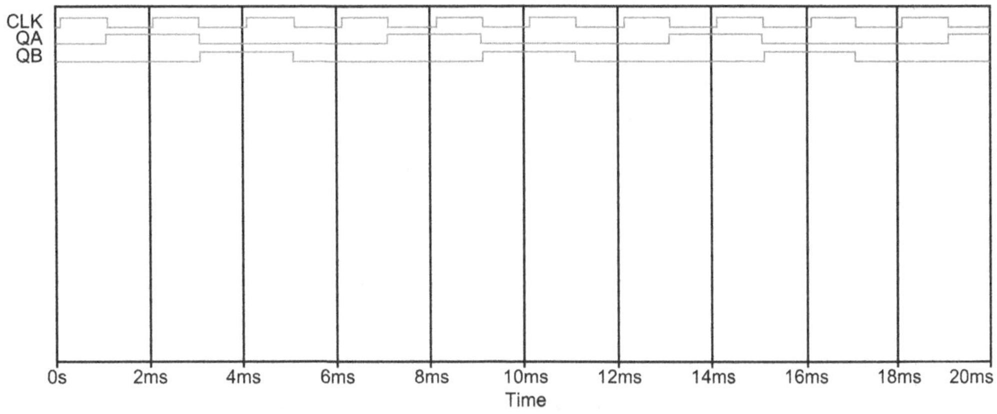

FIGURE 22.26
Output of counter.

Exercise 2

After the test bench simulation testing, any differences between the master and test bench designs, component additions, removals and change in component values can be highlighted.

1. In the master design, highlight **Digital Counter.dsn** and select **Tools > Test-Bench > Diff and Merge** (Figure 22.27).
2. In the SVS window, the test bench is on the left-hand side panel and the master design on the right-hand side panel. In the left panel, expand the **/Clk** as shown in Figure 22.28.

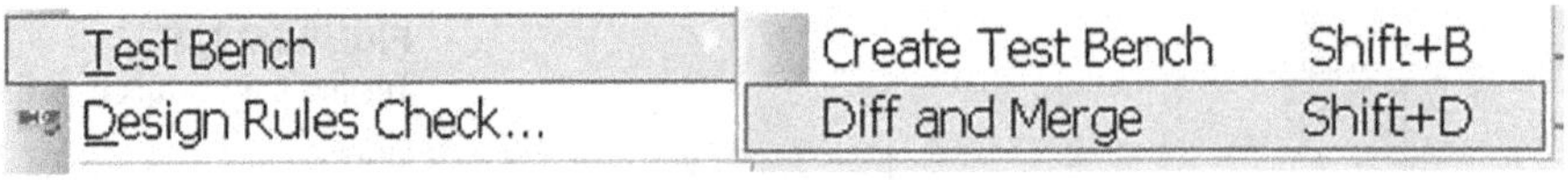

FIGURE 22.27
Comparing the differences between the master and test bench designs.

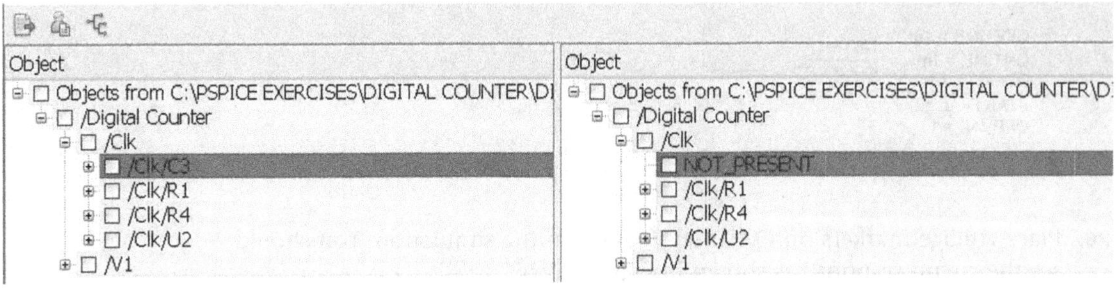

FIGURE 22.28
Highlighted differences between the master design and the Test_Clock test bench.

Yellow indicates the design differences between the Test_Clock test bench and the master design, which allows you to update any modified values to the master design.
The extra capacitor in the clock circuit has been highlighted in red and shown as NOT_PRESENT in the master design.

NOTE

You may have to save and close the Test_Benches before any differences are detected.

---

3. Expand **/Clk/R4** and you will see a Value box displayed for both the Test_Clock test bench and the master design. This allows you to decide which value of R4 to accept by checking the appropriate Value box.
   The modified value for R4 will be updated to the master design.
   Check the **/Clk/R4** box as shown in Figure 22.29. The BiasValue Power is a measurement and is not important for this exercise.
4. To update the master design with the new value for R4, click on the **Accept Left** icon; as there is no longer a difference between R4 in the test bench and the master design, R4 will not be displayed in the SVS window (Figure 22.30).

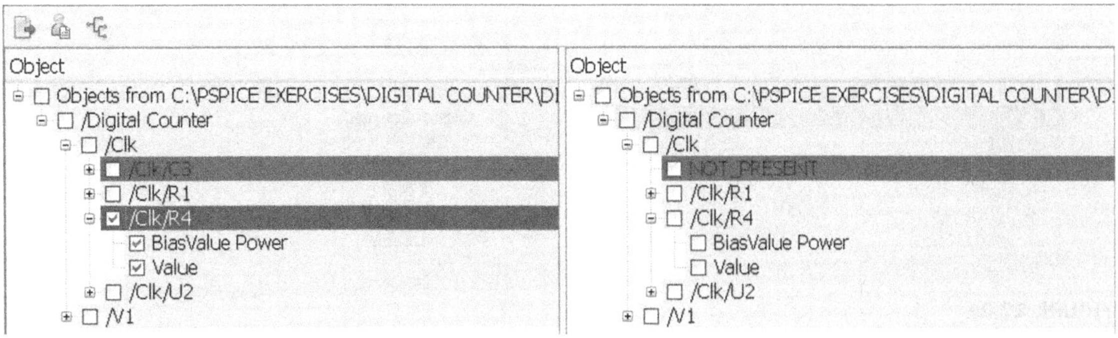

FIGURE 22.29
The modified value for R4 has been detected.

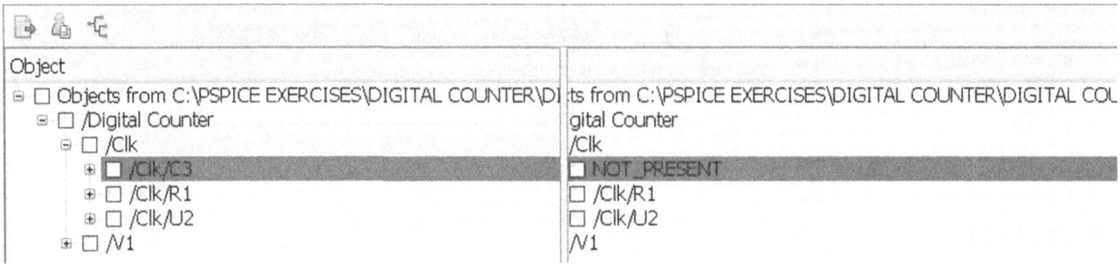

FIGURE 22.30
Differences for R4 no longer detected.

If you tried to update the extra capacitor, C3, the session log will display the warning message:

```
WARNING(ORCAP-37003): Could not add object '/Clk/C3' at
the target design, as this operation is not supported
```

5. Close the SVS utility and open the Test_Clock test bench design.
6. Open the **clock** schematic.
7. Delete the capacitor C3 and modify the value of C1 to 470n.
8. Save and close the clock schematic.
9. Open the master design and highlight the Digital Counter.dsn. Check for design differences: **Tools > Test Bench > Diff and Merge**.
10. The SVS window will show only value differences between the designs highlighted in yellow. Check the **/Digital Counter** box as in Figure 22.31 and click on the **Accept Left** icon .
11. The message window will appear, reporting no differences (Figure 22.32). The SVS window will appear as shown in Figure 22.33 with no entries.

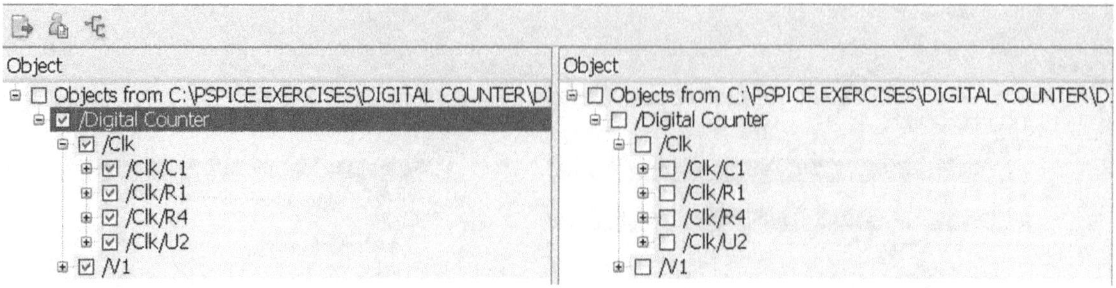

FIGURE 22.31
Selecting the test bench values to be updated.

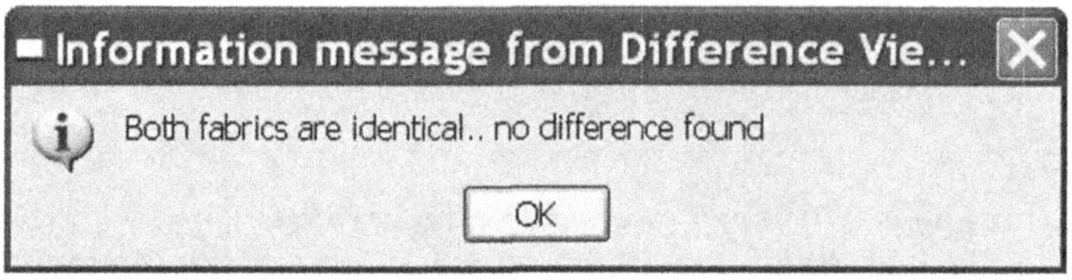

FIGURE 22.32
No design differences.

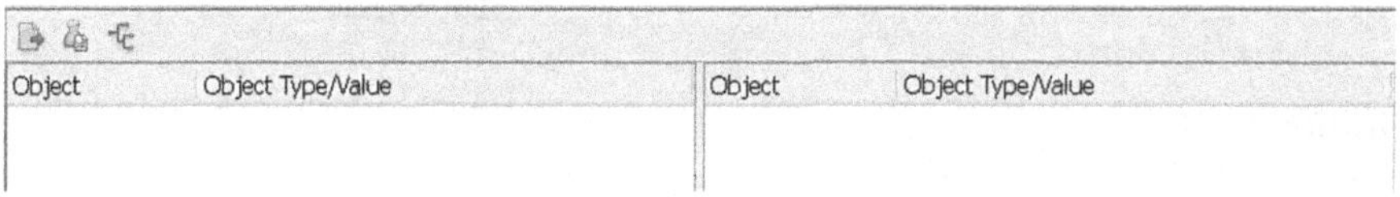

FIGURE 22.33
Clear SVS window.

NOTE

You could have selected the Value box for C1 as you did for R4, but if you have a large number of values to update, it is easier to select the whole test bench design to update. When you select the whole design, warning messages will be displayed in the session log, such as:

```
WARNING(ORCAP-37003): Could not add object '/Clk/R1' at the
target design, as this operation is not supported
```

which can be ignored.

PSpice Measurement Definitions

Bandwidth	Bandwidth of a waveform (you choose dB level)
Bandwidth_Bandpass_3dB	Bandwidth (3 dB level) of a waveform
Bandwidth_Bandpass_3dB_XRange	Bandwidth (3 dB level) of a waveform over a specified *X*-range
CenterFrequency	Center frequency (dB level) of a waveform
CenterFrequency_XRange	Center frequency (dB level) of a waveform over a specified *X*-range
ConversionGain	Ratio of the maximum value of the first waveform to the maximum value of the second waveform
ConversionGain_XRange	Ratio of the maximum value of the first waveform to the maximum value of the second waveform over a specified X-range
Cutoff_Highpass_3dB	High-pass bandwidth (for the given dB level)
Cutoff_Highpass_3dB_XRange	High-pass bandwidth (for the given dB level)
Cutoff_Lowpass_3dB	Low-pass bandwidth (for the given dB level)
Cutoff_Lowpass_3dB_XRange	Low-pass bandwidth (for the given dB level) over a specified range
DutyCycle	Duty cycle of the first pulse/period
DutyCycle_XRange	Duty cycle of the first pulse/period over a range
Falltime_NoOvershoot	Fall time with no overshoot
Falltime_StepResponse	Fall time of a negative-going step response curve
Falltime_StepResponse_XRange	Fall time of a negative-going step response curve over a specified range
GainMargin	Gain (dB level) at the first 180-degree out-of-phase mark
Max	Maximum value of the waveform

Max_XRange	Maximum value of the waveform within the specified range of *X*
Min	Minimum value of the waveform
Min_XRange	Minimum value of the waveform within the specified range of *X*
NthPeak	Value of a waveform at its *n*th peak
Overshoot	Overshoot of a step response curve
Overshoot_XRange	Overshoot of a step response curve over a specified range
Peak	Value of a waveform at its *n*th peak
Period	Period of a time domain signal
Period_XRange	Period of a time domain signal over a specified range
PhaseMargin	Phase margin
PowerDissipation_mW	Total power dissipation in milliwatts during the final period of time (can be used to calculate total power dissipation, if the first waveform is the integral of V(load)
Pulsewidth	Width of the first pulse
Pulsewidth_XRange	Width of the first pulse at a specified range
Q_Bandpass	Calculates *Q* (center frequency/ bandwidth) of a bandpass response at the specified dB point
Q_Bandpass_XRange	Calculates *Q* (center frequency/ bandwidth) of a bandpass response at the specified dB point and the specified range
Risetime_NoOvershoot	Rise time of a step response curve with no overshoot
Risetime_Step	Response rise time of a step response curve
Risetime_StepResponse_ XRange	Rise time of a step response curve at a specified range
SettlingTime	Time from <begin_x> to the time it takes a step response to settle within a specified band
SettlingTime_XRange	Time from <begin_x> to the time it takes a step response to settle within a specified band and within a specified range
SlewRate_Fall	Slew rate of a negative-going step response curve
SlewRate_Fall_XRange	Slew rate of a negative-going step response curve over an *X*-range
SlewRate_Rise	Slew rate of a positive-going step response curve

SlewRate_Rise_XRange	Slew rate of a positive-going step response curve over an *X*-range
Swing_XRange	Difference between the maximum and minimum values of the waveform within the specified range
XatNthY	Value of *X* corresponding to the nth occurrence of the given *Y*-value, for the specified waveform
XatNthY_NegativeSlope	Value of *X* corresponding to the nth negative slope crossing of the given *Y*-value, for the specified waveform
XatNthY_PercentYRange	Value of *X* corresponding to the nth occurrence of the waveform crossing the given percentage of its full *Y*-axis range; specifically, *n*th occurrence of Y=Ymin+(Ymax-Ymin)*Y_pct/100
XatNthY_Positive	Slope value of *X* corresponding to the *n*th positive slope crossing of the given *Y*-value, for the specified waveform
YatFirstX	Value of the waveform at the beginning of the *X*-value range
YatLastX	Value of the waveform at the end of the *X*-value range
YatX	Value of the waveform at the given *X*-value
YatX_PercentXRange	Value of the waveform at the given percentage of the *X*-axis range
ZeroCross	*X*-value where the *Y*-value first crosses zero
ZeroCross_XRange	*X*-value where the *Y*-value first crosses zero at the specified range

Index

0-9

A

B

J

K

L

M